T0073978

Studies in Computational Intelligence

Volume 785

Series editor

Janusz Kacprzyk, Polish Academy of Sciences, Warsaw, Poland
e-mail: kacprzyk@ibspan.waw.pl

The series "Studies in Computational Intelligence" (SCI) publishes new developments and advances in the various areas of computational intelligence—quickly and with a high quality. The intent is to cover the theory, applications, and design methods of computational intelligence, as embedded in the fields of engineering, computer science, physics and life sciences, as well as the methodologies behind them. The series contains monographs, lecture notes and edited volumes in computational intelligence spanning the areas of neural networks, connectionist systems, genetic algorithms, evolutionary computation, artificial intelligence, cellular automata, self-organizing systems, soft computing, fuzzy systems, and hybrid intelligent systems. Of particular value to both the contributors and the readership are the short publication timeframe and the world-wide distribution, which enable both wide and rapid dissemination of research output.

More information about this series at http://www.springer.com/series/7092

Leonardo Trujillo · Oliver Schütze
Yazmin Maldonado · Paul Valle
Editors

Numerical and Evolutionary Optimization – NEO 2017

 Springer

Editors
Leonardo Trujillo
Posgrado en Ciencias de la Ingeniería
Tecnológico Nacional de México - Instituto
  Tecnológico de Tijuana
Tijuana, Baja California
Mexico

Yazmin Maldonado
Posgrado en Ciencias de la Ingeniería
Tecnológico Nacional de México - Instituto
  Tecnológico de Tijuana
Tijuana, Baja California
Mexico

Oliver Schütze
Computer Science Department
Cinvestav-IPN
Mexico City, Mexico

Paul Valle
Posgrado en Ciencias de la Ingeniería
Tecnológico Nacional de México - Instituto
  Tecnológico de Tijuana
Tijuana, Baja California
Mexico

ISSN 1860-949X   ISSN 1860-9503   (electronic)
Studies in Computational Intelligence
ISBN 978-3-319-96103-3   ISBN 978-3-319-96104-0   (eBook)
https://doi.org/10.1007/978-3-319-96104-0

Library of Congress Control Number: 2018948219

This Springer imprint is published by the registered company Springer Nature Switzerland AG
The registered company address is: Gewerbestrasse 11, 6330 Cham, Switzerland

Preface

The Numerical and Evolutionary Optimization (NEO) workshop is a forum where researchers of two complimentary fields can meet and discuss issues related to one of the core problems of modern artificial intelligence, search and optimization. In particular, NEO integrates researchers that work with traditional numerical and mathematical optimization techniques, and those that develop bio-inspired and evolutionary algorithms. Moreover, the workshop also provides a forum for those working in specific application domains, where such techniques are known to be relevant or might provide new insights.

This volume comprises the post-proceeding of the fifth event of this series, the NEO 2017. The NEO 2017 returned to Tijuana for the second time (the first time in 2015), organized in collaboration with Tecnologico Nacional de Mexico/Instituto Tecnologico de Tijuana and CINVESTAV-IPN. The event was held from September 27 to September 29 in the city of Tijuana, Mexico. The event also had the first two industrial sponsors, IBM and TechData, as well as funding from CONACyT and CINVESTAV-IPN. For more details regarding the event, the venue and other information please visit the series webpage: http://neo.cinvestav.mx.

The event has grown since 2013, reaching 64 oral presentations and 7 invited speakers in 2017, covering topics from evolutionary computer vision, constraint handling, automatic design of electronic devices and many others that are relevant to the NEO attendees. This is the largest NEO to date, showing that this area of scientific discourse continues to grow, generating a lively conversation regarding the present and future of the manner in which search and optimization can be harnessed to solve the growing challenges of the modern world. Moreover, for the first time in the NEO history, a special one-day forum called NEO Engineering was held at the Instituto Tecnologico de Tijuana on September 26, to bring the NEO experience to undergraduate students in order to encourage the next generation of researchers in our field.

The presenters from the regular talks, as well as other members of the NEO community, were invited to submit chapters for the present post-proceedings volume. A total of 31 submissions underwent a thorough blind-review process, from which 13 chapters were chosen for publication. Moreover, three of the keynote speakers, Efren Mezura, Miguel Moreles and Esteban Tlelo, contributed with invited chapters.

The present volume contains sixteen chapters that are organized into four parts: Constraint Handling Techniques (Part I), Evolutionary and Genetic Computation (Part II), Optimal Control (Part III) and Real-World Applications (Part IV).

In Part I, three chapters are presented related to constraint handling techniques. The first chapter is an invited work from the laboratory of Efren Mezura that provides an extensive comparison of parameter control methods for differential evolution. The second chapter presents a descent method for efficient search in constrained multi-objective problems. The third and final chapter presents a constrained real-world application of a gradient descent method using the hypervolume indicator.

Part II contains two chapters on evolutionary and genetic computation. The first one deals with a problem in evolutionary computer vision, fitting multiple ellipses with multi-objective evolutionary search. The second chapter crosses over to the study of human interaction with evolutionary art by means of a kinect system.

Part III contains five chapters on optimal control. The first chapter applies control techniques to optimize supply chains. The NSGA-II algorithm is used to tune the parameters of a predictive controller in the second chapter. Similarly, controller tuning is also studied in the following chapter, in this case with the steepest descent method. The fourth chapter in this part deals with a self-tuning fuzzy controller, while the final chapter presents work on the optimal design of a sliding mode controller using differential evolution and the interior point method.

Part IV presents the final six chapters of this volume, dealing with real-world applications. The first chapter presents work on parameter estimation of a biological system with important biomedical applications, the second invited chapter of this volume written by Miguel Moreles. The next work is the invited chapter from Esteban Tlelo, where authors apply NSGA-II for the optimal sizing of voltage regulators. The third chapter presents a genetically optimized fuzzy system for classifying the quality of treated water. The fourth chapter deals with the optimal stabilization of civil structures that are exposed to external forces, such as earthquakes. The comparison of two methods for I/Q imbalance compensation in RF power amplifiers is presented in the fifth chapter. The final chapter of this part, and of the NEO 2017 volume, deals with the use of a linear programming approach to assess the impact of social networks in the hotel industry, an important financial activity in Mexico.

We express our gratitude to all invited speakers and attendees of the NEO 2017, you are the driving force behind this workshop and this post-proceeding volume. Thanks are also extended to all the authors and reviewers of contributed chapters, your hard work will make this volume a reference for future research.

September 2018   Leonardo Trujillo
                                                                      Oliver Schütze
                                                                     Yazmin Maldonado
                                                                          Paul Valle

Acknowledgements

To all of the researchers, authors and reviewers of the great work presented at NEO 2017 and this post-proceeding volume, without you there is no NEO! Special thanks are given to the invited speakers of NEO 2017, Dr. Efrén Mezura-Montes, Dr. Miguel Ángel Moreles, Dr. Michael Emmerich, Dr. Gustavo Olague, Dr. Esteban Tlelo Cuautle, Dr. Marcial Gonzalez and Ing. Leobardo Morales.

To the funding provided by CONACyT Fronteras de la Ciencia project FC-2015-2:944 "Aprendizaje evolutivo a gran escala", the CONACyT Basic Science project 285599 "Toma de decisiones multiobjetivo para sistemas altamente complejos", the CONACyT-DAAD project 207493 "Memetic strategies for indicator based multi-objective evolutionary algorithms", the TecNM project 6350.17-P "Planificación inteligente de recursos en sistemas reconfigurables", the TecNM project 6575.18-P "Ingeniería aplicada mediante el modelizado matemático para comprender la evolución del cáncer, la respuesta inmunológica y el efecto de algunos tratamientos" and the two industrial sponsors, IBM and TechData.

Finally, thanks are extended to the student organizers of the event, M.C. Perla Juárez and Ing. Rogelio Valdez for their invaluable participation, also M.C. Guadalupe Aĺarez, M.C. Darian Reyes, M.C. Carlos Goribar, Ing. Noelia Torres, M.C. Luis Muñoz, M.C. Itzel Gaytán, M.C. Jorge Alonso, Ing. César Bernal, Ing. Ernesto Aĺvarez, M.C. Victor López, M.C. Uriel Lopez and Ing. Carlos Dibene.

Contents

List of Contributors

Georgina Aguilar Instituto Tecnológico de Tijuana, Tijuana, Mexico

Luis T. Aguilar Instituto Politécnico Nacional—CITEDI, Tijuana, Mexico

Marco A. Alcaraz-Rodriguez Tecnológico Nacional de Mexico - Instituto Tecnológico de Tijuana, Tijuana, BC, Mexico

E. Allende-Chávez Instituto Tecnológico de Tijuana, Tecnológico Nacional de México, Tijuana, B.C., Mexico

Pablo M. Ayllon-Lorenzo Instituto Tecnológico de Tijuana, Tecnológico Nacional de México, Tijuana, Baja California, Mexico

Koen van der Blom LIACS, Leiden University, Leiden, The Netherlands

Sjonnie Boonstra Eindhoven University of Technology, Eindhoven, The Netherlands

Victor Hugo Carbajal-Gomez Universidad Autónoma de Tlaxcala, Tlaxcala, Mexico

Selene L. Cardenas-Maciel Tecnológico Nacional de México - Instituto Tecnológico de Tijuana, Tijuana, Baja California, México

M. A. Castro Tecnólogico Nacional de México, Instituto Tecnológico de La Paz, La Paz, Mexico

Nohe R. Cazarez-Castro Tecnológico Nacional de México - Instituto Tecnológico de Tijuana, Tijuana, Baja California, Mexico

Sergio Contreras-Hernandez Tecnológico Nacional de Mexico - Instituto Tecnológico de Tijuana, Tijuana, BC, Mexico

Luis N. Coria Tecnológico Nacional de Mexico - Instituto Tecnológico de Tijuana, Tijuana, BC, Mexico

Heriberto Cruz Hernández Computer Science Department, Cinvestav, Mexico City, Mexico

Cayetano Cruz Universidad de Extremadura, Badajoz, Spain

J. R. Cárdenas-Valdez Instituto Tecnológico de Tijuana, Tecnológico Nacional de México, Tijuana, B.C., Mexico

Manuel Cázares Research Institute for Economic and Social Growth, Mazatlán, Mexico

Michael Dellnitz Department of Mathematics, Paderborn University, Paderborn, Germany

Miguel Aurelio Duarte-Villaseñor Catedrático CONACYT en el Instituto Tecnológico de Tijuana, Tijuana, Mexico

Michael T. M. Emmerich LIACS, Leiden University, Leiden, The Netherlands

J. Enríquez-Zárate Facultad de Ingeniería y Ciencias Aplicadas, Universidad de los Andes, Las Condes, Santiago, Chile

Francisco Fernández de Vega Universidad de Extremadura, Badajoz, Spain

Luis Gerardo de la Fraga Computer Science Department, Cinvestav, Mexico City, Mexico

D. Garcia Tecnólogico Nacional de México, Instituto Tecnológico de La Paz, La Paz, Mexico

Mario García-Valdez Instituto Tecnológico de Tijuana, Tijuana, Mexico

Itzel G. Gaytan-Reyes Tecnológico Nacional de México-Instituto Tecnológico de Tijuana, Tijuana, Baja California, México

Bennet Gebken Department of Mathematics, Paderborn University, Paderborn, Germany

R. C. Gutiérrez-Urquídez Tecnológico Nacional de México, Instituto Tecnológico de Hermosillo, Hermosillo, Mexico

C. Hernández Centro de Investigación y de Estudios Avanzados del I.P.N. Departamento de Ingeniería Eléctrica, Sección de Computación, Mexico City, Mexico

Patricia Hernández University of Seville, Seville, Spain

C. Higuera Tecnólogico Nacional de México, Instituto Tecnológico de La Paz, La Paz, Mexico

Hèrm Hofmeyer Eindhoven University of Technology, Eindhoven, The Netherlands

S. A. Juárez-Cázares CITEDI, Instituto Politécnico Nacional, Tijuana, B.C., Mexico

David A. Lara-Ochoa Tecnológico Nacional de México-Instituto Tecnológico de Tijuana, Tijuana, Baja California, México

Jesus Lopez-Arredondo INAOE, Tonantzintla, Puebla, Mexico

F. R. López-Estrada Tecnológico Nacional de México, Instituto Tecnológico de Tuxtla Gutiérrez, Tuxtla Gutiérrez, Mexico

Armando Martinez-Graciliano Tecnológico Nacional de México-Instituto Tecnológico de Tijuana, Tijuana, Baja California, México

J. J. Merelo Computer Architecture and Technology, University of Granada, Granada, Spain

Efrén Mezura-Montes Artificial Intelligence Research Center, University of Veracruz, Centro, Xalapa, Veracruz, Mexico

J. A. Morales Tecnólogico Nacional de México, Instituto Tecnológico de La Paz, La Paz, Mexico

Miguel Angel Moreles Mathematics Department, Universidad de Guadalajara, CUCEI, Guadalajara, Mexico; CIMAT, Guanajuato, GTO, Mexico

Jose Angel Neria CIMAT, Guanajuato, GTO, Mexico

J. C. Nuñez-Pérez CITEDI, Instituto Politécnico Nacional, Tijuana, B.C., Mexico

J. A. Orrante-Sakanassi CONACYT-Tecnológico Nacional de México, Instituto Tecnológico de Hermosillo, Hermosillo, Mexico

Sebastian Peitz Department of Mathematics, Paderborn University, Paderborn, Germany

Joaquin Peña CIMAT, Guanajuato, GTO, Mexico

Pablo J. Prieto Tecnológico Nacional de México - Instituto Tecnológico de Tijuana, Tijuana, Baja California, Mexico

Octavio Ramos-Figueroa Artificial Intelligence Research Center, University of Veracruz, Centro, Xalapa, Veracruz, Mexico

María-Margarita Reyes-Sierra Artificial Intelligence Research Center, University of Veracruz, Centro, Xalapa, Veracruz, Mexico

O. M. Rodríguez-Elías Tecnológico Nacional de México, Instituto Tecnológico de Hermosillo, Hermosillo, Mexico

Y. Sandoval-Ibarra Instituto Tecnológico de Tijuana, Tecnológico Nacional de México, Tijuana, B.C., Mexico

J. Sandoval Tecnólogico Nacional de México, Instituto Tecnológico de La Paz, La Paz, Mexico

Oliver Schütze Computer Science Department, Cinvestav-IPN, México City, Mexico

Claudia N. Sánchez Facultad de Ingeniería, Universidad Panamericana, Aguascalientes, Aguascalientes, Mexico

Esteban Tlelo-Cuautle Instituto Nacional de Astrofísica, Óptica y Electrónica, Tonatzintla, Puebla, Mexico

L. Trujillo Tree-Lab, Posgrado en Ciencias de la Ingeniería, Instituto Tecnológico de Tijuana, Tijuana, B.C., Mexico

G. Valencia-Palomo Tecnológico Nacional de México, Instituto Tecnológico de Hermosillo, Hermosillo, Mexico

Hao Wang LIACS, Leiden University, Leiden, The Netherlands

Constraint Handling Techniques

Deterministic Parameter Control in Differential Evolution with Combined Variants for Constrained Search Spaces

Octavio Ramos-Figueroa[(✉)], María-Margarita Reyes-Sierra, and Efrén Mezura-Montes

Artificial Intelligence Research Center, University of Veracruz,
Sebastián Camacho 5, 91000 Centro, Xalapa, Veracruz, Mexico
oivatco.rafo@gmail.com, {mareyes, emezura}@uv.mx

Abstract. This chapter presents an empirical comparison of six deterministic parameter control schemes based on a sinusoidal behavior that are incorporated into a differential evolution algorithm called "Differential Evolution with Combined Variants" (DECV) to solve constrained numerical optimization problems. Besides, the feasibility rules and the ε-constrained method are adopted as constraint-handling techniques.

Two parameters are considered in this work, F (related with the mutation operator) and CR (related with the crossover operator). Two DECV versions (rand-best) and (best-rand) are assessed. From the above elements, 24 different variants are tested in 36 well-known benchmark problems (in 10 and 30 dimensions). Two performance measures used in evolutionary constrained optimization (successful percentage and average number of evaluations in successful runs) are adopted to evaluate the performance of each variant. Five experiments are proposed to compare (1) those variants with the feasibility rules, (2) those variants with the ε-constrained method, (3) the most competitive variants from the previous two experiments, (4) the convergence plots of those most competitive variants and (5) the significant statistical differences of feasible final results among variants.

The obtained results suggest that an increasing oscillation of F and CR values, starting around 0.5 and then moving between 0 and 1, is suitable for the (rand-best) DECV variant. In contrast, a decreasing oscillation of both parameter values is suitable for the (best-rand) DECV variant. The convergence behavior observed in the most competitive variants indicates the convenience of the increasing oscillation of both parameters, coupled with the rand-best DECV version, to promote a faster convergence. The ε-constrained method showed to be more competitive with this type of parameter control than the feasibility rules. Finally, no significant differences among variants were observed based on final feasible results.

Keywords: Parameter control · Differential evolution
Constrained numerical optimization · Constraint-Handling techniques

© Springer International Publishing AG, part of Springer Nature 2019
L. Trujillo et al. (Eds.): NEO 2017, SCI 785, pp. 3–28, 2019.
https://doi.org/10.1007/978-3-319-96104-0_1

1   Introduction

Throughout history, different numerical optimization problems of practical and theoretical importance have been identified within fields like Artificial Intelligence and Operations Research [1]. Mathematical programming methods are precisely designed to solve optimization problems. Such classic algorithms can be divided in two groups: (1) direct methods, e.g., the Hooke-Jeeves search method, the Nelder-Mead Simplex search method, among others [2, 3], and (2) gradient-based methods, e.g., the Cauchy and the Newton method, among others [4]. However, each day new and more complex problems are identified, and classic methods may not be effective enough to solve them.

In order to deal with complex numerical optimization problems, as an alternative option, approximate algorithms have been proposed. They are also known as meta-heuristic algorithms, i.e., strategies comprised by heuristics that guide the optimization problem solving. Such meta-heuristic algorithms show strong capabilities to solve search and optimization problems, and some of them are inspired by nature, i.e., biological processes are emulated [5].

There are different types of nature-inspired algorithms. Among the most known are: (1) swarm intelligence (SI), inspired by the complex collective behavior that emerges during the interaction of individuals in different social groups, usually of the same species, such as humans or biological organisms, to perform a particular task [6]; and (2) evolutionary algorithms (EAs), inspired by the biological evolution (reproduction, mutation, recombination, natural selection and survival of the fittest) [5]. Different algorithms have been developed under both approaches (SI and EAs) to solve a wide range of theoretical and real-world problems, showing a very competitive performance [7].

When an EA is designed, different aspects should be taken into consideration, such as selection and replacement strategies, and solution encoding as well. Moreover, suitable values for the parameters of an EA (e.g., population size, crossover and mutation rates) must be chosen. Parameter Setting (PS) is an important sub-area devoted to deal with the parameter fine-tuning required by EAs to provide competitive results. PS can be divided in two main branches: (1) parameter tuning problem (PT) and (2) parameter control (PC) [8, 9].

PT is given at the design stage, where high-quality parameter values are carefully chosen [8]. PT requires the EA to be run several times. Thus, the PT main weakness is the high computational cost required to identify appropriate parameter values for a certain EA, and such cost may be prohibitive for optimization problems where solution evaluations are particularly time-consuming. On the other hand, PC is given at the run stage. Thereby, parameter values may change as the search process progresses. Hence, a suitable parameter control scheme (PCS) should be employed for an efficient variation of parameter values [9]. The modification of parameter values over the run is considered the main virtue of PC. Nevertheless, getting a general PC which can deal with different types of optimization problems is its main weakness.

Different PCS have been proposed and they can be classified into (1) deterministic, (2) adaptive, and (3) self-adaptive. The definition of a fixed change pattern for parameter values is named as deterministic PC. Adaptive PC uses information of the search process to modify the parameter values. Finally, self-adaptive PC uses the same evolution process to evolve parameter values at the same time that solutions to the optimization problems are evolved [9].

Nowadays, one of the most popular EAs is Differential Evolution (DE), which was introduced in [10]. Since its emergence, this algorithm was so-well received, mainly because it is considered an easy yet straightforward search strategy. However, as well as other EAs, the parameter setting is considered as the key element in its success. Three main parameters are required by DE, (1) the population size NP, (2) the scaling factor of the mutation operator F, and (3) the crossover rate CR. Originally, DE was designed to solve unconstrained numerical optimization problems (UNOPs). The specialized literature indeed reports efforts to add PC to DE, e.g., the self-adaptive Differential Evolution algorithm (SaDE) [11] and jDE [12], both for unconstrained search spaces.

To deal with a constrained search space, EAs require the addition of a constraint-handling technique (CHT) [13]. According to the specialized literature, the selection of a suitable CHT is closely related to the performance reached by an EA for Constrained Numerical Optimization Problems (CNOPs) [14]. Penalty functions are very popular CHTs [15, 16], where a CNOP is transformed into an unconstrained one by incorporating a certain value to the objective function based on the amount of constraints violation in a given solution. Other popular CHTs are the feasibility rules introduced in [15] and the ε-constrained method presented in [16].

DE has been successfully adopted to solve CNOPs [13]. Among those DE-based EAs for CNOPs, the Differential Evolution with Combined Variants (DECV) was proposed in [17]. This EA is based on two well-known DE-variants (DE/rand/1/bin and DE/best/1/bin), where a switch between variants is based on an expected percentage of feasible solutions in the current population. An extension of SaDE proposed in [11] was presented in [18] to solve CNOPs. The authors modified the replacement criterion, such as feasible and infeasible individuals are differentiated using the set of feasibility rules above mentioned. In the same way, the jDE adaptive algorithm introduced in [12] was improved incorporating a CHT based on a penalty function.

On the other hand, the Diversity Differential Evolution (DDE) algorithm was proposed in [19]. Besides NP, CR and F, two additional parameters were used: (1) NO, which determines how many descendants can be generated by a target vector (potential solution of the problem to be solved); and (2) S_r, which is responsible to control the way vectors are selected. Thus, in that study, the use of a deterministic and a self-adaptive parameter control schemes were proposed. Thereby, F, CR and NO are self-adapted; while S_r, is controlled by a deterministic approach.

Another important adaptive algorithm is the Generalized Differential Evolution (GDE) with Exponential Weighting Moving Average (EWMA), proposed in [20]. EWMA is a PCS that performs parameter control adaptation by updating weighted average history of successful control parameter values. This algorithm has shown being suitable for single- and multi-objective optimization with or without constraints.

In [21], the authors presented a Memetic Differential Evolution for CNOPs, where DE was enhanced with a mathematical programming method called Powell´s conjugate direction as a local search operator and the ε-constrained method as a CHT. Such algorithm was able to quickly reach the feasible region of the search space and it also provided competitive feasible solutions.

An enhanced multi-operator differential evolution algorithm (E-MODE) was introduced in [22], where the initial population was divided into sub-populations, and each one of them was evolved with a different variation operator. Furthermore, a self-adaptive PCS was used for controlling F and CR parameters, while the sub-populations sizes were reduced with a linear reduction strategy. This algorithm showed a highly competitive performance when solving a wide set of test problems [22]. In [23] a framework with a mix of four CHTs was designed, where each CHT was employed by a certain sup-population, such that different combinations of mutation and crossover DE strategies were incorporated in each sub-population, and the parameter values of F and CR were updated over the run, between pre-defined intervals.

According to the above described literature review, DE-based EAs for CNOPs consider different mechanisms to improve their performance. Among them, PC is adopted. However, to the best of our knowledge, those works do not necessarily analyze the effects of different PCS combined with particular CHT in the performance of a DE-based EA for CNOPs.

This is precisely the motivation of this work, where a particular PCS, deterministic in this case, based on sinusoidal functions, already studied for UNOPs, is analyzed now in constrained search spaces to control two DE parameters (F and CR). The aim is to analyze the positive (or negative) effects of a given deterministic PCS when working with a particular CHT in the performance of an EA to solve CNOPs. Therefore, 24 variants of the algorithm are designed, based on versions of the Differential Evolution with Combined Variants (DECV) algorithm, two different CHT (the feasibility rules and the ε-constrained method) and six different deterministic PCS based on a sinusoidal function to adapt CR and/or F parameter(s).

From the above mentioned, the contribution of this work consists on analyzing the capacity of a deterministic PCS coupled with a CHT to improve the capabilities of a DE-based EA to solve CNOPs.

This chapter is organized as follows: the CNOP is formally defined in Sect. 2. After that, Sect. 3 introduces (1) DE, (2) the two CHT adopted in this work, (3) DECV, which is used as the search algorithm, and (4) the six configurations under analysis and used as PCS. Section 4 includes the experimental design, the results obtained and their corresponding discussion. Finally, Sect. 5 summarizes the conclusions and future paths of research.

2 Problem Statement

A CNOP, known also as the non-linear programming problem [13–23], without loss of generality, can be defined as to:

Find x which minimizes:

$$f(x) \tag{1}$$

subject to a set of constraints:

$$g_j(\boldsymbol{x}) \leq 0, i = 1, \ldots, m \tag{2}$$

$$h_k(\boldsymbol{x}) = 0, j = 1, \ldots, p \tag{3}$$

$$x_i^L \leq x_i \leq x_i^U, i = 1, \ldots, n \tag{4}$$

where x denotes a solution, comprised by a set of n decision variables as follows: $x = [x_1, x_2, \ldots, x_n]$. On the other hand, the m inequality constraints and p equality constraints are depicted by $g_j(x), i = 1, \ldots, m$ and $h_k(x), k = 1, \ldots, p$, respectively. Finally, the limits for each decision variable x_i are represented by $\left[x_i^L, x_i^U\right]$.

To deal with equality constraints a tolerance δ (a very small value) is adopted. Therefore, each equality constraint is transformed into an inequality constraint by using Eq. 5:

$$|h_k(\boldsymbol{x})| - \delta \leq 0 \tag{5}$$

3 Proposed Study

3.1 Differential Evolution Algorithm

Recalling from Sect. 1, DE has become one of the most popular EAs, because it has shown a high performance on solving a wide range of theoretical and real-world problems. Moreover, it is simple to code EA. Thereby, numerous DE variants have been proposed, giving rise to the emergence of DE/x/y/z notation, introduced to classify the DE variants, where x denotes the base vector, y is the number of difference vectors; and z specifies the type of crossover operator [10]. However, like other EAs, the performance of DE is strongly related to the setting of its three main parameters: (1) the population size "NP", (2) the scaling factor "F", and (3) the crossover rate "CR".

DE/rand/1/bin is the most popular DE variant, and the steps comprised by its search process are presented in Algorithm 1 [10]. As in most DE variants, the search starts with an initial population of vectors (set of potential solutions to the problem) randomly generated. Subsequently, the population is evolved as follows: for each vector in the population, known as target vector $\boldsymbol{x}_{G,i}$ at the reproduction step, three different vectors are randomly selected from the population (line 7 of Algorithm 1). After that, two vectors, $\boldsymbol{x}_{G,r1}$ and $\boldsymbol{x}_{G,r2}$ (out of the three randomly chosen) are used to compute a difference, which is scaled by the user-defined parameter F (line 11 of Algorithm 1). Next, the resulting vector is added to the third vector $\boldsymbol{x}_{G,r3}$ (known as base vector) to generate the so-called mutant vector $\boldsymbol{v}_{i,G}$. Besides, the crossover is performed simultaneously based on the user-defined parameter CR (line 10 of Algorithm 1). Thus, each element of the trial vector $\boldsymbol{u}_{G,i}$ (offspring vector) is taken from the target vector or the mutant vector according to the CR-value. Furthermore, the variable j_{rand} (line 8 of Algorithm 1) ensures that at least one element in the trial vector $\boldsymbol{u}_{G,i}$ is taken from the mutant vector. Based on that, exact copies of the target vector are avoided.

Algorithm 1. DE/rand/1/bin

1: **Begin**
2:    G=0
3:    Create a random initial population $x_{G,i}$ $\forall i, i = 1, ..., NP$
4:    Evaluate $f(x_{G,i})$ $x_{G,i}$ $\forall i, i = 1, ..., NP$
5:    **For** $G = 1$ to $MaxGen$ **Do**
6:        **For** $i = 1$ hasta NP **Do:**
7:            Select randomly r1 $\neq$ r2 $\neq$ r3 $\neq$ i
8:            j_{rand}= randint(1, D)
9:            **For** j = 1 to D **Do**
10:           **If(** $rand_j$ [0, 1) < CR or j = j_{rand}) **Then**
11:               $u_{G,i,j}$= $x_{G,r3,j}$ + $F(x_{G,r1,j} - x_{G,r2,j})$
12:           **Else:**
13:               $u_{G,i,j} = x_{G,i,j}$
14:           **End If**
15:       **End for**
16:       **If** $(f(u_{G,i}) \leq f(x_{G,i}))$
17:           $x_{G+1,i} = u_{G,i}$
18:       **Else:**
19:           $x_{G+1,i} = x_{G,i}$
20:       **End If**
21:    **End For**
22: **End For**
23: **End**

Another popular DE variant is DE/best/1/bin. In this variant, the mutation is guided by the best vector in the population of the current generation. Thus, the best vector $x_{G,best}$ is used as the base vector to generate each mutant vector. Thereby, a faster convergence is sought, i.e., a competitive solution can be found in less time because exploitation of promising areas of the search space is promoted. However, the risk of premature convergence may increase. In contrast, DE/rand/1/bin focuses more on the exploration of the search space [17].

In Fig. 1 it can be seen that the mutation process is performed according to parameter F, which establishes a step size, i.e., when F tends to 0, the mutant vector is generated closer to the base vector and exploitation is promoted. In contrast, when F has a value close to 1, the mutant vector is generated away from the base vector and exploration is promoted instead.

On the other hand, the crossover process is directed by parameter CR. Therefore, high CR values, i.e., close to 1, are used to generate a trial vector with the most of values taken from the mutant vector. Thus, the offspring is generated far away from the target vector, promoting exploration; while a lower CR value, i.e., close to 0, is used to generate an offspring close to its target vector, promoting exploitation.

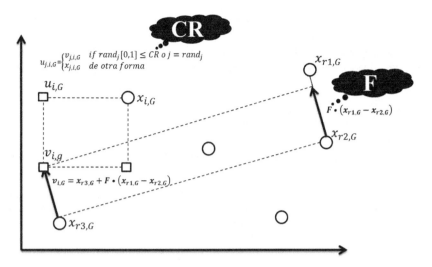

Fig. 1. DE mutation and crossover operators applied to trial vector i at generation G. $x_{i,G}$ is the target vector. $x_{r1,G} \neq x_{r2,G} \neq x_{r3,G} \neq x_{i,G}$ are three different vectors, randomly selected and different to the target vector. $v_{i,G}$ is the mutant vector. Finally, $u_{i,G}$ is the trial vector.

From the above, the fine-tuning of parameters F and CR helps to provide a good trade-off between exploration and exploitation, and it may lead to a success search. It is important to emphasize that the two parameters (F and CR) are somehow related. Hence, if one parameter is misconfigured, the performance of the other parameter may be affected.

3.2   Constraint-Handling Techniques

Recalling from Sect. 1, EAs, in their original versions, were designed to sample unconstrained search spaces. Therefore, they need the addition of a constraint-handling technique (CHT), which incorporates feasibility information into the EAs, like DE, to deal with a constrained search space. In this way, the EA is able to choose between feasible and infeasible solutions and promote diversity in this type of search space as well.

To get information about the feasibility of a solution, some value must be calculated. In this way, the constraint violation of each solution is computed by adding the amounts of all violated constraints. In such a way, the sum of constraint violation can be expressed as detailed in Eq. (6) [14]:

$$\varphi(\boldsymbol{x_i}) = \sum_j \max\left(0, g_j(\boldsymbol{x_i})\right)^2 + \sum_k (\max(0, |h_k(\boldsymbol{x_i})| - \delta))^2 \qquad (6)$$

where $\boldsymbol{x_i}$ is the solution evaluated, and the inequality and equality constraints are expressed with g_j and h_k, respectively. Below the two CHT studied in this work are described in detail.

The set of feasibility rules [15] is one of the most popular CHT. It is based on the sum of constraint violation described in Eq. (6). This CHT is widely used, mainly because is simple and easy to implement and no parameters are required. Solutions are compared considering the following three criteria:

1. When comparing two feasible solutions, the one with the best objective function is chosen.
2. When comparing a feasible solution and an infeasible solution, the feasible one is chosen.
3. When comparing two infeasible solutions, the one with the lowest sum of constraint violation (computed as in Eq. (6)) is chosen.

The ε-constrained method is another popular CHT [16]. In this technique, an ε-level comparison is defined as an order relation on a pair of objective function values and sums of constraints violation $(f(x_i), \varphi(x_i))$, as it can be seen in Eq. 7.

$$[f(x_1), \varphi(x_1)] < \varepsilon f(x_2), \varphi(x_2)) \iff \begin{cases} f(x_1) < f(x_2), \text{ if } \varphi(x_1), \varphi(x_2) \leq \varepsilon; \\ f(x_1) < f(x_2), \text{ if } \varphi(x_1), \varphi(x_2) = \varepsilon; \\ \varphi(x_1) < \varphi(x_2), otherwise \end{cases} \quad (7)$$

When $\varepsilon = 0$, the solutions are compared based on the constraints violation. Otherwise, if $\varepsilon = \infty$ the solution are compared using the objective function. The ε-level is updated with Eqs. 8 and 9.

$$\varepsilon(0) = \varphi(x_0) \quad (8)$$

$$\varepsilon(t) \begin{cases} \varepsilon(0) \left(1 - \frac{t}{Tc}\right)^{cp} & 0 < t < Tc; \\ 0 & t \geq Tc \end{cases} \quad (9)$$

where t is the current iteration, Tc is the maximum number of iterations using the ε-level comparison, x_θ is the top θ-th solution, $\theta = 0.2 * n$, and the ε tolerance reduction speed is controlled by cp.

3.3 Differential Evolution with Combined Variants for Constrained Numerical Optimization

Differential Evolution with Combined Variants (DECV) [17] is the algorithm chosen for this research. Such decision lies in the fact that it has provided competitive results when solving CNOPs. However, there is room for improvement in its performance and this research looks to analyze such positive effect with different PCS combined with two CHTs in its two original versions.

DECV conveniently combines two DE variants: DE/rand/1/bin and DE/best/1/bin. DECV starts with DE/rand/1/bin and changes to DE/best/1/bin when a certain population percentage has reached the feasible region. The moment of DE variant change is controlled by the Percentage Change Variant parameter (PCV). Therefore, the percentage of feasible solutions must be computed at each generation. Such value is compared against the PCV parameter and when the computed value surpasses PCV, the variant change is carried out. Originally, the feasibility rules are used as CHT in DECV. Such rules are used when the best vector between the target and its trial must be chosen. The steps comprised by the DECV algorithm are shown in Algorithm 2.

Algorithm 2. DECV

1: **Begin**
2:   G=0
3:   Create a random initial population $x_{G,i} \; \forall i, i = 1, ..., NP$
4:   Evaluate $f(x_{G,i}) \; x_{G,i} \; \forall i, i = 1, ..., NP$
5:   **For $G = 1$ to *MaxGen* Do**
6:       Compute the feasibilityPercentage
7:       **For $i = 1$ hasta NP Do**:
8:           **If** feasibilityPercentage $\geq$**PCV**:
9:               Select randomly r1 $\neq$r2 $\neq$i
10:          **Else:**
11:              Select randomly r1 $\neq$r2 $\neq$r3 $\neq$i
12:          **End If**
13:          j_{rand}= randint(1, D)
14:          **For** j = 1 to D **Do**
15:              **If($rand_j$ [0, 1) < *CR* or j = j_{rand}) Then**
16:                  **If** feasibilityPercentage $\geq$**PCV**:
17:                      $u_{G,i,j} = x_{G,best,j} + F(x_{G,r1,j} - x_{G,r2,j})$
18:                  **Else:**
19:                      $u_{G,i,j} = x_{G,r3,j} + F(x_{G,r1,j} - x_{G,r2,j})$
20:                  **End If**
21:              **Else:**
22:                  $u_{G,i,j} = x_{G,i,j}$
23:              **End If**
24:          **End for**
25:          If $f(u_{G,i}) \leq f(x_{G,i})$ (according to the feasibility rules)
26:              $x_{G+1,i} = u_{G,i}$
27:          **Else:**
28:              $x_{G+1,i} = x_{G,i}$
29:          **End If**
30:      **End For**
31: **End For**
32: **End**

Other DECV versions have been proposed [17], i.e., the DECV version where the DE variants order is swapped. Thus, the algorithm starts with DE/best/1/bin, and changes to DE/rand/1/bin after the percentage of feasible solutions surpasses the PCV value [17].

The two DECV versions mentioned above are studied in this research. To identify them, they will be referred as follows: (1) DECV (rand-best) is the algorithm which starts with DE/rand/1/bin and changes to DE/best/1/bin with PCV = 90%; and (2) DECV (best-rand) is the algorithm that starts with DE/best/1/bin and changes to DE/rand/1/bin with PCV = 10%.

3.4   Deterministic Parameter Control Scheme Based on a Sinusoidal Function

A deterministic PCS, which has shown competitive results in unconstrained search spaces [24], is studied here in constrained optimization. That is, the parameters in question are altered based on a particular rule, and the parameter values are generated in a deterministic way using a fixed function. Thus, search feedback is not used, and the time variation is employed to guide the parameter value changes.

In this study, six different deterministic PCS introduced in [24] are added to the two above mentioned DECV versions. The configuration adapts the F and/or CR parameter values based on a sinusoidal function. Thus, the initial parameter values are close to

Table 1. Deterministic PCS based on a sinusoidal function taken from [24]. *freq*: Frequency; *G*: current generation and *MaxGen*: maximum number of generations.

Name	Description	Behavior
Conf1	$F_G = \dfrac{1}{2} * \left(\sin(2\pi * freq * G) * \dfrac{G}{MaxGen} + 1\right)$ $CR_G = 0.9$	
Conf2	$F_G = \dfrac{1}{2} * \left(\sin(2\pi * freq * G) * \dfrac{G}{MaxGen} + 1\right)$ $CR_G = \dfrac{1}{2} * \left(\sin(2\pi * freq * G) * \dfrac{G}{MaxGen} + 1\right)$	
Conf3	$F_G = \dfrac{1}{2} * \left(\sin(2\pi * freq * G) * \dfrac{MaxGen - G}{MaxGen} + 1\right)$ $CR_G = \dfrac{1}{2} * \left(\sin(2\pi * freq * G) * \dfrac{MaxGen - G}{MaxGen} + 1\right)$	

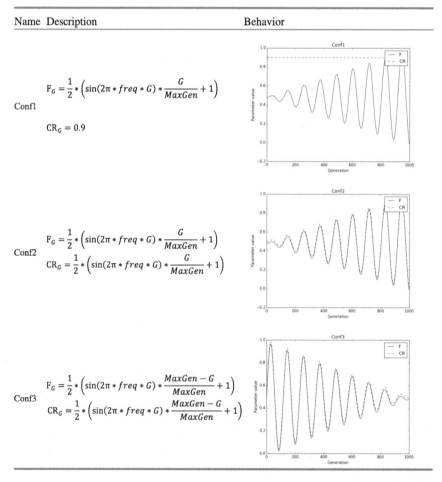

(continued)

Table 1. *(continued)*

Conf4

$$F_G = \frac{1}{2} * \left(\sin(2\pi * freq * G) * \frac{MaxGen - G}{MaxGen} + 1\right)$$

$$CR_G = 0.9$$

Conf5

$$F_G = 0.5$$

$$CR_G = \frac{1}{2} * \left(\sin(2\pi * freq * G) * \frac{G}{itMax} + 1\right)$$

Conf6

$$F_G = 0.5$$

$$CR_G = \frac{1}{2} * \left(\sin(2\pi * freq * G) * \frac{MaxGen - G}{MaxGen} + 1\right)$$

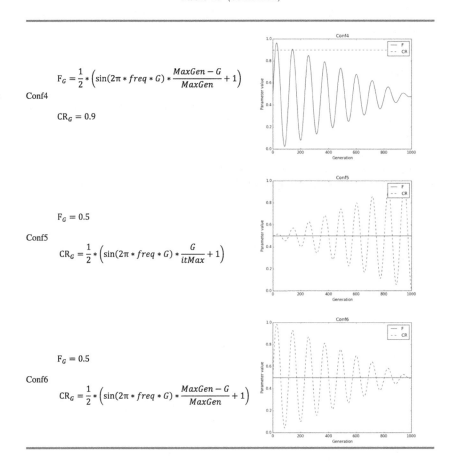

0.5; as the search process progresses, the sinusoidal function is employed to increase and decrease the parameter-values, oscillating between 0 and 1. That is to say, an increasing oscillatory behavior is presented.

On the other hand, there are other configurations where the parameters are initialized with values close to 0.5. After that, as the search advances, the parameter values present a remarked oscillation between 0 and 1; such oscillation decreases once the end of the search approaches and the values reach 0.5 again, i.e., a decreasing oscillatory behavior is presented.

The different deterministic PCS studied in this work, e.g., Conf1, Conf2, ..., Conf6, are shown in Table 1. It can be observed that in some schemes, e.g., Conf1, Conf4, Conf5 and Conf6, only one parameter is adapted and the other remains with a fixed value. In contrast, in the remaining two schemes, e.g., Conf2 and Conf3, both parameters are controlled by the sinusoidal behavior. Besides, Table 1 also contains the expected behavior of the six DPCSs, where those values F and CR can take at generation x are depicted in the y-axis. Thus, an increasing oscillatory behavior is presented by Conf1, Conf2 and Conf5, while a decreasing oscillatory behavior is indicated in Conf3, Conf4 and Conf6.

4 Experiments and Results

The experiments were performed based on the following four variants: (1) DECV (rand-best) with the feasibility rules, (2) DECV (best-rand) with the feasibility rules, (3) DECV (rand-best) with the ε-constrained method, and (4) DECV (best-rand) with the ε-constrained method. Each one of the six different deterministic PCS (Conf1, ..., Conf6) were added to each one of the four algorithm variants. Therefore, a total of 24 algorithm variants were assessed. Table 2 contains a detailed description of the 24 variants designed.

Table 2. Description of the 24 variants designed. "CHT" means constraint-handling technique. "Variant" indicates the DECV version. "DPCS" means deterministic parameter control scheme. "Algorithm" is the variant description.

CHT	Variant	Deterministic PCS	Algorithm
fr	rand-best	Conf1, ..., Conf6	fr-DECV (rand-best) Conf1, ..., Conf6
	best-rand	Conf1, ..., Conf6	fr-DECV (best-rand) Conf1, ..., Conf6
ε	rand-best	Conf1, ..., Conf6	ε-DECV (rand-best) Conf1, ..., Conf6
	best-rand	Conf1, ..., Conf6	ε-DECV (best-rand) Conf1, ..., Conf6

36 well-known constrained test problems [25] were solved in the experiments. A summary of their features is presented in Table 3 and further details can be found in [25]. 18 test functions have 10 Dimensions (10D) and the remaining 18 functions have 30 Dimensions (30D).

Performance measures taken from the specialized literature in constrained optimization with EAs were computed. Besides analyzing effects of a CHT with a particular PCS, the aim was to identify (if any) the variant with the most competitive performance. Such performance was measured by computing two measures proposed in [26]: (1) probability of convergence P (percentage of successful runs, where a successful run is that where a feasible solution in the neighborhood of the best feasible known solution is found, and in this work an error of 1×10^{-4} was adopted), and (2) average number of function evaluations in successful runs (AFES). 25 independent runs per function were performed to validate the results.

In this research, a test function was considered as "solved", when at least one of the 25 independent runs was successful. On the other hand, when an algorithm has the ability to solve several test problems with high P values, it was then considered as "robust".

Table 3. Details of 36 test problems [25], 18 with 10D and 18 with 30D. D is the number of decision variables, ρ is the estimated ratio between the feasible region and the search space, I is the number of inequality constraints, E is the number of equality constraints.

Problem/search range	Type of objective	Number of constraints		ρ	
		E	I	10D	30D
C01 $[0, 10]^D$	Non separable	0	2 Non separable	0.997689	1.000000
C02 $[-5.12, 5.12]^D$	separable	1 separable	2 separable	0.000000	0.000000
C03 $[-1000, 1000]^D$	Non separable	1 Non separable	0	0.000000	0.000000
C04 $[-50, 50]^D$	Separable	2 Non separable, 2 separable	0	0.000000	0.000000
C05 $[-600, 600]^D$	Separable	2 separable	0	0.000000	0.000000
C06 $[-600, 600]^D$	Separable	2 rotated	0	0.000000	0.000000
C07 $[-140, 140]^D$	Non separable	0	1 separable	0.505123	0.503725
C08 $[-140, 140]^D$	Non separable	0	1 rotated	0.379512	0.375278
C09 $[-500, 500]^D$	Non separable	1 separable	0	0.000000	0.000000
C10 $[-500, 500]^D$	Non separable	1 rotated	0	0.000000	0.000000
C11 $[-100, 100]^D$	Rotated	1 Non separable	0	0.000000	0.000000
C12 $[-1000, 1000]^D$	Separable	1 Non separable	1 separable	0.000000	0.000000
C13 $[-500, 500]^D$	Separable	0	2 separable, 1 non separable	0.000000	0.000000
C14 $[-1000, 1000]^D$	Non separable	0	3 separable	0.003112	0.006123
C15 $[-1000, 1000]^D$	Non separable	0	3 rotated	0.003210	0.006023
C16 $[-10, 10]^D$	Non separable	2 separable	1 separable, 1 non separable	0.000000	0.000000
C17 $[-10, 10]^D$	Non separable	1 separable	2 non separable	0.000000	0.000000
C18 $[-50, 50]^D$	Non separable	1 separable	1 separable	0.000010	0.000000

Those parameters which are not part of this study were tuned according to the suggested values in the specialized literature [10]. Those parameters values for fr-DECV and ε-DECV variants are shown in Tables 4 and 5, respectively.

Table 4. Parameter values used in *fr*-DECV (rand-best) Conf1, ..., Conf6 and *fr*-DECV (best-rand) Conf1, ..., Conf6 algorithms, "PCV" means percentage to change DE variant, "NP" is the population size, "MaxEval" is the maximum number of evaluations, "*freq*" is the oscillation frequency of sinusoidal functions.

Variant	PCV	Dimension	NP	MaxEval	freq
rand-best	90%	10D	90	200000	0.5
		30D	90	600000	0.5
best-rand	10%	10D	90	200000	0.5
		30D	90	600000	0.5

Table 5. Parameter values used in ε-DECV (rand-best) Conf1, ..., Conf6 and ε-DECV (best-rand) Conf1, ..., Conf6 algorithms. "PCV" means percentage to change DE variant, "NP" is the population size, "MaxEval" is the maximum number of evaluations, "*freq*" is the oscillation frequency of sinusoidal functions, "cp" is the descent speed of ε tolerance, and "Tc" is the percentage of iterations considering ε.

Variant	PCV	Dimension	NP	MaxEval	freq	cp	Tc
rand-best	90%	10D	90	200000	0.5	5	0.2
		30D	90	600000	0.5	5	0.2
best-rand	10%	10D	90	200000	0.5	5	0.2
		30D	90	600000	0.5	5	0.2

Five experiments were carried out:

(1) The 12 variants *fr*-DECV (rand-best) Conf1, ..., Conf6 and *fr*-DECV (best-rand) Conf1, ..., Conf6 were compared in the 36 test functions (with 10 and 30 dimensions) by computing the P measure.
(2) The 12 variants ε-DECV (rand-best) Conf1, ..., Conf6 and ε-DECV (best-rand) Conf1, ..., Conf6 were compared in the 36 test functions (with 10 and 30 dimensions) by computing P values.
(3) The two most competitive variants of the first experiment and the two most competitive variants of the second experiment were further compared by using P and AFES.
(4) The convergence behavior of the four variants of the third experiment was analyzed.
(5) The 24 variants were compared based on final feasible results by computing the 95%-confidence Kruskal-Wallis test and the Bonferroni post-hoc test.

4.1   Experiment 1

The results comparison based on the P measure of the six fr-DECV (rand-best) variants (one for each deterministic PCS) in 10D and 30D are shown in Tables 6 and 7, respectively. The most competitive configuration, out of the six compared, is located in the first column of the tables.

Table 6. Performance comparison based on the P measure of the six fr-DECV (rand-best) variants (Conf1, …, Conf6) in 10D.

Conf2	Conf1	Conf3	Conf4	Conf5	Conf6
wins (+)	5	4	7	4	4
losses (−)	1	2	0	2	2
ties (=)	12	12	11	12	12

Table 7. Performance comparison based on the P measure of the six fr-DECV (rand-best) variants (Conf1, …, Conf6) in 30D.

Conf2	Conf1	Conf3	Conf4	Conf5	Conf6
wins (+)	5	5	5	4	4
losses (−)	1	1	1	2	1
ties (=)	12	12	12	12	13

The number of wins reported in each column represents those test problems where a better P value was reached by the configuration located in the first column with respect to the configuration located in the corresponding column. The number of losses reported in each column represents those test problems were a worst P value was obtained by the configuration located in the first column with respect to the configuration located in the corresponding column. Finally, the number of ties indicates the test problems where similar P values were provided by the configuration in the corresponding column and the configuration located in the first column.

In the same way, the summary of results of the six fr-DECV (best-rand) variants is presented in Tables 8 and 9.

Table 8. Performance comparison based on the P measure of the six fr-DECV (best-rand) variants (Conf1, …, Conf6) in 10D.

Conf3	Conf1	Conf2	Conf4	Conf5	Conf6
wins (+)	5	4	7	5	6
losses (−)	2	2	1	2	1
ties (=)	11	12	10	11	11

Table 9. Performance comparison based on the P measure of the six *fr*-DECV (best-rand) variants (Conf1, ..., Conf6) in 30D.

Conf3	Conf1	Conf2	Conf4	Conf5	Conf6
wins (+)	4	3	5	4	4
losses (−)	2	2	1	2	2
ties (=)	12	13	12	12	12

From Tables 6 and 7 it can be observed that Conf2 provides a better performance. Such configuration is based on F and CR values starting with values close to 0.5 and an increasing oscillation between 0 and 1. Therefore, mutant vectors close and far from the base vector (oscillating F values) and trial vectors similar to its target vector and also similar to its mutant vector (oscillating CR values) seem to be more suitable for the feasibility rules CHT and a change from rand (exploration) to best (exploitation) DE variant.

It is quite interesting to note that, according to Tables 8 and 9, the opposite behavior promoted by Conf3 (decreasing oscillation starting with 0.5 for both parameters, F and CR) is suitable for the same CHT (feasibility rules) but with the opposite order between DECV versions (best first and rand later).

As a conclusion of this first experiment it was found that, for the feasibility rules which always favor feasible solutions over infeasible ones, an increasing oscillation of F and CR values is suitable for a switch between rand and best DE variants, and the opposite, i.e., decreasing oscillation of both parameter values, are suitable for a switch between best and rand DECV versions.

4.2   Experiment 2

The 12 variants now with the ε-constrained method as CHT were compared in a similar way as in Experiment 1. It is worth remarking that, unlike the feasibility rules, the ε-constrained method allows slightly infeasible solutions to be compared based on their objective function value, promoting then more diversity in the population (feasible solutions, but also infeasible solutions with good objective function values). The performance comparison summary of the six ε-DECV (rand-best) variants (Conf1, ..., Conf6) in 10D and 30D is presented in Tables 10 and 11, respectively. Meanwhile, the results of the six ε-DECV (best-rand) variants (Conf1, ..., Conf6) in 10D and 30D are presented in Tables 12 and 13, respectively.

Table 10. Performance comparison based on the P measure of the six ε-DECV (rand-best) variants (Conf1, ..., Conf6) in 10D.

Conf5	Conf1	Conf2	Conf3	Conf4	Conf6
wins (+)	8	7	7	13	6
losses (−)	7	6	7	4	4
ties (=)	3	5	4	1	8

Table 11. Performance comparison based on the P measure of the six ε-DECV (rand-best) variants (Conf1, ..., Conf6) in 30D.

Conf2	Conf1	Conf3	Conf4	Conf5	Conf6
wins (+)	6	4	6	6	6
losses (−)	2	3	2	1	3
ties (=)	10	11	10	11	9

Table 12. Performance comparison based on the P measure of the six ε-DECV (best-rand) variants (Conf1, ..., Conf6) in 10D.

Conf3	Conf1	Conf2	Conf4	Conf5	Conf6
wins (+)	9	7	9	7	7
losses (−)	7	4	6	5	5
ties (=)	2	7	9	6	6

Table 13. Performance comparison based on the P measure of the six ε-DECV (best-rand) variants (Conf1, ..., Conf6) in 30D.

Conf3	Conf1	Conf2	Conf4	Conf5	Conf6
wins (+)	7	5	7	6	6
losses (−)	1	3	2	2	2
ties (=)	10	10	9	10	10

It is worth noting that in Table 10 (10D test problems) the six configurations looked more competitive for the rand-best variant and it was hard to find one with a better performance (conf5, which fixes F = 0.5 and promotes an increasing oscillation for CR starting at 0.5). Moreover, in Table 11, Conf2 (F and CR values starting with values close to 0.5 and an increasing oscillation between 0 and 1) was the most competitive for the rand-best variant. On the other hand, in Tables 12 and 13 for the best-rand variant, the best configuration was conf3 (decreasing oscillation starting with 0.5 for both parameters, F and CR).

A finding of this second experiment about the ε-constrained method, which favors infeasible solutions with competitive objective function values to remain in the population, was the fact that, as in the case of Experiment 1, Conf3, which promotes a decreasing oscillation of F and CR values, was the most suitable for the best-rand variant. Another finding was the fact that an increasing oscillation of the CR parameter (as in Experiment 1) was more suitable for the rand-best variant. However, for 10D, fixing F = 0.5 provided better results, while for 30D the same increasing oscillation for F was more convenient.

4.3   Experiment 3

As a further assessment, the performances shown by the four most competitive algorithms of the previous two experiments by considering their P and AFES values are

Table 14. P and AFES values of the most competitive variants in 10D test problems. Best results are remarked in **boldface**.

Problem	Measure	fr-DECV (rand-best) Conf2	fr-DECV (best-rand) Conf3	ε-DECV (rand-best) Conf5	ε-DECV (best-rand) Conf3
c1	P	0.6	**0.92**	0.56	0.84
	AFES	4.22E+03	7.49E+03	4.59E+03	7.01E+03
c2	P	0	0	**0.24**	0
	AFES	0.00E+00	0.00E+00	6.65E+04	0
c3	P	0	0	**0.16**	0
	AFES	0.00E+00	0.00E+00	1.45E+05	0.00E+00
c4	P	**1**	0.72	**1**	**1**
	AFES	6.49E+04	8.58E+04	5.68E+04	7.00E+04
c5	P	0	0	0	0
	AFES	0.00E+00	0.00E+00	0.00E+00	0.00E+00
c6	P	0	0	0.08	0
	AFES	0.00E+00	0.00E+00	6.71E+04	0.00E+00
c7	P	0.8	**0.88**	0.76	**0.88**
	AFES	7.16E+04	5.22E+04	6.38E+05	5.07E+04
c8	P	0.32	0.24	0.24	**0.36**
	AFES	3.22E+04	4.41E+04	4.14E+04	4.49E+04
c9	P	0	0	**0.16**	0
	AFES	0.00E+00	0.00E+00	8.15E+04	0.00E+00
c10	P	0	0	0	0
	AFES	0.00E+00	0.00E+00	0.00E+00	0.00E+00
c11	P	**1**	**1**	0.96	**1**
	AFES	2.15E+04	2.12E+04	8.60E+04	5.90E+04
c12	P	0	0	**0.4**	0.32
	AFES	0.00E+00	0.00E+00	8.68E+04	1.07E+05
c13	P	0.08	0.24	0.72	**1**
	AFES	1.82E+04	1.69E+04	4.32E+04	4.60E+04
c14	P	0.44	0.64	**0.8**	**0.8**
	AFES	1.82E+05	1.74E+05	7.53E+04	7.91E+04
c15	P	0	0	0.04	**0.16**
	AFES	0.00E+00	0.00E+00	6.51E+04	6.52E+04
c16	P	0	0	**0.28**	0.24
	AFES	0.00E+00	0.00E+00	2.97E+04	3.06E+04
c17	P	0	0	0	0
	AFES	0.00E+00	0.00E+00	0.00E+00	0.00E+00
c18	P	0	0	**0.08**	0
	AFES	0.00E+00	0.00E+00	5.39E+04	0.00E+00

compared for 10D test problems in Table 14 and a summary based on the P values is provided in Table 15. A similar comparison for 30D test problems is provided in Table 16 with its summary in Table 17.

Table 15. Performance comparison based on the P measure of the four most competitive variants in 10D test problems.

ε− DECV (rand-best) Conf5	*fr*-DECV (rand-best) Conf2	*fr*-DECV (best-rand) Conf3	ε-DECV (best-rand) Conf3
wins (+)	10	11	7
losses (−)	4	3	6
ties (=)	4	4	5

The results in Tables 14, 15, 16 and 17 suggested that the algorithm with the most competitive performance was ε-DECV (rand-best) variant. Nevertheless, the configuration varied in 10D and in 30D, i.e., Conf5 was better in 10D, while Conf2 was better in 30D. Recalling from Table 1, Conf5 and Conf2 share the increasing oscillating behavior of CR. The difference is for parameter F, because Conf5 keeps it fixed at 0.5, while Conf2 starts it at 0.5 but promotes an increasing oscillating behavior between 0 and 1.

4.4   Experiment 4

Representative convergence plots for 10D and 30D test problems are presented in Figs. 2 and 3, respectively. In such plots, the behavior presented by each variant once the feasible zone was reached in test problem c07 is shown.

Based on such plots it can be observed that the behavior presented by each variant in 10D and 30D c07 test problem was very similar, i.e., the algorithms based on the DECV (rand-best) version showed a faster convergence than those based on DECV (best-rand) in both dimensions, regardless the CHT (the feasibility rules or the ε-constrained method) employed.

In the same way, it was found that Conf2, i.e., starting at 0.5 by F and CR, and an increasing oscillatory behavior between 0 and 1, regardless the CHT employed, provided the fastest convergence in the two fastest variants (*fr*-DECV (rand-best) in 10D and ε-DECV (rand-best) in 30D).

It is important to remark that the *fr*-DECV (rand-best) variant, despite not being the most competitive overall variant in this research, was benefited by Conf2 and the fastest convergence was promoted in 10D c07 test problem as shown in Fig. 3.

As a conclusion of this fourth experiment, it can be stated that the DECV version adopted, and not the CHT chosen, is responsible to promote a faster convergence, i.e., a (rand-best) version must be chosen if convergence must be reached in less time; otherwise a (best-rand) version must be adopted. Moreover, an incremental oscillatory behavior (Conf2 and even Conf5) was able to contribute to an even faster convergence in those two better variants based on convergence speed, regardless again of the CHT.

Table 16. P and AFES values of the most competitive variants in 30D test problems. Best results are remarked in **boldface**.

Problem	Measure	fr-DECV (rand-best) Conf2	fr-DECV (best-rand) Conf3	ε-DECV (rand-best) Conf2	ε-DECV (best-rand) Conf3
c1	P	0.04	0.04	**0.12**	0.08
	AFES	2.88E+04	7.31E+04	1.71E+04	1.00E+04
c2	P	0	0	0	0
	AFES	0.00E+00	0.00E+00	0.00E+00	0.00E+00
c3	P	0	0	0	0
	AFES	0.00E+00	0.00E+00	0.00E+00	0.00E+00
c4	P	**0.68**	0.36	0.36	0.64
	AFES	3.40E+05	4.31E+05	7.01E+04	3.13E+05
c5	P	0	0	0	0
	AFES	0.00E+00	0.00E+00	0.00E+00	0.00E+00
c6	P	0	0	0	0
	AFES	0.00E+00	0.00E+00	0.00E+00	0.00E+00
c7	P	**0.84**	0.68	0.68	0.6
	AFES	2.96E+05	2.18E+05	2.34E+05	2.92E+05
c8	P	0.48	**0.64**	0.56	0.44
	AFES	2.87E+05	2.55E+05	2.32E+05	2.81E+05
c9	P	0	0	0	0
	AFES	0.00E+00	0.00E+00	0.00E+00	0.00E+00
c10	P	0	0	0	0
	AFES	0.00E+00	0.00E+00	0.00E+00	0.00E+00
c11	P	**1**	**1**	**1**	**1**
	AFES	2.40E+04	2.14E+04	3.98E+04	3.58E+04
c12	P	**0.92**	0.64	0.04	0.48
	AFES	1.11E+05	1.85E+05	4.96E+05	2.02E+05
c13	P	0	0	**0.28**	0
	AFES	0.00E+00	0.00E+00	8.77E+04	0.00E+00
c14	P	0.56	0.6	**0.76**	0.56
	AFES	3.69E+05	3.94E+05	2.93E+05	2.76E+05
c15	P	0	0	**0.12**	0.04
	AFES	0.00E+00	0.00E+00	3.43E+05	3.57E+05
c16	P	0	0	**1**	**1**
	AFES	0.00E+00	0.00E+00	4.75E+04	4.39E+04
c17	P	0	0	0	0
	AFES	0.00E+00	0.00E+00	0.00E+00	0.00E+00
c18	P	0	0	0	0
	AFES	0.00E+00	0.00E+00	0.00E+00	0.00E+00

Table 17. Performance comparison based on the P measure of the four most competitive variants in 30D test problems.

ε-DECV (rand-best) Conf2	fr-DECV (rand-best) Conf2	fr-DECV (best-rand) Conf3	ε-DECV (best-rand) Conf3
wins (+)	6	5	6
losses (−)	3	2	2
ties (=)	9	11	10

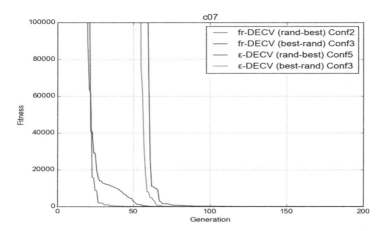

Fig. 2. Convergence plot for test problem c07 in 10D.

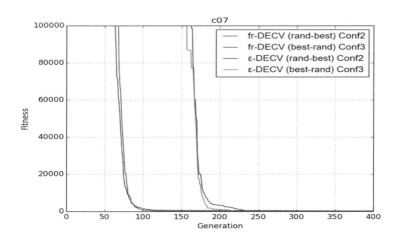

Fig. 3. Convergence plot for test problem c07 in 30D.

4.5   Experiment 5

The 24 variants designed were compared using the 95%-confidence Kruskal-Wallis test
with the Bonferroni post hoc test, based on final feasible objective function values in all
test problems. They were carried out by using the best feasible solution found by each
algorithm in those test functions where all 24 variants found at least one feasible
solution in a single execution. The results are shown in Figs. 4 and 5 for 10D and 30D,
respectively.

From Figs. 4 and 5 it can be observed that no significant differences based on final
feasible results among the 24 variants in 10D or 30D were found.

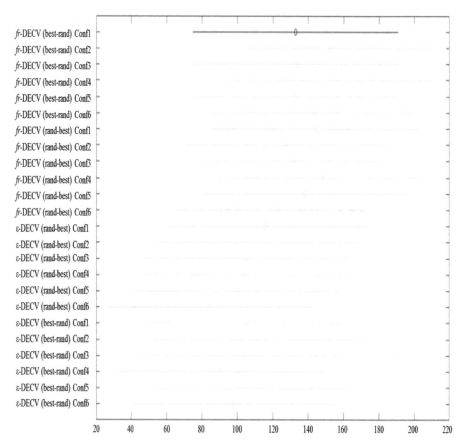

Fig. 4. 95%-confidence Kruskal-Wallis and Bonferroni post hoc tests for the 24 DECV variants
for CNOPs in 10D.

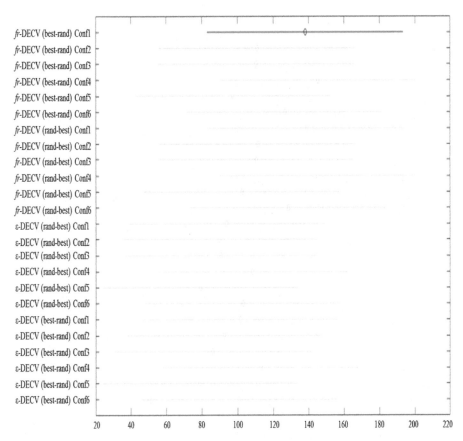

Fig. 5. 95%-confidence Kruskal-Wallis and Bonferroni post hoc tests for the 24 DECV variants for CNOPs in 30D.

5 Conclusions and Future Work

An analysis of the effects of six deterministic parameter control schemes based on a sinusoidal behavior with two constraint-handling techniques (feasibility rules and ε-constrained method) in the DECV algorithm to solve CNOPs was presented in this chapter. Two DECV versions (rand-best) and (best-rand) were considered as well. A total of 24 variants were assessed in 36 well-known benchmark problems (10D and 30D) by calculating the 95%-confidence Kruskal-Wallis and the Bonferroni post hoc tests, and by computing two performance measures based on the percentage of successful runs and the average number of evaluations in successful runs.

Five experiments were carried out (1) to compare the two DECV versions (rand-best) and (best-rand) with the six deterministic parameter control configurations with the feasibility rules, (2) to compare the two DECV versions (rand-best) and (best-rand) with the six deterministic parameter control configurations but now with the ε-constrained method, (3) to further compare the four best variants from the two previous experiments, (4) to analyze the convergence behavior of the four best variants, and (5) to compare the 24 variants designed using the best feasible solution found in each test function.

The overall results indicated that for both constraint-handling techniques an increasing oscillation of F and CR values is suitable for the (rand-best) DECV variants, while the opposite i.e., a decreasing oscillation of both parameter values, is suitable for (best-rand) DECV variant. Such finding is important because, for constrained search spaces, the rand DE variant (which is known for promoting exploration) is favored by F and CR values close to 0.5, while the best DE variant (which promotes exploitation) works better with more variations (between 0 and 1) in their F and CR values, according to the no-free-lunch theorems [27].

On the other hand, the remaining schemes (where one of the parameters is adapted and the other is fixed) had a less promising performance. In such schemes, the DPCS conf5 (where CR is adapted with an increasing oscillatory behavior and F is set at 0.5) showed the best performance, mainly coupled with the ε-DECV (rand-best) version in 10D test problems.

Regarding the most competitive variant, ε-DECV (rand-best) provided the most robust performance, mainly with an increasing oscillatory behavior for F and CR, both starting with a value of 0.5. This is relevant because the original DECV algorithm uses the feasibility rules as constraint-handling technique.

Finally, the convergence behavior analysis indicated that the increasing oscillatory behavior for F and CR, coupled with the DECV (rand-best) version, and regardless of the CHT, led to a faster convergence.

Derived from this research, the following future work is proposed: (1) controlling other related parameters like NP of DECV, or cp and Tc of the ε-constrained method, (2) analyzing the effect of the *freq* parameter required by the six configurations adopted in this work, (3) adding the increasing oscillating behavior for F and CR in other DE-based algorithms and other evolutionary approaches with similar properties (e.g., Particle Swarm Optimization, Artificial Bee Colony, etc.) for CNOPs, and (4) testing the opposite trends for F and CR based on the original six proposals analyzed in this chapter.

Acknowledgments. The first author acknowledges support from the Mexican Council for Science and Technology (CONACyT) through a scholarship to pursue graduate studies at the University of Veracruz. The third author acknowledges support from CONACyT through project No. 220522.

References

1. Garey, M.R., Johnson, D.S.: Computers and Intractability: A Guide to the Theory of np-Completeness (1979)
2. Schütze, O., Alvarado, S., Segura, C., Landa, R.: Gradient subspace approximation: a direct search method for memetic computing. Soft. Comput. **21**(21), 6331–6350 (2017)

3. Alvarado, S., Segura, C., Schütze, O.: The gradient subspace approximation as local search engine within evolutionary multi-objective optimization algorithms. Computación y Sistemas (to appear)
4. Deb, K.: Optimization for Engineering Design: Algorithms and Examples. Prentices-Hall of India, New Delhi (2000)
5. Eiben, A.E., Smith, J.E., et al.: Introduction to Evolutionary Computing, vol. 53. Springer, Berlin (2003)
6. Yang X.S.: Nature-Inspired Metaheuristic Algorithms. Luniver Press (2008)
7. Dréo, J., Pétrowski, A., Siarry, P., Taillard, E.: Metaheuristics for Hard Optimization: Methods and Case Studies. Springer Science & Business Media, Berlin (2006)
8. Smit, S.K., Eiben, A.E.: Comparing parameter tuning methods for evolutionary algorithms. In: IEEE Congress on Evolutionary Computation (CEC 2009), pp. 399–406 (2009)
9. Karafotias, G., Hoogendoorn, M., Eiben, A.E.: Parameter control in evolutionary algorithms: trends and challenges. IEEE Trans. Evol. Comput. **19**(2), 167–187 (2015)
10. Storn, R., Price, K.: Differential evolution – a simple and efficient heuristic for global optimization over continuous spaces. J. Global Optim. **11**(4), 341–359 (1997)
11. Qin, A.K., Huang, V.L., Suganthan, P.N.: Differential evolution algorithm with strategy adaptation for global numerical optimization. IEEE Trans. Evol. Comput. **13**(2), 398–417 (2009)
12. Brest, J., Greiner, S., Boskovic, B., Mernik, M., Zumer, V.: Self-adapting control parameters in differential evolution: a comparative study on numerical benchmark problems. IEEE Trans. Evol. Comput. **10**(6), 646–657 (2006)
13. Mezura-Montes, E., Coello-Coello, C.A.: Constraint-handling in nature-inspired numerical optimization: past, present and future. Swarm Evolut. Comput. **1**(4), 173–194 (2011)
14. Coello-Coello, C.A.: Theoretical and numerical constraint-handling techniques used with evolutionary algorithms: a survey of the state of the art. Comput. Methods Appl. Mech. Eng. **191**(11–12), 1245–1287 (2002)
15. Deb, K.: An efficient constraint handling method for genetic algorithms. Comput. Methods Appl. Mech. Eng. **186**(2–4), 311–338 (2000)
16. Takahama, T., Sakai, S.: Constrained optimization by the ε constrained differential evolution with gradient-based mutation and feasible elites. In: IEEE Congress on Evolutionary Computation (2006)
17. Mezura-Montes, E., Miranda-Varela, M.E., del Carmen Gómez-Ramón, R.: Differential evolution in constrained numerical optimization: an empirical study. Inf. Sci. **180**(22), 4223–4262 (2010)
18. Huang, V.L., Qin, A.K., Suganthan, P.N.: Self-adaptive differential evolution algorithm for constrained real-parameter optimization. In: IEEE Congress on Evolutionary Computation (CEC 2006), pp. 17–24 (2006)
19. Mezura-Montes, E., Palomeque-Ortiz, A.G.: Parameter control in differential evolution for constrained optimization. In: IEEE Congress on Evolutionary Computation (CEC 2009), pp. 1375–1382 (2009)
20. Kukkonen, S., Coello-Coello, C.A.: Applying exponential weighting moving average control parameter adaptation technique with generalized differential evolution. In: IEEE Congress on Evolutionary Computation (CEC 2016), pp. 4755–4762 (2016)
21. Domínguez-Isidro, S., Mezura-Montes, E., Leguizamón, G.: Memetic differential evolution for constrained numerical optimization problems. In: IEEE Congress on Evolutionary Computation (CEC 2013), pp. 2996–3003 (2013)
22. Elsayed, S., Sarker, R., Coello-Coello, C.A.: Enhanced multi-operator differential evolution for constrained optimization. In: IEEE congress on evolutionary computation (CEC 2016), pp. 4191–4198 (2016)

23. Mallipeddi, R., Suganthan, P.N.: Differential evolution with ensemble of constraint handling techniques for solving CEC 2010 benchmark problems. In: IEEE Congress on Evolutionary Computation (CEC 2010), pp. 1–8 (2010)
24. Draa, A., Bouzoubia, S., Boukhalfa, I.: A sinusoidal differential evolution algorithm for numerical optimization. Appl. Soft Comput. **27**, 99–126 (2015)
25. Suganthan, P.N., Mallipeddi, R.: Problem definitions and evaluation criteria for the CEC 2010 competition on constrained real-parameter optimization. In: Technical report, Nanyang Technological University, Singapore (2010)
26. Mezura-Montes, E., Cetina-Domínguez, O.: Empirical analysis of a modified artificial bee colony for constrained numerical optimization. Appl. Math. Comput. 10943–10973 (2012)
27. Wolpert, D.H., Macready, W.G.: No free lunch theorems for optimization. IEEE Trans. Evol. Comput. (1997)

A Descent Method for Equality and Inequality Constrained Multiobjective Optimization Problems

Bennet Gebken$^{(\boxtimes)}$, Sebastian Peitz, and Michael Dellnitz

Department of Mathematics, Paderborn University,
Warburger Str. 100, 33098 Paderborn, Germany
{bgebken,speitz}@math.upb.de,dellnitz@uni-paderborn.de

Abstract. In this article we propose a descent method for equality and inequality constrained multiobjective optimization problems (MOPs) which generalizes the steepest descent method for unconstrained MOPs by Fliege and Svaiter to constrained problems by using two active set strategies. Under some regularity assumptions on the problem, we show that accumulation points of our descent method satisfy a necessary condition for local Pareto optimality. Finally, we show the typical behavior of our method in a numerical example.

Keywords: Multiobjective optimization · Constrained optimization
Descent method · Active set strategy

1 Introduction

In many problems we face in reality, there are multiple objectives that have to be optimized at the same time. In production for example, one often wants to minimize the cost of a product but also maximize its quality. When the objectives we want to optimize are conflicting (like in the above example), classical scalar-valued optimization methods are not suited. In this case, there is no single point that is optimal for all objectives at the same time. This is the motivation for *multiobjective optimization*.

A general *multiobjective optimization problem* (MOP) consists of m objective functions $F_i : \mathcal{N} \to \mathbb{R}$, where $i \in \{1, ..., m\}$ and $\mathcal{N} \subseteq \mathbb{R}^n$. The solution of this problem is the *Pareto set*, containing all *Pareto optimal points*. Roughly speaking, a point is Pareto optimal if it can not be improved in all objective functions simultaneously. The goal in multiobjective optimization is to find the Pareto set of a given MOP.

For unconstrained MOPs (i.e. $\mathcal{N} = \mathbb{R}^n$), there are multiple ways to compute the Pareto set. A popular approach is to scalarize the MOP by weighting and summarizing the objective functions to find single optimal points (see, e.g., [10,21]). A different scalarization approach calculates an approximation of the whole Pareto front directly by optimizing a unary performance indicator on the

© Springer International Publishing AG, part of Springer Nature 2019
L. Trujillo et al. (Eds.): NEO 2017, SCI 785, pp. 29–61, 2019.
https://doi.org/10.1007/978-3-319-96104-0_2

set of all possible approximations (with a fixed number of points) (see [29, 30]). Another widely used approach is based on heuristic optimization and results in evolutionary methods [5, 6, 27]. In the case where the Pareto set is a connected manifold, it can be computed using continuation methods [26]. Some methods for scalar-valued problems can be generalized to MOPs. Examples are the steepest descent method [13], the Newton method [12] and the trust-region method [4]. Finally, set-oriented methods can be applied to compute a covering of the global Pareto set [8, 28].

There also exist gradient-based methods which can handle both unconstrained MOPs as well as certain classes of constraints. If the MOP is constrained to a closed and convex set, it is possible to use a projected gradient method [9]. For more general inequality constraints, it is possible to use the steepest descent method described in [13, Sect. 8]. For equality constrained MOPs where the feasible set is a (Riemannian) manifold, it is possible to use the steepest descent method described in [3]. Until recently, the consideration of MOPs with equality and inequality constraints was limited to heuristic methods [22] and scalarization methods (see, e.g., [11]). In [20], a continuation method for equality and box-constrained MOPs was proposed together with a constrained version of the Newton method [12]. In 2016, Fliege and Vaz proposed a method to compute the whole Pareto set of general equality and inequality constrained MOPs that is based on SQP-techniques [14]. Their method operates on a finite set of points in the search space and has two stages: In the first stage, the set of points is initialized and iteratively enriched by nondominated points and in the second stage, the set of points is driven to the actual Pareto set.

The goal of this article is to extend the steepest descent method in [13] to the constrained case by using two different *active set strategies* and an adjusted step length. In active set strategies, inequality constraints are either treated as equality constraints (if they are close to 0) or neglected. This approach essentially combines the descent method on manifolds [3] with the descent method for inequality constrained problems described in [13].

In contrast to [14], the descent methods we propose compute single Pareto optimal (or critical) points rather than an approximation to the whole Pareto set. However, similar to the approach in [8], it is possible to interpret our descent methods as discrete dynamical systems and use set-oriented methods (see, e.g., [7]) to calculate their global attractors which contain the Pareto set. It is also possible to use evolutionary approaches to globalize our method.

The outline of the article is as follows. In Sect. 2 we give a short introduction to multiobjective optimization and the steepest descent method for unconstrained MOPs. In Sect. 3 we begin by generalizing this method to equality constraints and then proceed with the equality and inequality constrained case. We prove convergence for both cases. In Sect. 4 we apply our method to an example to show its typical behavior and discuss ways to optimize it. In Sect. 5 we draw a conclusion and discuss future work.

2   Multiobjective Optimization

The goal of (unconstrained) continuous multiobjective optimization is to minimize an objective function

$$F : \mathbb{R}^n \to \mathbb{R}^m, \quad x = (x_1, ..., x_n) \mapsto (F_1(x), ..., F_m(x)).$$

Except in the case where all F_i have the same minima, the definition of optimality from scalar-valued optimization does not apply. This is due to the loss of the total order for $m \geq 2$. We thus introduce the notion of *dominance*.

Definition 1. Let $v, w \in \mathbb{R}^m$.

1. $v \leq w :\Leftrightarrow v_i \leq w_i \; \forall i \in \{1, ..., m\}$. Define $<$, $\geq$ and $>$ analogously.
2. v *dominates* w, if $v \leq w$ and $v_i < w_i$ for some $i \in \{1, ..., m\}$.

The dominance relation defines a partial order on $\mathbb{R}^m$ that is not total, so we generally can not expect to find a minimum (or infimum) of $F(\mathbb{R}^n)$ with respect to that order. Instead, the "minimum" of $F(\mathbb{R}^n)$ that we are looking for is given by a set, the so-called *Pareto set*, which is defined as follows.

Definition 2

1. $x \in \mathbb{R}^n$ is *locally Pareto optimal* if there is a neighborhood $U \subseteq \mathbb{R}^n$ of x such that

$$\nexists y \in U : F(y) \text{ dominates } F(x). \tag{1}$$

The set of locally Pareto optimal points is called the *local Pareto set*.
2. $x \in \mathbb{R}^n$ is *globally Pareto optimal* if (1) holds for $U = \mathbb{R}^n$. The set of globally Pareto optimal points is called the *global Pareto set*.
3. The *local (global) Pareto front* is the image of the local (global) Pareto set under F.

Similar to the unbounded case in scalar-valued optimization, the Pareto set can be empty, which means that the MOP has no solution. Note that the well-known Karush-Kuhn-Tucker (KKT) conditions from scalar-valued optimization can be generalized to multiobjective optimization (see [17], Theorem 4.1).

Minimizing F can now be defined as finding the Pareto set of F. The minimization of F on a subset $\mathcal{N} \subseteq \mathbb{R}^n$ is defined the same way by replacing $\mathbb{R}^n$ in Definition 2 with $\mathcal{N}$. For the constrained case, a point $x \in \mathbb{R}^n$ is called *feasible* if $x \in \mathcal{N}$. In this paper we consider the case where $\mathcal{N}$ is given by a number of equality and inequality constraints. Thus, the general MOP we consider is of the form

$$\begin{aligned} \min_{x \in \mathbb{R}^n} \quad & F(x), \\ \text{s.t.} \quad & H(x) = 0, \\ & G(x) \leq 0, \end{aligned} \tag{MOP}$$

where $F : \mathbb{R}^n \to \mathbb{R}^m$, $G : \mathbb{R}^n \to \mathbb{R}^{m_G}$ and $H : \mathbb{R}^n \to \mathbb{R}^{m_H}$ are continuously differentiable.

For unconstrained MOPs, Fliege and Svaiter have proposed a descent method in [13] which we will now briefly summarize. Starting at a given point $x_0 \in \mathbb{R}^n$, the method generates a sequence $(x_l)_l \in \mathbb{R}^n$ with

$$F(x_{l+1}) < F(x_l) \quad \forall l \geq 0.$$

As the first-order necessary condition for local Pareto optimality they use

$$\text{im}(DF(x)) \cap (\mathbb{R}^{<0})^m = \emptyset,$$

which is equivalent to

$$\nexists v \in \mathbb{R}^n : \nabla F_i(x)v < 0 \quad \forall i \in \{1, ..., m\},$$

where DF is the Jacobian matrix of F, $\text{im}(DF(x))$ is the image of the matrix $DF(x)$ and ∇F_i is the gradient of F_i. Points $x \in \mathbb{R}^n$ satisfying this condition are called *Pareto critical*. If x is not Pareto critical, then $v \in \mathbb{R}^n$ is called a *descent direction in* x if $\nabla F_i(x)v < 0$ for all $i \in \{1, ..., m\}$. Such a descent direction can be obtained via the following subproblem:

$$\min_{(v,\beta) \in \mathbb{R}^{n+1}} \quad \beta + \tfrac{1}{2}\|v\|^2, \qquad\qquad (\text{SP})$$
$$\text{s.t.} \quad\quad \nabla F_i(x)v \leq \beta \quad \forall i \in \{1, ..., m\}.$$

By $\alpha(x)$ and $v(x)$ we denote the optimal value and optimal v of (SP) for a fixed x. As shown in [13], problem (SP) has the following properties.

Lemma 1

1. (SP) *has a unique solution.*
2. *If x is Pareto critical, then $v(x) = 0$ and $\alpha(x) = 0$.*
3. *If x is not Pareto critical, then $\alpha(x) < 0$.*
4. $x \mapsto v(x)$ *and* $x \mapsto \alpha(x)$ *are continuous.*

Thus, x is Pareto critical iff $\alpha(x) = 0$. We want to use this descent method in a line search approach [23] and to this end choose a step length

$$t(x, v) = \max \left\{ s = \frac{1}{2^k} : k \in \mathbb{N}, \quad F(x + sv) < F(x) + \sigma s DF(x)v \right\}$$

for some $\sigma \in (0, 1)$. Using the descent direction and the step length described above, the sequence $(x_l)_l$ is calculated via the scheme

$$x_{l+1} = x_l + t(x_l, v(x_l))v(x_l) \qquad\qquad (2)$$

for $l \geq 0$. As a convergence result the following was shown.

Theorem 1 ([13]). *Let $(x_l)_l$ be a sequence generated by the descent method described above. Then every accumulation point of $(x_l)_l$ is Pareto critical.*

For $m = 1$ (i.e. scalar-valued optimization), this method is reduced to the method of steepest descent where the step length satisfies the Armijo rule (see, e.g., [23]). In what follows, we will generalize this approach to constrained MOPs.

3   Descent Methods for the Constrained Case

In this section we propose two descent methods for constrained MOPs and thereby extend the results of [13]. We first define a descent direction and a step length for MOPs with equality constraints (ECs) and then use two different active set strategies to incorporate inequality constraints (ICs). As will be seen in this section, in the generic constrained case the feasible set of (MOP) has the structure of a differentiable (Riemannian) manifold with a piecewise differentiable boundary. This will be used to show well-definedness and convergence of our methods. To this end, we need some concepts from differential geometry, for which we refer to [19]. We handle ECs similar to [3] where the exponential map from Riemannian geometry is used to locally obtain new feasible points along a geodesic in a given (tangent) direction. Since in general, geodesics are obtained by solving an ordinary differential equation, it is not efficient to evaluate the exponential map by calculating geodesics. Instead, we will use *retractions* [1] that can be thought of as first order approximations of the exponential map.

3.1   Equality Constraints

Our approach for handling ECs is to explore the feasible set given by them via predictor-corrector type methods [2]. In the predictor step, we choose a descent direction along which the ECs are only mildly violated, and perform a step in that direction with a proper step length. In the corrector step, the resulting point is mapped onto the set satisfying the ECs. To this end, we choose a descent direction lying in the tangent space of the set given by the ECs. To ensure the existence of these tangent spaces, we make the following assumptions on F and H.

(A1) $F : \mathbb{R}^n \to \mathbb{R}^m$ is C^2 (two times continuously differentiable).
(A2) $H : \mathbb{R}^n \to \mathbb{R}^{m_H}$ is C^2 with regular value 0 (i.e. $\mathrm{rk}(DH(x)) = m_H$ for all $x \in \mathbb{R}^n$ with $H(x) = 0$, where $rk(DH(x))$ is the rank of $DH(x)$).

The assumption (A2) is also known as the *linear independence constraint qualification* (LICQ) and is a commonly used constraint qualification (see, e.g., [23]). Let $\mathscr{H} := H^{-1}(\{0\})$ be the set of points satisfying the ECs. According to the Level Set Theorem ([24], Example 1.6.4), $\mathscr{H}$ is a closed, $(n - m_H)$-dimensional C^2-submanifold of $\mathbb{R}^n$. It is Riemannian with the inner product $(v, w) \mapsto v^T w$. The tangent space in $x \in \mathscr{H}$ is $T_x \mathscr{H} = \ker(DH(x))$ and the tangent bundle is

$$T\mathscr{H} = \bigcup_{x \in \mathscr{H}} \{x\} \times \ker(DH(x)),$$

where $\ker(DH(x))$ is the kernel of $DH(x)$. Consequently, if we consider (MOP) without ICs, we may also write

$$\min_{x \in \mathscr{H}} F(x), \tag{Pe}$$

Before generalizing the unconstrained descent method, we first have to extend the definition of Pareto criticality to the equality constrained case.

Definition 3. A point $x \in \mathcal{H}$ is *Pareto critical*, if

$$\nexists v \in \ker(DH(x)) : DF(x)v < 0.$$

This means that x is Pareto critical iff it is feasible and there is no descent direction in the tangent space of $\mathcal{H}$ in x. We will show later (in Lemma 6) that this is indeed a first-order necessary condition for local Pareto optimality. We now introduce a modification of the subproblem (SP) to obtain descent directions of F in the tangent space of $\mathcal{H}$.

$$\min_{(v,\beta)\in\mathbb{R}^{n+1}} \quad \beta + \tfrac{1}{2}\|v\|^2$$
$$\text{s.t.} \qquad \nabla F_i(x)v \leq \beta \quad \forall i \in \{1,...,m\}, \qquad \text{(SPe)}$$
$$\nabla H_j(x)v = 0 \quad \forall j \in \{1,...,m_H\}.$$

An equivalent formulation is

$$\min_{v\in\ker(DH(x))} \left(\max_i \nabla F_i(x)v + \frac{1}{2}\|v\|^2 \right) =: \alpha(x).$$

Since $v = 0$ (or $(v,\beta) = (0,0)$) is always a feasible point for this subproblem, we have $\alpha(x) \leq 0$ for all $x \in \mathcal{H}$. In the next three lemmas we generalize some results about the subproblem (SP) from the unconstrained case to (SPe).

Lemma 2. $x \in \mathcal{H}$ is Pareto critical iff $\alpha(x) = 0$.

Proof. Assume that x is not Pareto critical, so there is some $v \in \ker(DH(x))$ with $DF(x)v < 0$. In particular, we have $\max_i \nabla F_i(x)v < 0$. For $s > 0$ consider the scaled vector sv. Then we have the equivalence

$$\max_i \nabla F_i(x)sv + \frac{1}{2}\|sv\|^2 < 0 \quad \Leftrightarrow \quad s\max_i \nabla F_i(x)v + \frac{1}{2}s^2\|v\|^2 < 0$$

$$\Leftrightarrow \quad \max_i \nabla F_i(x)v + \frac{1}{2}s\|v\|^2 < 0.$$

As $\max_i \nabla F_i(x)v < 0$, there has to be some $s > 0$ so that the last inequality holds. This shows that there is at least one feasible point of (SPe) where the objective function of the subproblem is negative, so $\alpha(x) < 0$.

Now assume that $\alpha(x) < 0$. This means that there has to be some $v \in \ker(DH(x))$ with

$$\max_i \nabla F_i(x)v + \frac{1}{2}\|v\|^2 < 0.$$

In particular, $\max_i \nabla F_i(x)v < 0$, so $DF(x)v < 0$. Thus, x is not Pareto critical, which completes the proof. □

Lemma 2 shows that (SPe) can indeed be used to obtain a descent direction v in the tangent space if x is not Pareto critical. The next lemma will show that the solution of (SPe) is a unique convex combination of the projected gradients of F_i onto the tangent space of $\mathcal{H}$.

Lemma 3

1. (SPe) *has a unique solution.*
2. $v \in \mathbb{R}^n$ *solves* (SPe) *if there exist* $\lambda_i \geq 0$ *for* $i \in I(x,v)$ *so that*

$$v = -\sum_{i \in I(x,v)} \lambda_i \nabla_{\mathscr{H}} F_i(x), \qquad \sum_{i \in I(x,v)} \lambda_i = 1,$$

where

$$I(x,v) := \{i \in \{1, ..., m\} : \nabla F_i(x)v = \max_{i \in \{1,...,m\}} \nabla F_i(x)v\}.$$

$\nabla_{\mathscr{H}} F_i(x)$ *denotes the gradient of* F_i *as a function on* $\mathscr{H}$, *i.e.* $\nabla_{\mathscr{H}} F_i(x)$ *is the projection of* $\nabla F_i(x)$ *onto* $T_x \mathscr{H} = \ker(DH(x))$.

Proof. The first result follows from the strict convexity of the objective function in the equivalent formulation of (SPe). The second result follows from theoretical results about subdifferentials. A detailed proof is shown in [3, Lemma 4.1]. □

By the last lemma, the function $v : \mathscr{H} \to \ker(DH(x))$ which maps $x \in \mathscr{H}$ to the descent direction given by (SPe) is well-defined. The following lemma shows that it is continuous.

Lemma 4. *The maps* $v : \mathscr{H} \to \ker(DH(x))$ *and* $\alpha : \mathscr{H} \to \mathbb{R}$ *are continuous.*

Proof. The continuity of v is shown in [3, Lemma 5.1]. The continuity of α can be seen when decomposing α into the two following continuous maps:

$$\mathscr{H} \to \mathscr{H} \times \mathbb{R}^n, \quad q \mapsto (q, v(q)),$$

$$\mathscr{H} \times \mathbb{R}^n \to \mathbb{R}, \quad (q, w) \mapsto \max_i \nabla F_i(q)w + \frac{1}{2}\|w\|^2.$$

□

Similar to [13], it does not matter for the convergence theory if we take the exact solution of (SPe) or an inexact solution in the following sense.

Definition 4. An *approximate solution of* (SPe) *at* $x \in \mathscr{H}$ *with tolerance* $\gamma \in (0, 1]$ *is a* $v' \in \mathbb{R}^n$ *such that*

$$v' \in \ker(DH(x)) \quad \text{and} \quad \max_i \nabla F_i(x)v' + \frac{1}{2}\|v'\|^2 \leq \gamma \alpha(x).$$

If $x \in \mathscr{H}$ is not Pareto critical, we obviously still have $DF(x)v' < 0$ for all approximate solutions v'. Thus, approximate solutions are still descent directions in the tangent space. By setting the tolerance γ to 1, we again obtain the exact solution. For the relative error δ of the optimal value α of (SPe) we get

$$\delta(x) := \frac{|\max_i \nabla F_i(q)v' + \frac{1}{2}\|v'\|^2 - \alpha(x)|}{|\alpha(x)|} \leq \frac{|\gamma\alpha(x) - \alpha(x)|}{|\alpha(x)|} = 1 - \gamma.$$

By solving (SPe), we can now compute descent directions in the tangent space of $\mathscr{H}$ at a feasible point $x \in \mathscr{H}$. In order to show that our method generates a decreasing sequence of feasible points, we will need some properties of the corrector step. To this end, we introduce a retraction map on $\mathscr{H}$ in the following definition.

Definition 5. For $y \in \mathbb{R}^n$ consider the set

$$P_{\mathscr{H}}(y) := \{x \in \mathscr{H} : \|x - y\| = \min_{z \in \mathscr{H}} \|z - y\|\}.$$

According to [1, Lemma 3.1], for every $x \in \mathscr{H}$ there exists some $\epsilon > 0$ such that $P_{\mathscr{H}}(y)$ contains only one element for all $y \in U_\epsilon(x) := \{y \in \mathbb{R}^n : \|y - x\| < \epsilon\}$. We can thus define the map

$$\pi : \mathscr{V} \to \mathscr{H}, \quad y \mapsto P_{\mathscr{H}}(y)$$

in a neighborhood $\mathscr{V}$ of $\mathscr{H}$. It was also shown that this map is C^1. According to [1, Proposition 3.2], the map

$$R_\pi : T\mathscr{H} \to \mathscr{H}, \quad (x, v) \mapsto \pi(x + v)$$

is a C^1-*retraction on* $\mathscr{H}$, i.e. for all $x \in \mathscr{H}$ there is a neighborhood $\mathscr{U}$ of $(x, 0) \in T\mathscr{H}$ such that

1. the restriction of R_π onto $\mathscr{U}$ is C^1,
2. $R_\pi(q, 0) = q \quad \forall (q, 0) \in \mathscr{U}$,
3. $DR_\pi(q, \cdot)(0) = \mathrm{id}_{\ker(DH(q))} \quad \forall (q, 0) \in \mathscr{U}$.

(More precisely, it was shown that π is C^{k-1} and R_π is a C^{k-1}-retraction on $\mathscr{H}$ if $\mathscr{H}$ is C^k.) Here, $\mathrm{id}_{\ker(DH(q))}$ is the identity map on $\ker(DH(q))$. In all $y \in \mathbb{R}^n$ where π is undefined, we set $\pi(y)$ to be some element of $P_{\mathscr{H}}(y)$. This will not matter for our convergence results since we only need the retraction properties when we are near $\mathscr{H}$.

Using v and π, we can now calculate proper descent directions for the equality constrained case. As in all line search strategies, we furthermore have to choose a step length which assures that the new feasible point is an improvement over the previous one and that the resulting sequence converges to a Pareto critical point. To this end – for $x \in \mathscr{H}$ and $v \in \ker(DH(x))$ with $DF(x)v < 0$ given – consider the step length

$$t = \beta_0 \beta^k, \tag{3}$$

where

$$k := \min\{k \in \mathbb{N} : F(\pi(x + \beta_0 \beta^k v)) < F(x) + \sigma \beta_0 \beta^k DF(x)v\}$$

with $\beta_0 > 0$, $\beta \in (0, 1)$ and $\sigma \in (0, 1)$. To show that such a k always exists, we first require the following lemma.

Lemma 5. *Let* $n_1, n_2 \in \mathbb{N}$, $U \subseteq \mathbb{R}^{n_1}$ *open and* $f : U \to \mathbb{R}^{n_2}$ C^2. *Let* $(x_k)_k \in U$, $(v_k)_k \in \mathbb{R}^{n_1}$ *and* $(t_k)_k \in \mathbb{R}^{>0}$ *such that* $\lim_{k \to \infty} x_k = x \in U$, $\lim_{k \to \infty} v_k = v$ *and* $\lim_{k \to \infty} t_k = 0$. *Then*

$$\lim_{k \to \infty} \frac{f(x_k + t_k v_k) - f(x_k)}{t_k} = Df(x)v.$$

If additionally $Df(x)v < 0$, then for all $\sigma \in (0,1)$ there exists some $K \in \mathbb{N}$ such that

$$f(x_k + t_k v_k) < f(x_k) + \sigma t_k Df(x)v \quad \forall k \geq K.$$

Proof. The proof follows by considering the Taylor series expansion of f. □

With the last lemma we can now show the existence of the step length (3).

Lemma 6. *Let $v \in ker(DH(x))$ with $DF(x)v < 0$. Let $\beta_0 > 0$, $\beta \in (0,1)$ and $\sigma \in (0,1)$. Then*

$$\{k \in \mathbb{N} : F(\pi(x + \beta_0 \beta^k v)) < F(x) + \sigma \beta_0 \beta^k DF(x)v\} \neq \emptyset \qquad (4)$$

and there exists some $K \in \mathbb{N}$ such that this set contains all $k > K$. Particularly

$$\lim_{k \to \infty} \frac{F(\pi(x + \beta_0 \beta^k v)) - F(x)}{\beta_0 \beta^k} = DF(x)v.$$

Proof. We use Lemma 5 with $x_k = x$, $t_k = \beta_0 \beta^k$ and $v_k = \frac{\pi(x + \beta_0 \beta^k v) - x}{\beta_0 \beta^k}$. We have to show that

$$\lim_{k \to \infty} \frac{\pi(x + \beta_0 \beta^k v) - x}{\beta_0 \beta^k} = v.$$

Since R_π is a retraction, we have

$$\lim_{k \to \infty} \frac{\pi(x + \beta_0 \beta^k v) - x}{\beta_0 \beta^k} = \lim_{k \to \infty} \frac{R_\pi(x, \beta_0 \beta^k v) - R_\pi(x, 0)}{\beta_0 \beta^k} = DR_\pi(x, \cdot)(0)(v) = v.$$

□

The inequality in the set (4) is called the *Armijo inequality*. In particular, the last lemma shows that Pareto criticality (according to Definition 3) is a necessary condition for Pareto optimality since $\pi(x + \beta_0 \beta^k v) \to x$ for $k \to \infty$ and $F(\pi(x + \beta_0 \beta^k v)) < F(x)$ for k large enough if v is a descent direction.

The descent method for equality constrained MOPs is summarized in Algorithm 1. To obtain a feasible initial point $x_0 \in \mathcal{H}$, we evaluate π at the initial point $x \in \mathbb{R}^n$.

Similar to [3, Theorem 5.1], we have the following convergence result.

Theorem 2. *Let $(x_l)_l \in \mathcal{H}$ be a sequence generated by Algorithm 1. Then $(x_l)_l$ is either finite (and the last element of the sequence is Pareto critical) or each accumulation point of $(x_l)_l$ is Pareto critical.*

Proof. A sequence generated by Algorithm 1 can only be finite if $\alpha(x_l) = 0$, which means x_l is Pareto critical. Let $(x_l)_l$ from now on be infinite, so no element of the sequence is Pareto critical.

Algorithm 1. (Descent method for equality constrained MOPs)

Require: $x \in \mathbb{R}^n$, $\gamma \in (0,1]$, $\beta_0 > 0$, $\beta \in (0,1)$, $\sigma \in (0,1)$.
1: Compute $x_0 = \pi(x)$.
2: **for** $l = 0, 1, \ldots$ **do**
3:     Compute an approximate solution v_l of (SPe) at x_l with tolerance γ.
4:     **if** x_l is Pareto critical (i.e. $\alpha(x_l) = 0$) **then**
5:         Stop.
6:     **else**
7:         Compute t_l as in (3) for x_l and v_l with parameters β_0, β and σ.
8:         Set $x_{l+1} = \pi(x_l + t_l v_l)$.
9:     **end if**
10: **end for**

Let $\bar{x} \in \mathcal{H}$ be an accumulation point of $(x_l)_l$. By construction of Algorithm 1, each component of $(F(x_l))_l$ is monotonically decreasing. Therefore, since F is continuous, we have $\lim_{l \to \infty} F(x_l) = F(\bar{x})$. By our choice of the descent direction and the step length (3) we have

$$F(x_{l+1}) - F(x_l) < \sigma t_l DF(x)v_l < 0 \quad \forall l \in \mathbb{N}.$$

It follows that

$$\lim_{l \to \infty} t_l DF(x_l)v_l = 0.$$

Let $(l_s)_s \in \mathbb{N}$ be a sequence of indices so that $\lim_{s \to \infty} x_{l_s} = \bar{x}$. We have to consider the following two possibilities:

1. $\limsup_{l \to \infty} t_l > 0$
2. $\lim_{l \to \infty} t_l = 0$

Case 1: It follows that

$$\lim_{l \to \infty} DF(x_l)v_l = 0,$$

and particularly

$$\lim_{l \to \infty} \max_i \nabla F_i(x_l)v_l = 0.$$

Since v_l is an approximate solution for some $\gamma \in (0,1]$, we have

$$\max_i \nabla F_i(x_l)v_l + \frac{1}{2}\|v_l\|^2 \le \gamma \alpha(x_l) < 0 \quad \forall l \in \mathbb{N},$$

and it follows that

$$\lim_{l \to \infty} \left(\max_i \nabla F_i(x_l)v_l + \frac{1}{2}\|v_l\|^2 \right) = 0.$$

Since α is continuous, this means

$$0 = \lim_{s \to \infty} \alpha(x_{l_s}) = \alpha(\bar{x}),$$

and $\bar{x}$ is Pareto critical.

Case 2: It is easy to see that the sequence $(v_{l_s})_s$ of approximate solutions is bounded. Therefore, it is contained in a compact set and possesses an accumulation point $\bar{v}$. Let $(l_u)_u \in \mathbb{N}$ be a subsequence of $(l_s)_s$ with $\lim_{u \to \infty} v_{l_u} = \bar{v}$. Since all v_{l_u} are approximate solutions to (SPe), we have

$$\max_i \nabla F_i(x_{l_u}) v_{l_u} \leq \max_i \nabla F_i(x_{l_u}) v_{l_u} + \frac{1}{2} \|v_{l_u}\|^2 \leq \gamma \alpha(x_{l_u}) < 0.$$

Letting $u \to \infty$ and considering the continuity of α, we thus have

$$\max_i \nabla F_i(\bar{x}) \bar{v} \leq \max_i \nabla F_i(\bar{x}) \bar{v} + \frac{1}{2} \|\bar{v}\|^2 \leq \gamma \alpha(\bar{x}) \leq 0. \tag{5}$$

Since $\lim_{l \to \infty} t_l = 0$, for all $q \in \mathbb{N}$ there exists some $N \in \mathbb{N}$ such that

$$t_{l_u} < \beta_0 \beta^q \quad \forall u > N,$$

and therefore – as a consequence of the definition of the step length (3) – we have

$$F(\pi(x_{l_u} + \beta_0 \beta^q v_{l_u})) \not< F(x_{l_u}) + \sigma \beta_0 \beta^q DF(x_{l_u}) v_{l_u} \quad \forall u > N.$$

Thus, there has to be some $i \in \{1, ..., m\}$ such that

$$F_i(\pi(x_{l_u} + \beta_0 \beta^q v_{l_u})) \geq F_i(x_{l_u}) + \sigma \beta_0 \beta^q \nabla F_i(x_{l_u}) v_{l_u}$$

holds for infinitely many $u > N$. Letting $u \to \infty$ we get

$$F_i(\pi(\bar{x} + \beta_0 \beta^q \bar{v})) \geq F_i(\bar{x}) + \sigma \beta_0 \beta^q \nabla F_i(\bar{x}) \bar{v},$$

and therefore

$$\frac{F_i(\pi(\bar{x} + \beta_0 \beta^q \bar{v})) - F_i(\bar{x})}{\beta_0 \beta^q} \geq \sigma \nabla F_i(\bar{x}) \bar{v}. \tag{6}$$

There has to be at least one $i \in \{1, ..., m\}$ such that inequality (6) holds for infinitely many $q \in \mathbb{N}$. By Lemma 6 and letting $q \to \infty$, we have

$$\nabla F_i(\bar{x}) \bar{v} \geq \sigma \nabla F_i(\bar{x}) \bar{v}$$

and thus, $\nabla F_i(\bar{x}) \bar{v} \geq 0$. Combining this with inequality (5), we get

$$\max_i \nabla F_i(\bar{x}) \bar{v} = 0.$$

Therefore, we have

$$\gamma \alpha(\bar{x}) \geq \max_i \nabla F_i(\bar{x}) \bar{v} + \frac{1}{2} \|\bar{v}\|^2 = 0,$$

and $\alpha(\bar{x}) = 0$ by which $\bar{x}$ is Pareto critical. $\square$

To summarize, the above result yields the same convergence as in the unconstrained case.

Remark 1. Observe that the proof of Theorem 2 also works for slightly more general sequences than the ones generated by Algorithm 1. (This will be used in a later proof and holds mainly due to the fact that we consider subsequences in the proof of Theorem 2.)

Let $(x_l)_l \in \mathcal{H}$, $K \subseteq \mathbb{N}$ with $|K| = \infty$ such that

- $(F_i(x_l))_l$ is monotonically decreasing for all $i \in \{1, ..., m\}$ and
- the step from x_k to x_{k+1} was realized by an iteration of Algorithm 1 for all $k \in K$.

Let $\bar{x} \in \mathcal{H}$ such that there exists a sequence of indices $(l_s)_s \in K$ with $\lim_{s \to \infty} x_{l_s} = \bar{x}$. Then $\bar{x}$ is Pareto critical.

3.2   Equality and Inequality Constraints

In order to incorporate inequality constraints (ICs), we consider two different active set strategies in which ICs are either *active* (and hence treated as ECs) or otherwise neglected. An inequality constraint will be considered active if its value is close to zero. The first active set strategy is based on [13, Sect. 8]. There, the active ICs are treated as additional components of the objective function, so the values of the active ICs decrease along the resulting descent direction. This means that the descent direction points into the feasible set with respect to the ICs and the sequence moves away from the boundary. The second active set strategy treats active ICs as additional ECs. Thus, the active ICs stay active and the sequence stays on the boundary of the feasible set with respect to the ICs. Before we can describe the two strategies rigorously, we first have to define when an inequality is active. Furthermore, we have to extend the notion of Pareto criticality to ECs and ICs.

As in the equality constrained case, we have to make a few basic assumptions about the objective function and the constraints.

(A1) $F : \mathbb{R}^n \to \mathbb{R}^m$ is C^2.
(A2) $H : \mathbb{R}^n \to \mathbb{R}^{m_H}$ is C^2 with regular value 0.
(A3) $G : \mathbb{R}^n \to \mathbb{R}^{m_G}$ is C^2.

By (A2), $\mathcal{H} := H^{-1}(\{0\})$ has the same manifold structure as in the equality constrained case. For the remainder of this section we consider MOPs of the form

$$\begin{aligned} \min_{x \in \mathcal{H}} \quad & F(x), \\ \text{s.t.} \quad & G_i(x) \leq 0 \quad \forall i \in \{1, ..., m_G\}, \end{aligned} \tag{P}$$

which is equivalent to (MOP). The set of feasible points is given by $\mathcal{N} := \mathcal{H} \cap G^{-1}((\mathbb{R}^{\leq 0})^{m_G})$ and can be thought of as a manifold with a boundary (and corners). In order to use an active set strategy we first have to define the active set.

Definition 6. For $\epsilon \geq 0$ and $x \in \mathcal{H}$ set

$$I_\epsilon(x) := \{i \in \{1, ..., m_G\} : G_i(x) \geq -\epsilon\}.$$

The component functions of G with indices in $I_\epsilon(x)$ are called *active* and $I_\epsilon(x)$ is called the *active set in x (with tolerance ϵ)*. A boundary $G_i^{-1}(\{0\})$ is called *active* if $i \in I_\epsilon(x)$.

In other words, the i-th inequality being active at $x \in \mathcal{N}$ means that x is close to the i-th boundary $G_i^{-1}(\{0\})$. Before defining Pareto criticality for problem (P), we need an additional assumption on G and H. Let

$$L(x) := \{v \in \ker(DH(x)) : \nabla G_i(x)v \leq 0 \quad \forall i \in I_0(x)\}$$

be the *linearized cone at x*.

(A4) The interior of $L(x)$ is non-empty for all $x \in \mathcal{N}$, i.e.

$$L^\circ(x) = \{v \in \ker(DH(x)) : \nabla G_i(x)v < 0 \quad \forall i \in I_0(x)\} \neq \emptyset.$$

For example, this assumption eliminates the case where $G^{-1}((\mathbb{R}^{\leq 0})^{m_G})$ has an empty interior. We will use the following implication of (A4):

Lemma 7. *(A4) holds iff for all $\delta > 0$, $x \in \mathcal{N}$ and $v \in \ker(DH(x))$ with*

$$\nabla G_i(x)v \leq 0 \quad \forall i \in I_0(x),$$

there is some $w \in U_\delta(v) \cap \ker(DH(x))$ with

$$\nabla G_i(x)w < 0 \quad \forall i \in I_0(x).$$

Proof. Assume that there is some $v \in L(x)$ and $\delta > 0$ such that

$$\nexists \epsilon \in U_\delta(0) \cap \ker(DH(x)) : \nabla G_i(x)(v + \epsilon) < 0 \quad \forall i \in I_0(x).$$

It follows that

$$\forall \epsilon \in U_\delta(0) \cap \ker(DH(x)) \ \exists j \in I_0(x) : \nabla G_j(x)\epsilon \geq 0,$$

and thus,

$$\forall w \in \ker(DH(x)) \ \exists j \in I_0(x) : \nabla G_j(x)w \geq 0.$$

So we have $L^\circ(x) = \emptyset$. On the other hand, the right-hand side of the equivalence stated in this lemma obviously can not hold if $L^\circ(x) = \emptyset$, which completes the proof. $\qquad\square$

We will now define Pareto criticality for MOPs with ECs and ICs. Similar to scalar-valued optimization, Pareto critical points are feasible points for which there exists no descent direction pointing into or alongside the feasible set $\mathcal{N}$.

Definition 7. A point $x \in \mathcal{N}$ is *Pareto critical*, if

$$\nexists v \in \ker(DH(x)) : DF(x)v < 0 \quad \text{and} \quad \nabla G_i(x)v \leq 0 \quad \forall i \in I_0(x).$$

By Lemma 7, the condition in the last definition is equivalent to

$$\nexists v \in \ker(DH(x)) : DF(x)v < 0 \quad \text{and} \quad \nabla G_i(x)v < 0 \quad \forall i \in I_0(x).$$

We will show in Lemma 8 that Pareto criticality is indeed a first-order necessary condition for Pareto optimality for the MOP (P). We will now look at the first active set strategy.

3.2.1   Strategy 1 – Active Inequalities as Additional Objectives

If we consider the active constraints in the subproblem (SPe) as additional components of the objective function F, then we obtain the following subproblem. For given $x \in \mathcal{N}$ and $\epsilon > 0$:

$$\alpha_1(x, \epsilon) := \min_{(v,\beta) \in \mathbb{R}^{n+1}} \quad \beta + \tfrac{1}{2}\|v\|^2$$

$$\text{s.t.} \quad \begin{aligned} \nabla F_i(x)v &\leq \beta && \forall i \in \{1, ..., m\}, \\ \nabla G_l(x)v &\leq \beta && \forall l \in I_\epsilon(x), \\ \nabla H_j(x)v &= 0 && \forall j \in \{1, ..., m_H\}. \end{aligned} \qquad \text{(SP1)}$$

Due to the incorporated active set strategy, the solution v and the optimal value $\alpha_1(x, \epsilon)$ (which now additionally depends on the tolerance ϵ) of this subproblem do in general not depend continuously on x.

Remark 2. The following two results for (SPe) can be translated to (SP1).

1. By Lemma 2 and (A4) we obtain that a point x is Pareto critical iff $\alpha_1(x, 0) = 0$. Furthermore, a solution of (SP1) is a descent direction of F if the corresponding optimal value α_1 is negative.
2. Using Lemma 3 with the objective function extended by the active inequalities, we obtain uniqueness of the solution of (SP1).

Using (SP1) instead of (SPe) and modifying the step length yields Algorithm 2 as a descent method for the MOP (P). Similar to the equality constrained case, we first have to calculate a feasible point to start our descent method. We have modified the step length such that in addition to the Armijo condition, the algorithm verifies whether the resulting point is feasible with respect to the inequality constraints. In order to show that this algorithm is well-defined, we have to ensure that this step length always exists.

Lemma 8. *Let $x \in \mathcal{N}$, $\beta_0 > 0$, $\beta \in (0,1)$, $\sigma \in (0,1)$, $\epsilon \geq 0$ and $v \in \ker(DH(x))$ with*

$$DF(x)v < 0 \quad \text{and} \quad \nabla G_i(x)v < 0 \quad \forall i \in I_\epsilon(x).$$

Algorithm 2. (Descent method for equality and inequality constrained MOPs, Strategy 1)

Require: $x \in \mathbb{R}^n$, $\beta_0 > 0$, $\beta \in (0,1)$, $\epsilon > 0$, $\sigma \in (0,1)$.
1: Compute some $x_0 \in \mathcal{N}$ with $\|x - x_0\| = \min\{\|x - x_0\| : x_0 \in \mathcal{N}\}$.
2: **for** $l = 0, 1, ...$ **do**
3:     Identify the active set $I_\epsilon(x_l)$.
4:     Compute the solution v_l of (SP1) at x_l with ϵ.
5:     **if** x_l is Pareto critical (i.e. $\alpha_1(x_l, \epsilon) = 0$) **then**
6:         Stop.
7:     **else**
8:         Compute the step length t_l as in (3) for x_l and v_l with parameters β_0, β and σ.
9:         **if** $\pi(x_l + t_l v_l) \notin \mathcal{N}$ **then**
10:             Compute the smallest k so that

$$F(\pi(x_l + \beta_0 \beta^k v_l)) < F(x_l) + \sigma \beta_0 \beta^k DF(x_l)v_l$$

            and $\pi(x_l + \beta_0 \beta^k v_l) \in \mathcal{N}$.
11:             Choose some $t_l \in [\beta_0 \beta^k, \beta_0 \beta^{k_l}]$, so that $\pi(x_l + t_l v_l) \in \mathcal{N}$ and the Armijo condition hold.
12:         **end if**
13:         Set $x_{l+1} = \pi(x_l + t_l v_l)$.
14:     **end if**
15: **end for**

Then there exists some $K \in \mathbb{N}$ such that

$$F(\pi(x + \beta_0 \beta^k v)) < F(x) + \sigma \beta_0 \beta^k DF(x)v \quad and \quad \pi(x + \beta_0 \beta^k v) \in \mathcal{N} \quad \forall k > K.$$

Particularly, Pareto criticality is a first-order necessary condition for local Pareto optimality.

Proof. Lemma 6 shows the existence of some $K \in \mathbb{N}$ for which the first condition holds for all $k > K$. We assume that the second condition is violated for infinitely many $k \in \mathbb{N}$. Then we have $G(x) \leq 0$ and

$$G_j(\pi(x + \beta_0 \beta^k v)) > 0 \tag{7}$$

for arbitrarily large $k \in \mathbb{N}$ and some $j \in \{1, ..., m_G\}$. Since G and π are continuous, we get $G_j(x) = 0$ and $j \in I_\epsilon(x)$. Using inequality (7) combined with Lemma 6 with $F = G_j$, we get

$$0 \leq \lim_{k \to \infty} \frac{G_j(\pi(x + \beta_0 \beta^k v)) - G_j(x)}{\beta_0 \beta^k} = \nabla G_j(x)v.$$

This contradicts our prerequisites. $\qquad \square$

Since the step length always exists, we know that Algorithm 2 generates a sequence $(x_l)_l \in \mathcal{N}$ with $F(x_{l+1}) < F(x_l)$. We will now prove a convergence result of this sequence (Theorem 3). The proof is essentially along the lines of [13, Sect. 8] and is based on the observation that the step lengths in the algorithm can not become arbitrarily small if $\alpha_1(x, \epsilon) < \rho < 0$ holds (for some $\rho < 0$) for all x in a compact set. To prove this, we have to show the existence of a positive lower bound for step lengths violating the two requirements in Step 10 of Algorithm 2. To this end, we first need the following technical result.

Lemma 9. *Let $f : \mathbb{R}^n \to \mathbb{R}$ be a continuously differentiable function so that $\nabla f : \mathbb{R}^n \to \mathbb{R}^n$ is Lipschitz continuous with a constant $L > 0$. Then for all $x, y \in \mathbb{R}^n$ we have*

$$f(y) \leq f(x) + \nabla f(x)(y - x) + \frac{1}{2}L\|y - x\|^2.$$

Proof. The proof follows by considering

$$f(y) = f(x) + \nabla f(x)(y - x) + \int_0^1 (\nabla f(x + t(y - x)) + \nabla f(x))(y - x)dt$$

and by using the Lipschitz continuity of ∇f to get an upper bound for the integral. □

The next lemma will show that there exists a lower bound for the step lengths that violate the first requirement in Step 10 of Algorithm 2.

Lemma 10. *Let $\mathcal{K} \subseteq \mathbb{R}^n$ be compact, $\mathcal{V} \subseteq \mathbb{R}^n$ be a closed sphere around 0, $\delta > 0$ and $\sigma \in (0, 1)$. Then there is some $T > 0$ so that*

$$F(\pi(x + tv)) \leq F(x) + t\sigma DF(x)v \qquad (8)$$

holds for all $x \in \mathcal{K} \cap \mathcal{N}$, $v \in \mathcal{V}$ with $DF(x)v \leq 0$ and

$$t \in \left[0, \min\left(T, -\frac{2(1 - \sigma)\max_i(\nabla F_i(x)v) + \delta}{L^3\|v\|^2}\right)\right],$$

where L is a Lipschitz constant on $\mathcal{K}$ for π and ∇F_i, $i \in \{1, ..., m\}$.

Proof. Since π and all ∇F_i are continuously differentiable and $\mathcal{K}$ is compact, there is a (common) Lipschitz constant L (cf. [16, Propositions 2 and 3]). Thus, by Lemma 9 we have

$$F_i(\pi(x + tv)) = F_i(x + (\pi(x + tv) - x))$$

$$\leq F_i(x) + \nabla F_i(x)(\pi(x + tv) - x) + \frac{1}{2}L\|\pi(x + tv) - x\|^2.$$

This means that (8) holds if for all $i \in \{1, ..., m\}$ we have

$$F_i(x) + \nabla F_i(x)(\pi(x + tv) - x) + \frac{1}{2}L\|\pi(x + tv) - x\|^2 \leq F_i(x) + t\sigma\nabla F_i(x)v,$$

which is equivalent to

$$t\nabla F_i(x)\left(\frac{\pi(x+tv)-x}{t}-\sigma v\right)\leq-\frac{1}{2}L\|\pi(x+tv)-x\|^2. \tag{9}$$

Since π is Lipschitz continuous we have

$$-\frac{1}{2}L\|\pi(x+tv)-x\|^2\geq-\frac{1}{2}L^3t^2\|v\|^2,$$

and (9) holds if

$$t\nabla F_i(x)\left(\frac{\pi(x+tv)-x}{t}-\sigma v\right)\leq-\frac{1}{2}L^3t^2\|v\|^2,$$

which is equivalent to

$$
\begin{aligned}
t&\leq-\frac{2\nabla F_i(x)(\frac{\pi(x+tv)-x}{t}-\sigma v)}{L^3\|v\|^2}\\
&=-\frac{2(1-\sigma)\nabla F_i(x)v+2\nabla F_i(x)(\frac{\pi(x+tv)-x}{t}-v)}{L^3\|v\|^2}
\end{aligned}
\tag{10}
$$

for all $i\in\{1,...,m\}$. Since $\mathscr{K}$ and $\mathscr{V}$ are compact and $\lim\limits_{t\to0}\frac{\pi(x+tv)-x}{t}=v$ (cf. the proof of Lemma 6), we know by continuity that for all $\delta'>0$, there exists some $T'>0$ such that

$$\sup_{x\in\mathscr{K}\cap\mathscr{N},v\in\mathscr{V},t\in(0,T']}\left\|\frac{\pi(x+tv)-x}{t}-v\right\|<\delta'.$$

Since all $\|\nabla F_i\|$ are continuous on $\mathscr{K}$ and therefore bounded, by the Cauchy-Schwarz inequality there exists some $T>0$ such that

$$\left|2\nabla F_i(x)\left(\frac{\pi(x+tv)-x}{t}-v\right)\right|<\delta$$

for all $i\in\{1,...,m\}$, $x\in\mathscr{K}\cap\mathscr{N}$, $v\in\mathscr{V}$ and $t\in[0,T]$. Combining this with inequality (10) completes the proof.   □

The following lemma shows that there is a lower bound for the second requirement in Step 10 of Algorithm 2. This means that if we have a direction v pointing inside the feasible set given by the active inequalities, then we can perform a step of a certain length in that direction without violating any inequalities.

Lemma 11. *Let $\mathscr{K}\subseteq\mathbb{R}^n$ be compact, $\mathscr{V}\subseteq\mathbb{R}^n$ be a closed sphere around 0, $\delta>0$ and L be a Lipschitz constant of G_i and ∇G_i on $\mathscr{K}$ for all i. Let $x\in\mathscr{K}\cap\mathscr{N}$, $v\in\mathscr{V}$, $\epsilon>0$ and $\rho<0$, such that*

$$\nabla G_i(x)v\leq\rho\quad\forall i\in I_\epsilon(x).$$

Then there exists some $T>0$ such that

$$\pi(x+tv)\in\mathscr{N}\quad\forall t\in\left[0,\min\left(T,\frac{\epsilon}{L(\|v\|+\delta)},-\frac{2(\rho+\delta)}{L(\|v\|+\delta)^2}\right)\right].$$

Proof. Since G_i and all ∇G_i are continuously differentiable, there exists a (common) Lipschitz constant L on $\mathscr{K}$. For $i \notin I_\epsilon(x)$ we have

$$|G_i(\pi(x + tv)) - G_i(x)| \leq L\|\pi(x + tv) - x\| = tL\left\|\frac{\pi(x + tv) - x}{t} - v + v\right\|$$

$$\leq tL\left(\left\|\frac{\pi(x + tv) - x}{t} - v\right\| + \|v\|\right) \leq tL(\delta + \|v\|)$$

for $t \leq T$ with some $T > 0$, where the latter estimation is similar to the proof of Lemma 10. Since

$$tL(\delta + \|v\|) \leq \epsilon \Leftrightarrow t \leq \frac{\epsilon}{L(\delta + \|v\|)},$$

we have

$$|G_i(\pi(x + tv)) - G_i(x)| \leq \epsilon \quad \forall t \in \left[0, \min\left(T, \frac{\epsilon}{L(\delta + \|v\|)}\right)\right],$$

which (due to $G_i(x) < -\epsilon$) results in

$$G_i(\pi(x + tv)) \leq 0 \quad \forall t \in \left[0, \min\left(T, \frac{\epsilon}{L(\delta + \|v\|)}\right)\right].$$

For $i \in I_\epsilon(x)$ we apply Lemma 9 to get

$$G_i(\pi(x + tv)) \leq G_i(x) + \nabla G_i(x)(\pi(x + tv) - x) + \frac{1}{2}L\|\pi(x + tv) - x\|^2$$

$$\leq t\nabla G_i(x)\left(\left(\frac{\pi(x + tv) - x}{t} - v\right) + v\right) + \frac{1}{2}Lt^2\left\|\left(\frac{\pi(x + tv) - x}{t} - v\right) + v\right\|^2. \tag{11}$$

Since all ∇G_i are bounded on $\mathscr{K}$, the first term can be estimated by

$$t\nabla G_i(x)\left(\left(\frac{\pi(x + tv) - x}{t} - v\right) + v\right)$$

$$= t\nabla G_i(x)\left(\frac{\pi(x + tv) - x}{t} - v\right) + t\nabla G_i(x)v$$

$$\leq t(\delta + \nabla G_i(x)v) \leq t(\delta + \rho)$$

for $t < T$ with some $T > 0$ (again like in the proof of Lemma 10). For the second term we have

$$\frac{1}{2}Lt^2\left\|\left(\frac{\pi(x + tv) - x}{t} - v\right) + v\right\|^2 \leq \frac{1}{2}Lt^2\left(\left\|\frac{\pi(x + tv) - x}{t} - v\right\| + \|v\|\right)^2$$

$$\leq \frac{1}{2}Lt^2(\delta + \|v\|)^2$$

for $t < T$. Combining both estimates with (11), we obtain

$$G_i(\pi(x + tv)) \leq t(\delta + \rho) + \frac{1}{2}Lt^2(\delta + \|v\|)^2.$$

Therefore, we have $G_i(\pi(x + tv)) \leq 0$ if

$$t(\delta + \rho) + \frac{1}{2}Lt^2(\delta + \|v\|)^2 \leq 0$$

$$\Leftrightarrow \quad t \leq -\frac{2(\rho + \delta)}{L(\delta + \|v\|)^2}.$$

Combining the bounds for $i \in I_\epsilon(x)$ and $i \notin I_\epsilon(x)$ (i.e. taking the minimum of all upper bounds for t) completes the proof. $\qquad\square$

As mentioned before, a drawback of using an active set strategy is the fact that this approach naturally causes discontinuities for the descent direction at the boundary of our feasible set. But fortunately, $\alpha_1(\cdot, \epsilon)$ is still upper semi-continuous, which is shown in the next lemma. We will later see that this is sufficient to prove a convergence result for the sequence generated by Algorithm 2.

Lemma 12. *Let* $(x_l)_l$ *be a sequence in* $\mathcal{N}$ *with* $\lim_{l \to \infty} x_l = \bar{x}$. *Then*

$$\limsup_{l \to \infty} \alpha_1(x_l, \epsilon) \leq \alpha_1(\bar{x}, \epsilon).$$

Proof. Let I' be the set of indices of the ICs which are active for infinitely many elements of $(x_l)_l$, i.e.

$$I' := \bigcap_{k \in \mathbb{N}} \bigcup_{l \geq k} I_\epsilon(x_l).$$

We first show that $I' \subseteq I_\epsilon(\bar{x})$. If $I' = \emptyset$ we obviously have $I' \subseteq I_\epsilon(\bar{x})$. Therefore, let $I' \neq \emptyset$ and $j \in I'$. Then there has to be a subsequence $(x_{l_u})_u$ of $(x_l)_l$ such that $j \in I_\epsilon(x_{l_u})$ for all $u \in \mathbb{N}$. Thus,

$$G_j(x_{l_u}) \geq -\epsilon \quad \forall u \in \mathbb{N}.$$

Since $(x_{l_u})_u$ converges to $\bar{x}$ and by the continuity of G_j, we also have

$$G_j(\bar{x}) \geq -\epsilon,$$

hence $j \in I_\epsilon(\bar{x})$ and consequently, $I' \subseteq I_\epsilon(\bar{x})$.
Let $(x_{l_k})_k$ be a subsequence of $(x_l)_l$ with

$$\lim_{k \to \infty} \alpha_1(x_{l_k}, \epsilon) = \limsup_{l \to \infty} \alpha_1(x_l, \epsilon).$$

Since there are only finitely many ICs, we only have finitely many possible active sets, and there has to be some $I^s \subseteq \{1, ..., m_G\}$ which occurs infinitely many times in $(I_\epsilon(x_{l_k}))_k$. W.l.o.g. assume that $I^s = I_\epsilon(x_{l_k})$ holds for all $k \in \mathbb{N}$. By Lemma 4, the map $\alpha_1(\cdot, \epsilon)$ is continuous on the closed set $\{x \in \mathcal{N} : I_\epsilon(x) = I^s\}$ (by extending the objective function F with the ICs in I^s). Thus, $\lim_{k \to \infty} \alpha_1(x_{l_k}, \epsilon)$ is the optimal value of a subproblem like (SP1) at $\bar{x}$, except we take I^s as the active set. This completes the proof since $\alpha_1(\bar{x}, \epsilon)$ is the optimal value of (SP1) at $\bar{x}$ with the actual active set $I_\epsilon(\bar{x})$ at $\bar{x}$ and $I^s \subseteq I' \subseteq I_\epsilon(\bar{x})$ holds. (The optimal value can only get larger when additional conditions in $I_\epsilon(\bar{x}) \setminus I^s$ are considered.) $\qquad\square$

Lemmas 10, 11 and 12 now enable us to prove the following convergence result.

Theorem 3. *Let $(x_l)_l$ be a sequence generated by Algorithm 2 with $\epsilon > 0$. Then*

$$\alpha_1(\bar{x}, \epsilon) = 0$$

for all accumulation points $\bar{x}$ of $(x_l)_l$ or, if $(x_l)_l$ is finite, the last element $\bar{x}$ of the sequence.

Proof. The case where $(x_l)_l$ is finite is obvious (cf. Step 5 in Algorithm 2), so assume that $(x_l)_l$ is infinite. Let $\bar{x}$ be an accumulation point of $(x_l)_l$. Since each component of $(F(x_l))_l$ is monotonically decreasing and F is continuous, $(F(x_l))_l$ has to converge, i.e.

$$\lim_{l \to \infty} (F(x_{l+1}) - F(x_l)) = 0.$$

We therefore have

$$\lim_{l \to \infty} t_l DF(x_l) v_l = 0. \tag{12}$$

Let $(x_{l_u})_u$ be a subsequence of $(x_l)_l$ with $\lim_{u \to \infty} x_{l_u} = \bar{x}$.

We now show the desired result by contradiction. Assume $\alpha_1(\bar{x}, \epsilon) < 0$. By Lemma 12 we have

$$\limsup_{u \to \infty} \alpha_1(x_{l_u}, \epsilon) < 0.$$

Let $\rho < 0$ so that $\alpha_1(x_{l_u}, \epsilon) \leq \rho$ for all $u \in \mathbb{N}$. By definition of α_1 we therefore have

$$\nabla F_i(x_{l_u}) v_{l_u} \leq \rho \quad \text{and} \quad \nabla G_j(x_{l_u}) v_{l_u} \leq \rho \tag{13}$$

for all $u \in \mathbb{N}$, $i \in \{1, ..., m\}$, $j \in I_\epsilon(x_{l_u})$. Since $(\|v_{l_u}\|)_u$ is bounded, there is some C_v with $\|v_{l_u}\| \leq C_v$ for all $u \in \mathbb{N}$. Due to the convergence of $(x_{l_u})_u$, all elements of the sequence are contained in a compact set. By Lemma 10, all step lengths in

$$\left[0, \min\left(T, -\frac{2(1-\sigma)\rho + \delta}{L^3 C_v^2}\right)\right]$$

satisfy the Armijo condition for arbitrary $\delta > 0$ and proper $T > 0$. With $\delta = -(1-\sigma)\rho$ and proper $T_1 > 0$, all step lengths in

$$\left[0, \min\left(T_1, -\frac{(1-\sigma)\rho}{L^3 C_v^2}\right)\right]$$

satisfy the Armijo condition. By Lemma 11 we have $x_{l_u} + t v_{l_u} \in \mathcal{N}$ for all step lengths t in

$$\left[0, \min\left(T, \frac{\epsilon}{L(C_v + \delta)}, -\frac{2(\rho + \delta)}{L(C_v + \delta)^2}\right)\right]$$

for arbitrary $\delta > 0$ and proper $T > 0$. With $\delta = -\frac{1}{2}\rho$ and a properly chosen $T_2 > 0$, the last interval becomes

$$\left[0, \min\left(T_2, \frac{\epsilon}{L(C_v - \frac{1}{2}\rho)}, -\frac{\rho}{L(C_v - \frac{1}{2}\rho)^2}\right)\right].$$

We now define

$$t' := \min\left(T_1, -\frac{(1-\sigma)\rho}{L^3 C_v^2}, T_2, \frac{\epsilon}{L(C_v - \frac{1}{2}\rho)}, -\frac{\rho}{L(C_v - \frac{1}{2}\rho)^2}\right).$$

Observe that t' does not depend on the index u of $(x_{l_u})_u$ and since $\epsilon > 0$, we have $t' > 0$. Since all $t \in [0, t']$ satisfy the Armijo condition and $x_{l_u} + t v_{l_u} \in \mathcal{N}$ for all u, we know that t_{l_u} has a lower bound.

By equality (12) we therefore have

$$\lim_{l \to \infty} DF(x_l)v_l = 0,$$

which contradicts (13), and we have $\alpha_1(\bar{x}, \epsilon) = 0$. □

We conclude our results about Strategy 1 with two remarks.

Remark 3. Unfortunately, it is essential for the proof of the last theorem to choose $\epsilon > 0$, so we can not expect the accumulation points to be Pareto critical (cf. Remark 2). But since

$$\alpha_1(x, 0) = 0 \quad \Rightarrow \quad \alpha_1(x, \epsilon) = 0 \quad \forall \epsilon > 0,$$

Theorem 3 still shows that accumulation points satisfy a necessary optimality condition, it is just not as strict as Pareto criticality. To obtain convergence, one could decrease ϵ more and more during execution of Algorithm 2. For the unconstrained case, this was done in [13, Algorithm 2], and indeed results in Pareto criticality of accumulation points. This result indicates that the accumulation points of our algorithm are close to Pareto critical points if we choose small values for ϵ.

Remark 4. Observe that similar to the proof of Theorem 2, the proof of Theorem 3 also works for slightly more general sequences. (This will be used in a later proof and holds mainly due to the fact that we consider subsequences in the proof Theorem 3.)

Let $(x_l)_l \in \mathcal{H}$, $K \subseteq \mathbb{N}$ with $|K| = \infty$ so that

- $(F_i(x_l))_l$ is monotonically decreasing for all $i \in \{1, ..., m\}$ and
- the step from x_k to x_{k+1} was realized by an iteration of Algorithm 2 for all $k \in K$.

Let $\bar{x} \in \mathcal{H}$ such that there exists a sequence of indices $(l_s)_s \in K$ with $\lim_{s \to \infty} x_{l_s} = \bar{x}$. Then $\alpha_1(\bar{x}, \epsilon) = 0$.

3.2.2   Strategy 2 – Active Inequalities as Equalities

We will now introduce the second active set strategy, where active inequalities are considered as equality constraints when calculating the descent direction. To be able to do so, we have to impose an additional assumption on the ICs.

(A5) The elements in

$$\{\nabla H_1(x), ..., \nabla H_{m_H}(x)\} \cup \{\nabla G_i(x) : i \in I_0(x)\}$$

are linear independent for all $x \in \mathcal{N}$.

This extends (A2) and ensures that the set

$$\mathcal{H}_I := \mathcal{H} \cap \bigcap_{i \in I} G_i^{-1}(\{0\})$$

is either empty or a C^2-submanifold of $\mathbb{R}^n$. For $\mathcal{H}_I \neq \emptyset$ define

$$\pi_I : \mathbb{R}^n \to \mathcal{H}_I$$

as the "projection" onto $\mathcal{H}_I$ (cf. Definition 5). Treating the active ICs as ECs in (SPe), we obtain the following subproblem for a given $x \in \mathcal{N}$.

$$
\alpha_2(x) := \min_{(v,\beta) \in \mathbb{R}^{n+1}} \quad \beta + \tfrac{1}{2}\|v\|^2
$$
$$
\text{s.t.} \quad
\begin{aligned}
\nabla F_i(x)v &\leq \beta & \forall i \in \{1, ..., m\}, \\
\nabla H_j(x)v &= 0 & \forall j \in \{1, ..., m_H\}, \\
\nabla G_l(x)v &= 0 & \forall l \in I_0(x).
\end{aligned}
\qquad \text{(SP2)}
$$

The following properties of (SPe) can be transferred to (SP2).

Remark 5

1. By Lemma 2, x is Pareto critical in $\mathcal{H}_{I_0(x)}$ iff $\alpha_2(x) = 0$.
2. By Lemma 3 (with extended ECs) we obtain uniqueness of the solution of (SP2).

Unfortunately, α_2 can not be used as a criterion to test for Pareto criticality in $\mathcal{N}$. $\alpha_2(x) = 0$ only means that the cone of descent directions of F, the tangent space of the ECs and the tangent space of the active inequalities have no intersection. On the one hand, this occurs when the cone of descent directions points outside the feasible set which means that x is indeed Pareto critical. On the other hand, this also occurs when the cone points inside the feasible set. In this case, x is not Pareto critical. Consequently, α_2 can only be used to test for Pareto criticality with respect to $\mathcal{H}_{I_0(x)}$, i.e. the constrained MOP where the active ICs are actually ECs (and there are no other ICs). The following lemma shows a simple relation between Pareto criticality in $\mathcal{N}$ and in $\mathcal{H}_{I_0(x)}$.

Lemma 13. *A Pareto critical point x in $\mathcal{N}$ is Pareto critical in $\mathcal{H}_{I_0(x)}$.*

Proof. By definition, there is no $v \in \ker(DH(x))$ such that

$$DF(x)v < 0 \quad \text{and} \quad \nabla G_i(x) \leq 0 \quad \forall i \in I_0(x).$$

For $w \in \mathbb{R}^n$ let $w^\perp$ denote the orthogonal complement of the linear subspace of $\mathbb{R}^n$ spanned by w. In particular, there exists no

$$v \in \ker(DH(x)) \cap \bigcap_{i \in I_0(x)} \nabla G_i(x)^\perp = T_x(\mathscr{H}_{I_0(x)})$$

such that $DF(x)v < 0$. Therefore, x is Pareto critical in $\mathscr{H}_{I_0(x)}$. $\qquad\square$

The fact that α_2 can not be used to test for Pareto criticality (in $\mathscr{N}$) poses a problem when we want to use (SP2) to calculate a descent direction. For a general sequence $(x_n)_n$, the active set $I_0(x)$ can change in each iteration. Consequently, active ICs can also become inactive. However, using (SP2), ICs can not become inactive. By Remark 5 and Theorem 2, simply using Algorithm 1 (with (SP2) instead of (SPe)) will result in a point $x \in \mathscr{N}$ that is Pareto critical in $\mathscr{H}_I$ for some $I \subseteq \{1, ..., m_G\}$, but not necessarily Pareto critical in $\mathscr{N}$. This means we can not just use (SP2) on its own. To solve this problem, we need a mechanism that deactivates active inequalities when appropriate. To this end, we combine Algorithms 1 and 2 and introduce a parameter $\eta > 0$. If in the current iteration we have $\alpha_2 > -\eta$ after solving (SP2), we calculate a descent direction using (SP1) and use the step length in Algorithm 2. Otherwise, we take a step in the direction we calculated with (SP2). The entire procedure is summarized in Algorithm 3.

Since Algorithm 3 generates the same sequence as Algorithm 2 if we take η large enough, it can be seen as a generalization of Algorithm 2. To obtain the desired behavior of "moving along the boundary", one has to consider two points when choosing η: On the one hand, η should be small enough such that active boundaries that indeed possess a Pareto critical point are not activated and deactivated too often. On the other hand, η should be large enough so that the sequence actually moves along the boundary and does not "bounce off" too early. Additionally, if it is known that all Pareto critical points in $\mathscr{H}_{I_0(x)}$ are also Pareto critical in $\mathscr{N}$, $\eta < 0$ can be chosen such that Strategy 1 is never used during execution of Algorithm 3.

In order to show that Algorithm 3 is well-defined, we have to prove existence of the step length.

Lemma 14. *Let $x \in \mathscr{N}$, $\beta_0 > 0$, $\beta \in (0,1)$, $\sigma \in (0,1)$ and*

$$v \in \ker(DH(x)) \cap \bigcap_{i \in I_0(x)} \nabla G_i(x)^\perp \quad \text{with} \quad DF(x)v < 0.$$

Then there is some $l \in \mathbb{N}$ so that

$$F(\pi_{I_0(x)}(x + \beta_0 \beta^k v)) < F(x) + \sigma \beta_0 \beta^k DF(x)v \quad \forall k > l$$

and

$$\pi_{I_0(x)}(x + \beta_0 \beta^k v) \in \mathscr{N} \quad \forall k > l.$$

Algorithm 3. (Descent method for equality and inequality constrained MOPs, Strategy 2)

Require: $x \in \mathbb{R}^n$, $\beta_0 > 0$, $\beta \in (0,1)$, $\epsilon > 0$, $\eta > 0$, $\sigma \in (0,1)$.

1: Compute some x_0 with $\|x - x_0\| = \min\{\|x - x_0\| : x_0 \in \mathcal{N}\}$.

2: **for** $l = 0, 1, \ldots$ **do**

3:     Identify the active set $I_0(x_l)$.

4:     Compute the solution v_l of (SP2) at x_l.

5:     **if** $\alpha_2(x_l) > -\eta$ **then**

6:         Compute t_l and v_l as in Algorithm 2. If $\alpha_1(x_l, \epsilon) = 0$, stop.

7:         Set $x_{l+1} = \pi(x_l + t_l v_l)$.

8:     **else**

9:         Compute

$$k_l := \min\{k \in \mathbb{N} : F(\pi_{I_0(x_l)}(x_l + \beta_0 \beta^k v_l)) < F(x) + \sigma \beta_0 \beta^k DF(x_l)v_l\}.$$

    and set $t_l := \beta_0 \beta^{k_l}$.

10:         **if** $\pi(x_l + t_l v_l) \notin \mathcal{N}$ **then**

11:             Compute some $t < t_l$ such that

$$\pi_{I_0(x_l)}(x_l + t v_l) \in \mathcal{N} \quad \text{and} \quad |I_0(\pi_{I_0(x_l)}(x_l + t v_l))| > |I_0(x_l)|$$

            and the Armijo condition holds. Set $t_l = t$.

12:         **end if**

13:         Set $x_{l+1} = \pi_{I_0(x_l)}(x_l + t_l v_l)$.

14:     **end if**

15: **end for**

Proof. Using Lemma 8 with the active ICs at $I_0(x)$ as additional ECs and $\epsilon = 0$. $\square$

Remark 6. In theory, it is still possible that Step 11 in Algorithm 3 fails to find a step length that satisfies the conditions, because the Armijo condition does not have to hold for all $t < t_l$. In that case, this problem can be avoided by choosing smaller values for β and β_0.

Since Algorithm 3 is well-defined, we know that it generates a sequence $(x_l)_l$ in $\mathcal{N}$ with $F(x_{l+1}) < F(x_l)$ for all $l \geq 0$. The following theorem shows that sequences generated by Algorithm 3 converge in the same way as sequences generated by Algorithm 2. Since Algorithm 3 can be thought of as a combination of Algorithm 1 with modified ECs and Algorithm 2, the idea of the proof is a case analysis of how a generated sequence is influenced by those algorithms. We call Step 5 and Step 10 in Algorithm 3 *active in iteration i*, if the respective conditions of those steps are met in the i-th iteration.

Theorem 4. *Let $(x_l)_l$ be a sequence generated by Algorithm 3 with $\epsilon > 0$. Then*

$$\alpha_1(\bar{x}, \epsilon) = 0$$

for all accumulation points $\bar{x}$ of $(x_l)_l$ or, if $(x_l)_l$ is finite, the last element $\bar{x}$ of the sequence.

Proof. If $(x_l)_l$ is finite, the stopping criterion in Step 6 was met, so we are done. Let $(x_l)_l$ be infinite, $\bar{x}$ an accumulation point and $(x_{l_s})_s$ a subsequence of $(x_l)_l$ so that $\lim_{s \to \infty} x_{l_s} = \bar{x}$. Consider the following four cases for the behavior of Algorithm 3:

Case 1: In the iterations where $l = l_s$ with $s \in \mathbb{N}$, Step 5 is active infinitely many times. Then the proof follows from Theorem 3 and Remark 4.

Case 2: In the iterations where $l = l_s$ with $s \in \mathbb{N}$, Step 5 and 10 are both only active a finite number of times. W.l.o.g. they are never active. Since the power set $\mathscr{P}(\{1, ..., m_G\})$ is finite, there has to be some $I \in \mathscr{P}(\{1, ..., m_G\})$ and a subsequence $(x_{l_u})_u$ of $(x_{l_s})_s$ so that $I_0(x_{l_u}) = I$ for all $u \in \mathbb{N}$. In this case, the proof of Theorem 2 and Remark 1 (with modified ECs) shows that $\lim_{u \to \infty} \alpha_2(x_{l_u}) = 0$, so Step 5 has to be active for some iteration with $l = k_u$, which is a contradiction. Thus Case 2 can not occur.

Case 3: In the iterations where $l = l_s$ with $s \in \mathbb{N}$, Step 5 is only active a finite number of times and $\limsup_{s \to \infty} t_{l_s} > 0$. W.l.o.g. Step 5 is never active and $\lim_{s \to \infty} t_{l_s} > 0$. As in Case 2 there has to be a subsequence $(x_{l_u})_u$ of $(x_{l_s})_s$ and some I so that $I_0(x_{l_u}) = I$ for all $u \in \mathbb{N}$. However, the first case in the proof of Theorem 2 (with modified ECs) yields $\lim_{u \to \infty} \alpha_2(x_{l_u}) = 0$, so Step 5 has to be active at some point which is a contradiction. Consequently, Case 3 can not occur either.

Case 4: In the iterations where $l = l_s$ with $s \in \mathbb{N}$, Step 5 is only active a finite amount of times (w.l.o.g. never), Step 10 is active an infinite amount of times (w.l.o.g. always) and $\lim_{s \to \infty} t_{l_s} = 0$. We have

$$\|\bar{x} - x_{l_s+1}\| = \|\bar{x} - x_{l_s} + x_{l_s} - x_{l_s+1}\| \leq \|\bar{x} - x_{l_s}\| + \|x_{l_s} - x_{l_s+1}\|.$$

For $s \in \mathbb{N}$ the first term gets arbitrarily small since we have $\lim_{s \to \infty} x_{l_s} = \bar{x}$ by assumption. As in Case 2, we can assume w.l.o.g. that $I_0(x_{l_s}) = I$. Then the second term becomes arbitrarily small since $\lim_{s \to \infty} t_{l_s} = 0$, $(\|v_{l_s}\|)_s$ is bounded and the projection $\pi_{I_0(x_{l_s})} = \pi_I$ is continuous. Thus $(x_{l_s+1})_s$ is another sequence that converges to $\bar{x}$. Since Step 10 is always active for $l = l_s$, we have

$$I_0(x_{l_s+1}) \geq I_0(x_{l_s}) + 1 \quad \forall s \in \mathbb{N}. \tag{14}$$

If the prerequisites of Case 1 hold for $(x_{l_s+1})_s$, the proof is complete. So w.l.o.g. assume that $(x_{l_s+1})_s$ satisfies the prerequisites of Case 4. Now consider the sequence $(x_{l_s+2})_s$. Using the same argument as above, we only have to consider Case 4 for this sequence and we obtain

$$I_0(x_{l_s+2}) \geq I_0(x_{l_s}) + 2 \quad \text{for infinitely many } s \in \mathbb{N}.$$

(Note that this inequality does not have to hold for all $s \in \mathbb{N}$ since it is possible that we only consider a subsequence in Case 4.) If we continue this procedure, there has to be some $k \in \mathbb{N}$ such that the sequence $(x_{l_s+k})_s$ converges to $\bar{x}$ but does not satisfy the prerequisites of Case 4, since there is only a finite amount of ICs and by inequality (14) the amount of active ICs increases in each step of our procedure. Thus, $(x_{l_s+k})_s$ has to satisfy the prerequisites of Case 1, which completes the proof. □

4   Numerical Results

In this section we will present and discuss the typical behavior of our method using an academic example. Since Algorithm 2 is a special case of Algorithm 3 (when choosing η large enough), we will from now on only consider Algorithm 3 with varying η. Consider the following example for an inequality constrained MOP.

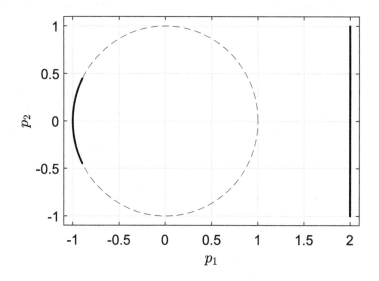

Fig. 1. The unit circle (dashed) and the set of Pareto critical points (thick).

Example 1. Let

$$F : \mathbb{R}^2 \to \mathbb{R}^2, \quad \begin{pmatrix} x_1 \\ x_2 \end{pmatrix} \mapsto \begin{pmatrix} (x_1 - 2)^2 + (x_2 - 1)^2 \\ (x_1 - 2)^2 + (x_2 + 1)^2 \end{pmatrix}$$

and

$$G : \mathbb{R}^2 \mapsto \mathbb{R}, \quad \begin{pmatrix} x_1 \\ x_2 \end{pmatrix} \mapsto -x_1^2 - x_2^2 + 1.$$

Consider the MOP

$$\min_{x \in \mathbb{R}^2} \quad F(x),$$
$$\text{s.t.} \quad G(x) \leq 0.$$

The set of feasible points is $\mathbb{R}^2$ without the interior of the unit circle (cf. the dashed line in Fig. 1). The set of Pareto critical points is

$$\left\{ \begin{pmatrix} \cos(t) \\ \sin(t) \end{pmatrix} : t \in [\pi - \theta, \pi + \theta] \right\} \cup \left\{ \begin{pmatrix} 2 \\ -1 + s \end{pmatrix} : s \in [0, 2] \right\},$$

with $\theta = \arctan(\frac{1}{2})$ and the line between $(2, 1)^T$ and $(2, -1)^T$ being globally Pareto optimal. Figure 1 shows the set of Pareto critical points.

As parameters for Algorithm 3 we choose

$$\beta = \frac{1}{2}, \quad \beta_0 = \frac{1}{10}, \quad \epsilon = 10^{-4}$$

and $\eta \in \{1, \infty\}$. (The parameters for the Armijo step length are set to relatively small values for better visibility of the behavior of the algorithm.) Figure 2 shows a sequence generated by Algorithm 3 with $\eta = \infty$ (i.e. Strategy 1 (Algorithm 2)). When the sequence hits the boundary, the IC G is activated and in the following subproblem, G is treated as an additional objective function. Minimizing the active ICs results in leaving the boundary. During the first few steps on the boundary, all possible descent directions are almost orthogonal to the gradients ∇F_1 and ∇F_2. Consequently, the Armijo step lengths are very short. The effect of this is that the sequence hits the boundary many times before the descent directions allow for more significant steps. (Note that this behavior is amplified by our choice of parameters for the Armijo step length.) After the sequence has passed the boundary, the behavior is exactly as in the unconstrained steepest descent method.

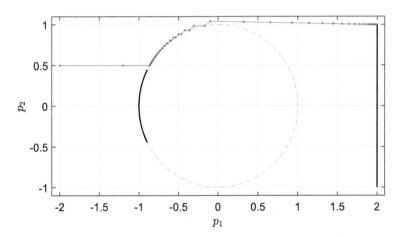

Fig. 2. Algorithm 3 with $\eta = \infty$ starting in $(-2, 0.5)^T$ for Example 1.

Figure 3 shows a sequence generated by Algorithm 3 with $\eta = 1$ which means that active ICs are often treated as ECs instead of objectives. When the sequence first hits the boundary, the IC becomes active and in the following iteration, a solution of (SP2) is computed. Since the descent direction has to lie in the tangent space of $G^{-1}(\{0\})$ at the current boundary point and there is no descent direction of acceptable quality in this subspace, the optimal value of (SP2) is less than $-\eta$. Consequently, the solution of (SP2) is discarded and (SP1) is solved to compute a descent direction. This means that the sequence tries to leave the boundary and we obtain the same behavior as in Fig. 2. This occurs as long as the optimal value of (SP2) on the boundary is less than $-\eta$. If it is greater than $-\eta$ (i.e. close to zero), the sequence "sticks" to the boundary as our method treats the active inequality as an equality. This behavior is observed as long as the optimal value of (SP2) is larger than $-\eta$. Then the algorithm will again use (SP1) which causes the sequence to leave the boundary and behave like the unconstrained steepest descent method from then on.

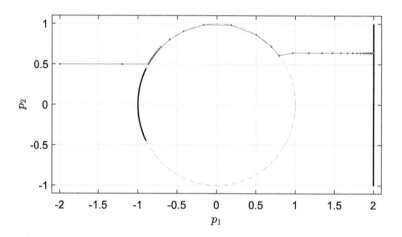

Fig. 3. Algorithm 3 with $\eta = 1$ starting in $(-2, 0.5)^T$ for Example 1.

The above example shows that it makes sense to use Algorithm 3 with a finite η since a lot less steps are required on the boundary (and in general) than with $\eta = \infty$. The fact that the behavior with $\eta = 1$ is similar to $\eta = \infty$ when the boundary is first approached (i.e. it trying to leave the boundary) indicates that in this example, it might be better to choose a smaller value for η. In general, it is hard to decide how to choose a good η as it heavily depends on the shape of the boundaries given by the ICs.

Note that in the convergence theory of the descent method in Sect. 3, we have only focused on convergence and have not discussed efficiency. Therefore, there are a few things to note when trying to implement this method efficiently.

Remark 7

1. If Step 5 in Algorithm 3 is active, two subproblems will be solved in one iteration. A case where this is obviously very inefficient is when the sequence

converges to a point $\bar{x}$ that is nowhere near a boundary. This can be avoided by only solving (SP2) if $I_\epsilon(x_l) \neq \emptyset$ and only solving (SP1) if $I_\epsilon(x_l) = \emptyset$ or $\alpha_2(x_l) > -\eta$. In order to avoid solving two subproblems in one iteration entirely, one could consider using the optimal value of the subproblem solved in the previous iteration of the algorithm as an indicator to decide which subproblem to solve in the current iteration. But one has to keep in mind that by doing so, it is slightly more difficult to globalize this method since it does no longer exclusively depend on the current point.

2. For each evaluation of the projection π onto the manifold given by the ECs and active ICs, the problem

$$\min_{y \in \mathbb{R}^n} \quad \|x - y\|^2,$$
$$\text{s.t.} \quad H(y) = 0,$$
$$G_i(y) = 0 \quad \forall i \in I,$$

for some $I \subseteq \{1, ..., m_G\}$ needs to be solved. This is an n-dimensional optimization problem with a quadratic objective function and nonlinear equality constraints. The projection is performed multiple times in Steps 9 and 11 and once in Step 13 of Algorithm 3 and thus has a large impact on the computational effort. In the convergence theory of Algorithms 2 and 3, we have only used the fact that R_π (cf. Definition 5) is a retraction and not the explicit definition of π. This means that we can exchange R_π by any other retraction and obtain the same convergence results. By choosing a retraction which is faster to evaluate, we have a chance to greatly improve the efficiency of our algorithm. An example for such a retraction (for $m_H = 1$) is the map $\psi : T\mathcal{H} \to \mathcal{H}$ which maps (x, tv) with $x \in \mathcal{H}$, $t \in \mathbb{R}$ and $v \in T_x\mathcal{H}$ to

$$x + tv + s\nabla H(x).$$

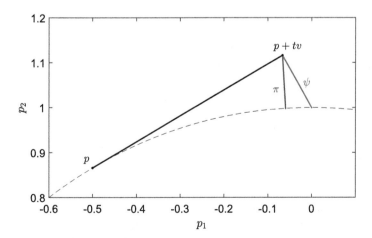

Fig. 4. ψ (red) and the projection π (blue) on the boundary of the unit circle.

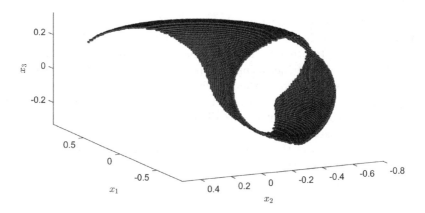

Fig. 5. Pareto set of an MOP that is constrained to the surface of the torus, calculated via the subdivision algorithm [7] combined with our descent method.

Here, $s \in \mathbb{R}$ is the smallest root (by absolute value) of

$$\mathbb{R} \to \mathbb{R}, \quad s \mapsto H(x + tv + s\nabla H(x)).$$

In general, such a map is easier to evaluate than π. Figure 4 shows the behavior of ψ for $H(x) := x_1^2 + x_2^2 - 1$. For smaller values of $t\|v\|$, these two maps differ even less. If one can show that this map is indeed a retraction (and could possibly be generalized for $m_H > 1$), this would be a good alternative to π.

5   Conclusion and Future Work

5.1   Conclusion

In this article we propose a descent method for equality and inequality constrained MOPs that is based on the steepest descent direction for unconstrained MOPs by Fliege and Svaiter [13]. We begin by incorporating equality constraints using an approach similar to the steepest descent direction on general Riemannian manifolds by Bento et al. [3] and show Pareto criticality of accumulation points. We then treat inequalities using two different active set strategies. The first one is based on [13] and treats active inequalities as additional objective functions. The second one treats active inequalities as additional equality constraints. Since for the second strategy we require a mechanism to deactivate active inequalities when necessary, we merge both strategies to one algorithm (Algorithm 3) and introduce a parameter η to control how both strategies interact. We show convergence in the sense that accumulation points of sequences generated by that algorithm satisfy a necessary optimality condition for Pareto optimality. Finally, the typical behavior of our method is shown using an academic example and some of its numerical aspects are discussed, in particular the choice of the parameter η. In combination with set-oriented methods or evolutionary algorithms, this approach can be used to significantly accelerate the computation of global Pareto sets of constrained MOPs.

5.2   Future Work

There are several possible future directions which could help to improve both theoretical results as well as the numerical performance.

- For the unconstrained steepest descent method in [13], Fukuda and Drummond present a stronger convergence result than Pareto criticality of accumulation points in [15]. It might be interesting to investigate possible similar extensions for our method.
- Being generalizations of the steepest descent method for scalar-valued problems, the unconstrained steepest descent method for MOPs as well as the constrained descent method presented here only use information about the first-order derivatives. In [12], a generalization of Newton's method to MOPs is proposed which also uses second-order derivatives. Extending our active set approach to those Newton-like descent methods could help to significantly accelerate the convergence rate.
- As explained at the end of Sect. 4, the computation of the projection onto the manifold can be relatively expensive. The proposed alternative map or similar maps may significantly reduce the computational effort. In order to use such maps without losing the convergence theory, the retraction properties (cf. Definition 5) need to be shown.
- Since the method presented in this article only calculates single Pareto critical points, globalization strategies and nondominance tests have to be applied to compute the global Pareto set. Since the dynamical system which stems from our descent direction is discontinuous on the boundary of the feasible set, this is not trivial. Although the subdivision algorithm in [7] can be used for this (with minor modifications), it may be possible to develop more efficient techniques that specifically take the discontinuities into account. As a first step, we tested the subdivision algorithm, combined with our descent method, on the equality constrained MOP given in [25], Example S4. Figure 5 shows the promising result, which coincides with the result in [25].
- Evolutionary algorithms are a popular approach to solve MOPs. It would be interesting to see if our method can be used in a hybridization approach similar to how the unconstrained steepest descent method was used in [18].

References

1. Absil, P.-A., Malick, J.: Projection-like retractions on matrix manifolds. SIAM J. Optim. Soc. Ind. Appl. Math. **22**(1), 135–158 (2012)
2. Allgower, E.L., Georg, K.: Numerical Continuation Methods: An Introduction. Springer, Heidelberg (1990)

3. Bento, G.C., Ferreira, O.P., Oliveira, P.R.: Unconstrained steepest descent method for multicriteria optimization on Riemannian manifolds. J. Optim. Theor. Appl. **154**, 88–107 (2012)

4. Carrizo, G.A., Lotito, P.A., Maciel, M.C.: Trust region globalization strategy for the nonconvex unconstrained multiobjective optimization problem. Math. Program. **159**(1), 339–369 (2016)

5. Coello Coello, C.A., Lamont, G., van Veldhuizen, D.: Evolutionary Algorithms for Solving Multi-objective Problems. Springer, New York (2007)

6. Deb, K.: Multi-objective Optimization Using Evolutionary Algorithms. Wiley, New York (2001)

7. Dellnitz, M., Hohmann, A.: A Subdivision Algorithm for the Computation of Unstable Manifolds and Global Attractors. Numerische Mathematik **75**(3), 293–317 (1997)

8. Dellnitz, M., Schütze, O., Hestermeyer, T.: Covering pareto sets by multilevel subdivision techniques. J. Optim. Theor. Appl. **124**(1), 113–136 (2005)

9. Drummond, L.G., Iusem, A.: A projected gradient method for vector optimization problems. Comput. Optim. Appl. **28**(1), 5–29 (2004)

10. Ehrgott, M.: Multicriteria Optimization. Springer, Heidelberg (2005)

11. Eichfelder, G.: Adaptive Scalarization Methods in Multiobjective Optimization. Springer, Heidelberg (2008)

12. Fliege, J., Drummond, L.M.G., Svaiter, B.F.: Newton's method for multiobjective optimization. SIAM J. Optim. **20**(2), 602–626 (2009)

13. Fliege, J., Svaiter, B.F.: Steepest descent methods for multicriteria optimization. Math. Methods Oper. Res. **51**(3), 479–494 (2000)

14. Fliege, J., Vaz, A.I.F.: A method for constrained multiobjective optimization based on SQP techniques. SIAM J. Optim. **26**(4), 2091–2119 (2016)

15. Fukuda, E.H., Drummond, L.M.G.A.: A survey on multiobjective descent methods. Pesquisa Operacional **34**, 585–620 (2014)

16. Hildebrandt, S.: Analysis 2. Springer, Heidelberg (2003)

17. Hillermeier, C.: Nonlinear Multiobjective Optimization: A Generalized Homotopy Approach. Birkhäuser, Basel (2001)

18. Lara, A., Sanchez, G., Coello Coello, C.A., Schütze, O.: HCS: a new local search strategy for memetic multiobjective evolutionary algorithms. IEEE Trans. Evol. Comput. **14**(1), 112–132 (2010)

19. Lee, J.: Introduction to Smooth Manifolds. Springer, New York (2012)

20. Martín, A., Schütze, O.: Pareto Tracer: a predictor-corrector method for multiobjective optimization problems. Eng. Optim. **50**(3), 516–536 (2018)

21. Miettinen, K.: Nonlinear Multiobjective Optimization. Springer, New York (1998)

22. Mo, H., Xu, Z., Xu, L., Wu, Z., Ma, H.: Constrained multiobjective biogeography optimization algorithm. Sci. World J. **2014** (2014)

23. Nocedal, J., Wright, S.: Numerical Optimization. Springer, New York (2006)

24. Rudolph, G., Schmidt, M.: Differential Geometry and Mathematical Physics Part 1. Springer, Dordrecht (2013)

25. Schütze, O.: Set oriented methods for global optimization. Ph.D. thesis, University of Paderborn, Paderborn, Germany (2004)

26. Schütze, O., Dell'Aere, A., Dellnitz, M.: On continuation methods for the numerical treatment of multi-objective optimization problems. In: Practical Approaches to Multi-Objective Optimization, number 04461 in Dagstuhl Seminar Proceedings. Internationales Begegnungs- und Forschungszentrum für Informatik (IBFI), Schloss Dagstuhl, Germany (2005)

27. Schütze, O., Mostaghim, S., Dellnitz, M., Teich, J.: Covering pareto sets by multi-level evolutionary subdivision techniques. In: International Conference on Evolutionary Multi-criterion Optimization (EMO), pp. 118–132 (2003)
28. Schütze, O., Witting, K., Ober-Blöbaum, S., Dellnitz, M.: Set oriented methods for the numerical treatment of multiobjective optimization problems. In: EVOLVE - A Bridge Between Probability, Set Oriented Numerics and Evolutionary Computation. Studies in Computational Intelligence, vol. 447, pp. 187–219. Springer, Heidelberg (2013)
29. Sosa Hernández, V.A., Schütze, O., Emmerich, M.: Hypervolume maximization via set based Newton's method. In: EVOLVE - A Bridge Between Probability, Set Oriented Numerics, and Evolutionary Computation V, pp. 15–28. Springer, Cham (2014)
30. Zitzler, E.: Evolutionary algorithms for multiobjective optimization: methods and applications (1999)

Evaluating Memetic Building Spatial Design Optimisation Using Hypervolume Indicator Gradient Ascent

Koen van der Blom[1(✉)], Sjonnie Boonstra[2], Hao Wang[1], Hèrm Hofmeyer[2], and Michael T. M. Emmerich[1]

[1] LIACS, Leiden University, Niels Bohrweg 1, 2333 CA Leiden, The Netherlands
{k.van.der.blom,h.wang,m.t.m.emmerich}@liacs.leidenuniv.nl
[2] Eindhoven University of Technology, P.O. Box 513,
5600 MB Eindhoven, The Netherlands
{s.boonstra,h.hofmeyer}@tue.nl

Abstract. In traditional, single objective, optimisation local optima may be found by gradient search. With the recently introduced hypervolume indicator (HVI) gradient search, this is now also possible for multi-objective optimisation, by steering the whole Pareto front approximation (PFA) in the direction of maximal improvement. However, so far it has only been evaluated on simple test problems. In this work the HVI gradient is used for the real world problem of building spatial design, where the shape and layout of a building are optimised. This real world problem comes with a number of constraints that may hamper the effectiveness of the HVI gradient. Specifically, box constraints, and an equality constraint which is satisfied by rescaling. Moreover, like with regular gradient search, the HVI gradient may overstep an optimum. Therefore, step size control is also investigated. Since the building spatial designs are encoded in mixed-integer form, the use of gradient search alone is not sufficient. To navigate both discrete and continuous space, an evolutionary multi-objective algorithm (EMOA) and the HVI gradient are used in hybrid, forming a so-called memetic algorithm. Finally, the effectiveness of the memetic algorithm using the HVI gradient is evaluated empirically, by comparing it to an EMOA without a local search method. It is found that the HVI gradient method is effective in improving the PFA for this real world problem. However, due to the many discrete subspaces, the EMOA is able to find better solutions than the memetic approach, albeit only marginally.

Keywords: Multi-objective optimisation · Memetic algorithm
Hypervolume indicator gradient · Building spatial design

1 Introduction

Many engineering experts from different disciplines cooperate in the process of building design. Whenever these experts change the design their limited

© Springer International Publishing AG, part of Springer Nature 2019
L. Trujillo et al. (Eds.): NEO 2017, SCI 785, pp. 62–86, 2019.
https://doi.org/10.1007/978-3-319-96104-0_3

knowledge about the other disciplines may lead to a negative impact on the quality of the design with respect to these other disciplines. Automated multi-objective building spatial design optimisation can play a part in alleviating this issue. In [4,6] the authors introduced representations for building spatial design for the optimisation of structural and thermal performance. Following this in [2,3] the authors developed problem specific operators for evolutionary algorithms to ensure that only feasible designs are considered. Furthermore, parameter tuning is employed to maximise the performance of the algorithm [2]. However, finding exact optima remains a challenge.

The ultimate goal in optimisation is finding the global optimum. Convergence to the global optimum is quick for convex and continuous optimisation problems when exact methods like gradient search are used. In complex functions however, such exact methods may get stuck in local optima. Exploring multiple local optima requires different methods, such as evolutionary algorithms for example. However, heuristic methods like evolutionary algorithms may be slower to lock in on an exact (local) optimum. As such, a combination of these methods could provide advantages over either of the individual methods. Hybrids of such heuristic and exact methods are called memetic algorithms [21]. Here, a combination of evolutionary search and gradient search is proposed.

Gradient search, like other exact optimisation methods, has traditionally been used for the single-objective case. In fact, until recently gradients were only defined for single points. In the generalisation of gradient methods for multi-objective optimisation the challenge arises that typically a set of points is considered together, the so-called Pareto front. One way to measure the quality of a Pareto front approximation is the hypervolume indicator (HVI). In [11] the authors describe the HVI gradient, which allows a set of points to be moved towards the Pareto front, and to be distributed well across the Pareto front. Although the HVI gradient has been tested on benchmark functions [12], and improvements to its computation [11], as well as to the navigation of dominated points [26,27] have been proposed, it was never tested on real world problems.

Other gradient approaches for multi-objective optimisation exist as well. The key difference is the use of set gradients (in case of the HVI gradient), as opposed to single point gradients [13,22] and gradients that are used for the computation of bounds on subspaces [8]. Other ideas to use gradients in memetic search have been proposed in the literature, such as continuation methods that locally extend Pareto fronts by steps along tangent planes [20,23]. Moreover, directed search has been proposed, which steers points in a desired direction, either across or towards the PF. Such methods also use gradients in order to construct these directions in the decision space [24]. The HVI gradient is favoured here since it updates the Pareto front approximation as a whole and local optimality has been verified [12].

In this work the HVI gradient is applied to the real world problem of building spatial design for the first time. The combination of an evolutionary multi-objective algorithm (EMOA) and the HVI gradient results in a memetic multi-objective (MEMO) algorithm. Specifically, here the $\mathscr{S}$-metric (also known as HVI) selection EMOA (SMS-EMOA [10]), with specialised operators [2] for the building spatial design optimisation problem is used. For the HVI gradient component the hypervolume indicator gradient ascent multi-objective optimisation (HIGA-MO) approach from [26] is employed.

To summarise, key points of this work are as follows. The HVI gradient is used in a memetic setting for the first time. Since local optimality of the HVI gradient has been verified [12] it is an excellent candidate to explore the potential of local search for the considered problem. In addition, it may provide guarantees with regard to local convergence in the continuous subspace. Furthermore, the HVI gradient method is also subjected to constraints and a mixed-integer search space for the first time. These new challenges should provide insight into the effectiveness of the HVI gradient in more complex search spaces.

The remainder of this work is structured as follows. Section 2 introduces the problem of building spatial design and its representation. Next, Sect. 3 reviews the basics of multi-objective optimisation, and briefly describes the principles of the hypervolume indicator. In Sect. 4 the hypervolume indicator (HVI) gradient and its use in algorithms are discussed. Section 5 starts by describing a local search algorithm (based on the HVI gradient) and a global search algorithm (based on SMS-EMOA) for the building spatial design problem, and then considers how they can be combined into a memetic algorithm. Following this, Sect. 6 describes the experimental setup for the evaluation of the different algorithms. This is naturally followed by Sect. 7, where the results are analysed. Finally, Sect. 8 discusses the work as a whole, and summarises the resulting conclusions.

2   Problem

This section briefly introduces the problem of building spatial design optimisation in its first subsection. Following that the supercube representation, which encodes the problem in mixed-integer form, is introduced.

2.1   Building Spatial Design

Early stage building design consists of design decisions that have a profound influence on the many disciplines involved, and as such on their quality. By its very nature the shape and the layout of the building, i.e. the building spatial design, are normally decided early in the design process. The building spatial design includes not only the exterior building shape, but also the partitioning of spaces within the building.

Since the building spatial design influences numerous qualities of a building, it is also something worth optimising. Here, the goal is to optimise the building spatial design for two aspects. First, optimal structural performance is desirable, as it will minimise material use. For this the stiffness of a building's structural design has to be maximised. Here compliance is used as a fitting measure for a structure's stiffness. Compliance has an inverse relationship with the stiffness, and therefore has to be minimised. Second, the energy efficiency, which is here measured by the required heating and cooling energy of a building during operational hours. Efficient energy use may reduce both upkeep costs and the environmental impact. This too, has to be minimised. Both of these objectives are evaluated through simulation. Details of the simulation process are discussed later in Subsect. 6.1.

2.2 Supercube Representation

Algorithmic optimisation of building spatial designs requires a numerical representation. Here the representation from [4,6] is used, and will be briefly reintroduced on a conceptual level in the following.

In the basis the representation consists of a supercube (Fig. 1, left), which encompasses the different spaces that define the building. This supercube is a superstructure, i.e. the relevant subset of the design space. It should be noted that *space* here refers to a section of the building, somewhat similar to a room. While, *design space* refers to the collection of candidate solutions in the used representation.

The supercube is partitioned in cells by width, depth, and height indices ($i \in \{1, \ldots, N_w\}, j \in \{1, \ldots, N_d\}, k \in \{1, \ldots, N_h\}$ respectively). Each cell may then be set to the active or inactive state for a specific space $\ell \in \{1, \ldots, N_{spaces}\}$ by its corresponding binary variable $b_{i,j,k}^{\ell}$. A space may consist of multiple cells, allowing spaces to be elongated in one dimension such that they neighbour spaces that are shortened in that same dimension. Depending on which of the cells are active, and which are not, buildings with different shapes may be carved out of the supercube. Moreover, the different rows, columns, and beams in the supercube may be stretched and shrunk by their corresponding continuous variables $w_{i \in \{1, \ldots, N_w\}}, d_{j \in \{1, \ldots, N_d\}}, h_{k \in \{1, \ldots, N_h\}}$, which results in a more diverse set of possible shapes. An example of a spatial design that could be composed from the supercube is included on the right of Fig. 1.

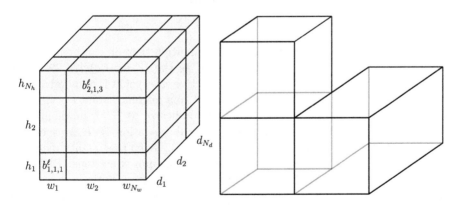

Fig. 1. Schematic of the supercube representation (left); and an example spatial design (right)

Naturally, this representation may lead to infeasible designs when it is not restricted by constraints, for example floating spaces could occur. On the binary variables the following constraints are defined: (C1) Each cell must have support below it, in the form of either another cell, or the ground. (C2) Each space must have at least one active cell assigned to it. (C3) Each cell is active for no more than one space. (C4 + C5) Each space forms a cuboid (3D rectangle) out of the active cells assigned to it (mathematically this is described in two parts). For the interested reader these constraints are described in mixed-integer nonlinear programming form in [4,6].

Further, a number of constraints are considered on the continuous subspace as well. All continuous variables are restricted between a lower bound, and an upper bound with box-constraints. In addition, a constant volume V_0 is maintained during optimisation to allow a fair comparison between different spatial designs. This is achieved by iteratively scaling the design according to the scaling functions given in [3], until the volume is within a cubic millimetre of the desired value.

3   Multi-objective Optimisation and the Hypervolume Indicator

Loosely speaking, in multi-objective optimisation problems (MOPs) the goal is to search for the candidate decision vector $\mathbf{x} = [x_1, \ldots, x_d]$ that optimises a tuple of objective functions $\mathbf{y} = \mathbf{f}(\mathbf{x}) := [f_1(\mathbf{x}), \ldots, f_m(\mathbf{x})]$, simultaneously. Without loss of generality, it is assumed that each objective function $f_i : \mathbb{R}^d \to \mathbb{R}$ has to be *minimised*.

Given that the different functions will rarely have common optimal values for $\mathbf{x}$, the outcome of a MOP is usually a Pareto front of solutions, with differing values for $\mathbf{y}$. Particularly in *continuous* MOPs, the Pareto front's efficient set [9] is approximated by a finite set (of size μ): $X = \{\mathbf{x}_1, \ldots, \mathbf{x}_\mu\} \subset \mathbb{R}^d$. The corresponding Pareto front approximation $Y = \{\mathbf{y}_1, \ldots, \mathbf{y}_\mu\} \subset \mathbb{R}^m$ is the image of X under $\mathbf{f}$, namely $\mathbf{y}_i = \mathbf{f}(\mathbf{x}_i)$, $i = 1, 2, \ldots, \mu$.

The quality of a Pareto front, or an approximation thereof, can be measured with the hypervolume indicator (also known as the $\mathscr{S}$-metric) [28,29]. This metric measures the volume (or area in bi-objective cases) that is dominated by a set of points in the objective space, with respect to a reference point $\mathbf{r} \in \mathbb{R}^m$. For notational brevity we denote the hypervolume indicator as H for mathematical use, while otherwise using the more expressive abbreviation HVI. As such the hypervolume indicator for Y can be found by $H(Y)$. Given this quality measure, it is also possible to compare different Pareto front approximations (PFAs) to each other. However, it should be noted that this measure is entirely dependent on how the reference point is chosen. In other words, the ranking of PFAs is determined in part by the value used for the reference point.

4 Hypervolume Indicator Gradient Ascent Multi-objective Optimisation

In this section, first the hypervolume indicator (HVI) gradient is introduced in its general form. This is followed by subsections for normalisation of the HVI gradient, step size adaptation for the HVI gradient, and finally update rules for the considered points.

4.1 Hypervolume Indicator Gradient

To give a proper derivation of the hypervolume indicator (HVI) gradient over approximation sets, it is proposed to use the so-called *set-oriented* approach: By concatenating the vectors in X the μd-vector $\mathbf{X} = [\mathbf{x}_1^\top, \ldots, \mathbf{x}_\mu^\top]^\top \in \mathbb{R}^{\mu d}$ is defined. Note that the restriction to $\mathbb{R}$ here is intentional, the gradients will be taken exclusively for the real subspace of the considered problem, while the discrete subspace remains constant. Likewise a μm-vector $\mathbf{Y} = [\mathbf{y}_1^\top, \ldots, \mathbf{y}_\mu^\top]^\top \in \mathbb{R}^{\mu m}$ can be defined for the objective values. Furthermore, the following mapping can be introduced: $\mathbf{F} : \mathbb{R}^{\mu d} \to \mathbb{R}^{\mu m}, \mathbf{X} \mapsto \mathbf{Y}$. Using the mapping $\mathbf{F}$, the HVI can be related to the decision space: $\mathscr{H}_\mathbf{F}(\mathbf{X}) := H(\mathbf{F}(\mathbf{X})) = H(\mathbf{Y})$. Note that this is simply the definition of a more concise symbol for the same concept.

The full HVI gradient can then be expressed as in Eq. 1, and represents the direction of steepest improvement of the HVI for the entire Pareto Front Approximation (PFA).

$$\nabla \mathscr{H}_\mathbf{F}(\mathbf{X}) = \left[\frac{\partial \mathscr{H}_\mathbf{F}(\mathbf{X})}{\partial \mathbf{x}^{(1)}}^\top, \ldots, \frac{\partial \mathscr{H}_\mathbf{F}(\mathbf{X})}{\partial \mathbf{x}^{(\mu)}}^\top \right]^\top .\tag{1}$$

Subsequently, Eq. 2 defines subgradients for each point of the PFA. Note that although subgradients are computed for individual points, their combination is not merely the direction of maximal improvement for each point, but for the whole set.

$$\frac{\partial \mathscr{H}_\mathbf{F}(\mathbf{X})}{\partial \mathbf{x}^{(i)}} = \left[\frac{\partial \mathscr{H}_\mathbf{F}}{\partial x_1^{(i)}}, \ldots, \frac{\partial \mathscr{H}_\mathbf{F}}{\partial x_d^{(i)}} \right]^\top .\tag{2}$$

Each subgradient can then be computed as:

$$\frac{\partial \mathscr{H}_\mathbf{F}}{\partial x_j^{(i)}}(\mathbf{X}) = \sum_{k=1}^m \frac{\partial H}{\partial y_k^{(i)}}(\mathbf{Y}) \times \frac{\partial f_k(\mathbf{x}^{(i)})}{\partial x_j^{(i)}} .\tag{3}$$

For the $m = 2$ case, when the indices are given by the ascending order of the first objective f_1, the first term of the summation can be expressed as follows:

$$\frac{\partial H}{\partial y_1^{(i)}} = y_2^{(i)} - y_2^{(i-1)} , \qquad \frac{\partial H}{\partial y_2^{(i)}} = y_1^{(i)} - y_1^{(i+1)} .$$

Or, for a greater number of objective functions:

$$\forall_{k=1,\ldots,m} : \frac{\partial H}{\partial y_k^{(i)}} = \prod_{\ell \neq k} y_\ell^{(i)} - y_\ell^{(i+)} , \quad k = 1, 2, \ldots, m ,$$

where $i+$ indicates the next greater value in objective ℓ, or the reference point if no such value exists. For maximisation this would be $i-$, the next smaller value. Note that strictly greater (smaller) values are taken, since any points that are equivalent in some objective, should also move the same in that objective.

Applying the HVI gradient is only possible if the gradient can be computed. Since for many problems, like the one considered in this work, the analytical expression of derivatives is not available, numerical computation of the gradient is considered as alternative. As such, the second term of the summation in Eq. 3 may be computed numerically according to the finite difference method. For a small number h, the approximation reads,

$$\frac{\partial f_k(\mathbf{x}^{(i)})}{\partial x_j^{(i)}} = \frac{f_k(\mathbf{x}^{(i)} + \mathbf{e}_j h) - f_k(\mathbf{x}^{(i)})}{h} .$$

Note that $\mathbf{e}_j$ is the j-th standard basis in $\mathbb{R}^d$. Here $h = 0.01$ is chosen, which equates to a change of $10\,\mathrm{mm}$ (millimetre) in the building spatial design. This value was chosen such that it both represents a meaningful change to the design, and it is small enough such that it provides a sufficient accuracy to approximate the gradient. Moreover, it was ensured that the employed simulator for the evaluation of a solution's quality was sufficiently sensitive. In other words, that it gave different objective values for changes of this size.

4.2 Normalisation

A limitation of the HVI gradient method is the so-called *creepiness* behaviour, as analysed in [25]. Creepiness refers to how the points move towards the Pareto front in a suboptimal way. When the differences between subgradients are large, the steps taken by the points are largely unbalanced, leading to a non-uniform convergence to the Pareto front. To avoid creepiness, the subgradients are normalised according to Eq. 4 before using them to update the original points.

$$G_{norm} = \left(\frac{\frac{\partial \mathcal{H}_\mathbf{F}}{\partial \mathbf{x}^{(1)}}}{\left\| \frac{\partial \mathcal{H}_\mathbf{F}}{\partial \mathbf{x}^{(1)}} \right\|}, \ldots, \frac{\frac{\partial \mathcal{H}_\mathbf{F}}{\partial \mathbf{x}^{(\mu)}}}{\left\| \frac{\partial \mathcal{H}_\mathbf{F}}{\partial \mathbf{x}^{(\mu)}} \right\|} \right), \quad \text{where} \quad \left\| \frac{\partial \mathcal{H}_\mathbf{F}}{\partial \mathbf{x}^{(i)}} \right\| = \sqrt{\sum_{j=1}^{d} \left(\frac{\partial \mathcal{H}_\mathbf{F}}{\partial x_j^{(i)}} \right)^2} . \quad (4)$$

4.3 Step Size Adaptation

In [27], the authors mentioned that the normalised subgradients may lead to oscillatory (even divergent) behaviour. To mitigate this effect, the step-size adaptation that has been proposed in [26] is adopted as follows. In Eq. 5 $\langle \cdot, \cdot \rangle$ stands for the dot product in $\mathbb{R}^d$. For each search point, I is calculated by the inner

product of the normalised HVI subgradients in two consecutive iterations. This
is used to find whether the step size should be increased, for positive values, or
decreased otherwise. The subscript t on the subgradient is used to indicate the
iteration.

$$I_t^{(i)} = \left\langle \left(\frac{\partial \mathcal{H}_{\mathbf{F}}(\mathbf{X})}{\partial \mathbf{x}^{(i)}} \right)_{t-1}, \left(\frac{\partial \mathcal{H}_{\mathbf{F}}(\mathbf{X})}{\partial \mathbf{x}^{(i)}} \right)_t \right\rangle, \quad i = 1, \ldots, \mu, \quad t = 1, 2, \ldots . \quad (5)$$

Since the inner product may vary largely between generations, the stabilisation
is achieved by taking the cumulative p of this value, over t generations with
exponential decay (Eq. 6). The accumulation coefficient $0 < c < 1$ controls how
much new information will be incorporated.

$$p_t^{(i)} \leftarrow (1 - c) \times p_{t-1}^{(i)} + c \times I_t^{(i)}, \qquad i = 1, \ldots, \mu, \qquad t = 1, 2, \ldots . \quad (6)$$

Given the cumulative inner product the value of the step size $\sigma_{t+1}^{(i)}$ for the next
time step can be found according to Eq. 7. The parameter α controls the rate of
change for updates to the step size.

$$\sigma_{t+1}^{(i)} = \begin{cases} \sigma_t^{(i)} \times \alpha & \text{if } p_t^{(i)} < 0, \\ \sigma_t^{(i)} & \text{if } p_t^{(i)} = 0, \\ \sigma_t^{(i)}/\alpha & \text{if } p_t^{(i)} > 0. \end{cases} \qquad 0 < \alpha < 1 . \quad (7)$$

The parameters $c = 0.7$ and $\alpha = 0.8$ are used here as they were used in [26], but
should ideally be tuned for the specific problem. Both $\left(\frac{\partial \mathcal{H}_{\mathbf{F}}(\mathbf{X})}{\partial \mathbf{x}^{(i)}} \right)_{t-1}$ and $p_{t-1}^{(i)}$
are initialised to zeros, such that the starting position is neutral.

4.4 Update

Which points are updated, and when, has a large influence on how the Pareto
front approximation (PFA) changes. The obvious choice is to move the points on
the current PFA. However, what to do with dominated points is not immediately
obvious. In [18] the authors suggested to move dominated points according to
the there defined direction of Lara.

   In the bi-objective case, this direction is defined as the sum of normalised
gradients from two objective functions. It guarantees that dominated decision
points move into the dominance cone [27]. However, such a method only considers
the movement of single points, instead of a set of search points, and does not
generalise naturally to higher dimensions.

   Alternatively, in [26], the authors suggested to move all points, including the
dominated points, according to the HVI gradient. In order to do this, the whole
population is partitioned by the so-called nondominated sorting procedure [7],
resulting in multiple subsets (fronts) of nondominated solutions. Subsequently,
the HVI gradient is well-defined on each front by ignoring other fronts that
dominate it. Since both approaches require the same number of evaluations the

exact method from [26] is used here, as shown in Eq. 8. Given that the numerical computation of the gradients requires a large number of evaluations (equal to the number of continuous decision variables) investigating alternatives that use fewer, or no, evaluations could be a promising future direction.

In this work, the step size parameter σ is initialised to $0.0025 \times (ub_r - lb_r)$ according to practical usage of the algorithm. Here ub_r and lb_r refer to the upper and lower bounds of continuous decision variable r respectively. The gradient-based update is as follows,

$$\mathbf{x}_j^{(i)} \leftarrow \mathbf{x}_j^{(i)} + \sigma^{(i)} \frac{\partial \mathcal{H}_{\mathbf{F}}(\mathbf{X})}{\partial \mathbf{x}_j^{(i)}} , \qquad i = 1,\ldots,\mu , \qquad j = 1,\ldots,d . \qquad (8)$$

5   Algorithms

Each subsection here describes one of the considered algorithms. First the HIGA-MO-SC approach, as adapted from the standard HIGA-MO [26] algorithm. Second the SMS-EMOA-SC algorithm, previously introduced in [2]. Finally, a combination of the two in the form of a memetic algorithm, MEMO-SC, is considered.

5.1   HIGA-MO-SC

In full, the HVI gradient method adapted to the context of building spatial design is described in Algorithm 1. There HIGA-MO [26] is adjusted to work with the supercube representation and forms the HIGA-MO-SC algorithm. An initial population is generated with the specialised initialisation procedure introduced in [2]. Following this, the population is sorted according to nondominated sorting [7]. Each front is then updated separately as follows. First, the HVI gradient is computed for the continuous subspace as previously described in Sect. 4.1. Second, the HVI gradients are normalised using Eq. 4. Third, step sizes are updated by employing Eqs. 5, 6 and 7. Finally, the old points are replaced with new points generated according to the normalised HVI gradients and the updated step sizes.

Note that HIGA-MO-SC as used here differs from HIGA-MO from [26] on two points. First, the step sizes are updated before moving points, rather than after. As a result the gradient information of the current iteration is immediately taken into account. Second, and most significantly, here gradients are numerically approximated. As a result they require a number of function evaluations equal to the number of continuous decision variables.

Discussion on whether to call this algorithm memetic or not is possible, since the HVI gradient operates on a population level. Here, it is important to note that there are two equivalent views on the HVI gradient. One view is, that the HVI gradient consists of the gradients of the hypervolume contributions. In that sense, a point can locally improve by increasing its hypervolume contribution. If this is performed simultaneously for all points, the second view, the effect is equivalent to following the set gradient of the HVI. A detailed discussion is provided in [11].

Algorithm 1. HIGA-MO-SC

1: **input:** $\mu, \lambda, \sigma, c, \alpha, h$
2: **output:** PFA based on all evaluated solutions
3: Initialise population X of μ parents as in [2]
4: **while** Stop condition not met **do**
5:     **while** $X \neq \varnothing$ **do**
6:         $X_{nds} \leftarrow \text{NDS}_1(X) \triangleright$ Where NDS_1 returns the first front after nondominated sorting
7:         $X \leftarrow X \setminus X_{nds}$
8:         Compute the HVI gradient for X_{nds} according to Sect. 4.1
9:         Normalise HVI gradient of X_{nds} according to Eq. 4
10:         Update step size of X_{nds} according to Eqs. 5, 6 and 7
11:         Move X_{nds} according to Eq. 8
12:         $X' \leftarrow X' \cup X_{nds}$
13:     **end while**
14:     $X \leftarrow X'$
15: **end while**

5.2   SMS-EMOA-SC

The SMS-EMOA SuperCube (SMS-EMOA-SC) algorithm was previously developed by the authors in [2,3] to navigate the heavily constrained landscape represented by the supercube. In [3] it was shown by the authors that specialised operators become essential to the building spatial design problem as soon as building designs beyond the trivially small cases are considered. Standard algorithms such as NSGA-II [7] and SMS-EMOA [10] simply do not handle the constraints well, and waste a large amount of time on infeasible solutions.

Through the use of specialised initialisation and mutation operators SMS-EMOA-SC as described in Algorithm 2 considers only feasible solutions. In discrete space the initialisation operator generates random building spatial designs composed of cuboid spaces consisting of a random number of cells, within the restrictions of the supercube representation. The continuous variables are initialised uniformly at random within their bounds. Either discrete mutations are applied with probability $MT = 0.4993$, or continuous mutations with probability $1 - MT$. Mutation in discrete space works by extending or contracting existing spaces to change their shape, while ensuring that these changes finally lead to another feasible design. Since this mutation procedure can consist of multiple steps, it is possible to move into infeasible regions, and then back to feasible space. As a result, disconnected feasible areas can be reached as well. For mutation of the continuous variables, polynomial mutation is applied with probability $MC = 0.4381$. Both MT and MC are used with values as found by parameter tuning in [2], although somewhat different objective functions were considered there.

Note that mutation is applied either in discrete space, or in continuous space. When mutations are applied on the discrete variables a design may get a significantly altered shape. As a result the optimal settings for the continuous variables change, and mutating them at the same time may have little meaning. Further, all designs are rescaled in the continuous domain (as described in [3]) to the same volume to be able to make a sensible comparison between them. Therefore, any

Algorithm 2. SMS-EMOA-SC

1: **input:** μ, MT, MC
2: **output:** PFA based on all evaluated solutions
3: Initialise population X of μ parents as in [2]
4: **while** Stop condition not met **do**
5:    $\mathbf{x}' \leftarrow$ A uniform random individual from X
6:    **if** $U(0,1) \leq MT$ **then**   ▷ Where $U(0,1)$ returns a uniform random number
7:       **if** $U(0,1) \leq 0.5$ **then**
8:          n_steps $\leftarrow 1$ ▷ Local move
9:       **else**
10:          n_steps $\leftarrow 3$ ▷ Explorative move
11:       **end if**
12:       Mutate binary variables in $\mathbf{x}'$ with n_steps as in [2]
13:    **else**
14:       Apply polynomial mutation to each continuous variable in $\mathbf{x}'$ with proba-
       bility MC
15:    **end if**
16:    Rescale the continuous variables of $\mathbf{x}'$ until the design reaches the desired spatial
       volume
17:    $X \leftarrow$ Select μ individuals from $X \cup \mathbf{x}'$
18: **end while**

---

changes in the discrete domain automatically also result in changes in the continuous domain. Finally, mutations in discrete space may – chosen uniformly at random – consist either of a single step, to make a local move, or of three steps, to make an explorative move.

5.3   MEMO-SC

Algorithm 3 shows how the SMS-EMOA-SC [2,3] and the HIGA-MO [26] algorithms are combined into a new memetic algorithm. The evaluation budget is split between the two approaches according to a given fraction $frac = 0.5$ to be used for global search. Aside from this, the behaviour is the same as for the separate algorithms.

Although different hybridisation strategies are possible, here a relay-hybrid is chosen. This is favoured here over an alternate-hybrid for various reasons. Applying the relatively expensive HVI gradient at earlier stages of the optimisation process may result in costly updates to points in suboptimal discrete subspaces. Furthermore, optimising the points in low quality discrete subspaces may even impede finding better solutions in overlapping discrete subspaces. For instance, only 10% of the solutions in some subspace A may be able to improve over the Pareto front (PF) of subspace B. As such, the further the search is away from the PF of B, the more likely it is that a newly discovered solution of the higher quality subspace A is accepted into the population by the evolutionary algorithm. Despite these possible issues, evaluating alternatives to the considered

Algorithm 3. MEMO-SC

 1: **input:** $\mu, MT, MC, \lambda, \sigma, c, \alpha, h$
 2: **output:** PFA based on all evaluated solutions
 3: Initialise population X of μ parents as in [2]
 4: **while** Stop condition not met **do**
 5:     **if** $eval \geq eval_{max} \times frac$ **then**
 6:         Generate a new population as in Algorithm 1
 7:     **else**
 8:         Generate a new population as in Algorithm 2
 9:     **end if**
10: **end while**

---

relay-hybrid, with appropriate consideration for the noted pitfalls, may still be worth investigating in future work.

6   Experiments

For the comparison of the three algorithms (SMS-EMOA-SC, HIGA-MO-SC, and MEMO-SC) two objectives are considered as discussed in Subsect. 6.1. Following that, Subsect. 6.2 describes the experimental setup.

6.1   Objective Functions

In this work two objectives are considered for the building spatial design problem, related to two disciplines: structural design, and building physics. For both objectives measurements are taken through simulations [5,6]. Settings for each of the simulation models are described briefly in the following.

6.1.1   Structural Design

The Structural Design (SD) objective for a given building spatial design is obtained by taking the total strain energy, here defined as compliance, in N mm (newton millimetre) from a Finite Element (FE) analysis that has been performed on an SD model of that spatial design. An SD model is obtained by means of a design grammar, i.e. a set of design rules that add discipline specific details to a building spatial design. Specifically, the SD grammar adds structural aspects – like structural components, loads, and constraints – to the spatial design [6].

The following SD grammar has been defined for the studies in this work: To every surface in the spatial design a concrete slab is added with thickness $t = 150$ mm (millimetre), elasticity modulus $E = 30000$ Nmm^{-2} (newton per square millimetre), and Poisson's ratio $v = 0.3$. Furthermore, each edge of a surface will be constrained if both endpoints of that edge have an equal z-coordinate that is at or below zero (i.e. ground level). Next, a live load case

$p_{live} = 5.0\,\mathrm{kNm}^{-2}$ (kilo newton per square metre) in $-z$-direction is applied on each concrete slab with a surface normal oriented vertically. Finally, wind load cases are applied, with for each wind load case three load types: $p_{w,p} = 1.0\,\mathrm{kNm}^{-2}$ for pressure; $p_{w,s} = 0.8\,\mathrm{kNm}^{-2}$ for suction; and $p_{w,sh} = 0.4\,\mathrm{kNm}^{-2}$ for shear. Four wind load cases are defined; in positive and negative x- and y-direction respectively. The load types are assigned to all external surfaces in the building spatial design (except to the ground floor surface). This is carried out according to the orientation of the external surface normal vector with respect to the wind direction vector: pressure if they are opposing; suction if they have the same orientation; and shear if they are perpendicular to each other.

FE analysis starts with meshing all the components into finite elements and nodes. Here a structural component is divided into ten elements along every dimension, which results in 10^n elements for n-dimensional components. For each load case, loads and boundary conditions are then applied to nodes, and stiffness relations between the nodes are obtained via finite element formulations. The discretised structural design is formulated as a sparse linear system, which is then solved by the simplicial-LLT solver from the C++ library Eigen [15]. For each load case, the strain energy for each element can be computed once the system has been solved. Finally, the objective is then easily computed as the sum of strain energies over all elements, over all load cases. Note that here, for each element, the strain energy is calculated by $\mathbf{u}^\top \mathbf{K} \mathbf{u}$, where $\mathbf{u}$ is the displacement vector of an element and $\mathbf{K}$ is its stiffness matrix.

6.1.2   Building Physics

The building physics (BP) objective is computed as the sum of heating and cooling energy in kW h (kilo watt hour) that is required to keep the air of all spaces of the building spatial design within a certain temperature range during a given simulation time period. The BP design grammar adds thermal related aspects – like volumes of air, thermal separations (e.g. walls and floors), temperature set points, and temperature profiles – to the building spatial design.

The BP grammar starts by defining temperature profiles for the weather and the ground. The ground temperature is set to be constant at $T_g = 10\,^\circ\mathrm{C}$. The temperature data of the weather is obtained from real world measured data by KNMI at De Bilt, The Netherlands [16]. Two periods are simulated, three full hot summer days starting 1976, July 2, and three full cold winter days starting 1978, December 30. The grammar initialises all spaces of the building spatial design with their volume, and assigns a heat capacity $C_s = 3600\,\mathrm{J\,K^{-1}\,m^{-3}}$ (joule per kelvin per cubic metre), a heating set point $T_h = 18\,^\circ\mathrm{C}$, a cooling set point $T_c 20\,^\circ\mathrm{C}$, a heating power $Q_h = 100\,\mathrm{W\,m^{-3}}$ (watt per cubic metre), a cooling power $Q_c = 100\,\mathrm{W\,m^{-3}}$, and a ventilation rate of one air change per hour. Subsequently, the thermal separations are added, with their heat conduction properties and their connections to the volumes and temperature profiles: All surfaces in the building spatial design are assigned a concrete slab with thickness $t = 150\,\mathrm{mm}$, density $\varrho = 2400\,\mathrm{kgm^{-3}}$ (kilogram per cubic metre), specific heat capacity $C = 850\,\mathrm{JK^{-1}\,kg^{-1}}$ (joule per kelvin per kilogram), and thermal conductivity $\lambda = 1.8\,\mathrm{W\ K^{-1}\,m^{-1}}$ (watt per kelvin per metre). Additionally, each external surface is assigned insulation on the outside with thickness $t = 150\,\mathrm{mm}$, density $\varrho = 60\,\mathrm{kgm^{-3}}$, specific heat capacity $C = 850\,\mathrm{JK^{-1}\,kg^{-1}}$, and thermal conductivity $\lambda = 0.04\mathrm{W\ K^{-1}\,m^{-1}}$. A warm-up period is defined for each simulation period, starting to run backwards from four days after the beginning of the actual simulation period and ending when the start of the period is reached.

For the simulation, the BP model is first abstracted as a Resistor-Capacitor (RC) network [17], where each volume or separation is modelled by a temperature point called a state. Between each temperature point a resistance is modelled, and a grounded capacitor is attached to each temperature point. The heat flux through the capacitors and resistors in the RC-network can be described by a set of first order ordinary differential equations (ODEs) [6]. This system is solved using time steps of 15 minutes using the error controlled explicit Runge-Kutta-Dopri5 solver by odeint [1]. The simulated heating or cooling of spaces is controlled at each time step by first predicting the energy demand for that time step with the system of ODEs. Then the predicted heating or cooling demand is accepted if it is lower than the available power in a space, if not it is set to the available power. All heating and cooling energies are summed over all spaces and time steps to finally yield the BP objective.

Note that in comparison with earlier work [2–4], the objectives were obtained differently. For the BP objective now realistic heating and cooling performance is used like in [5], instead of only a measure of the outer surface area. And for the SD objective, the number of wind load cases has been reduced to four and the magnitudes of the loads have changed to the values as mentioned above. Moreover, error control has been introduced in the solver of the BP simulations to prevent possible erroneous results compared to [5].

6.2   Setup

A number of aspects of the proposed approach are evaluated empirically. Specifically, a comparison is made between the three described methods: SMS-EMOA-SC, HIGA-MO-SC, and MEMO-SC. Moreover, two versions of both the HIGA-MO-SC and the MEMO-SC algorithms are considered, one with gradient step size adaptation and one without.

Note that due to the mixed-integer nature of the problem the pure HIGA-MO-SC approach cannot be expected to be competitive with the other methods. It is considered here solely to study the behaviour of the HVI gradient on the constrained landscape of this real world problem, and the value of step size adaptation. It may also be used to show that the exploration of the discrete subspace, which HIGA-MO-SC lacks, is essential to find high quality solutions, but this is not new information.

A problem with a supercube size 3333 is considered here. Meaning the supercube has three cells in with, depth, and height dimensions, and also consists of three spaces. Although in earlier work [4] it was found that for a mid-sized supercube like this constraint navigation is still reasonably simple, this problem size already consists of nine continuous variables. For the hypervolume indicator

(HVI) gradient, including numerically computing the gradient (nine evaluations, one per continuous variable), this means each new point requires ten evaluations. Since the focus of this study is on analysing the behaviour of the HVI gradient, and not on constraint navigation, the problem size is considered to be sufficient here.

Given a $3 \times 3 \times 3$ supercube, 27 discrete variables, and nine continuous variables exist: three each for width, depth, and height. The continuous variables for width and depth are bounded in $]0.5, 20]$, while those for height are bounded in $]3, 20]$. This ensures all spaces in the building are sufficiently large for human occupation. During optimisation these variables are rescaled such that the volume of the building spatial designs is maintained at $V_0 = 300 \, \text{cm}^3$ (cubic metre), like in [5].

For these experiments each algorithm is executed 35 times with an evaluation budget of 10000. The MEMO-SC approaches are set to switch halfway (i.e. $frac = 0.5$), and thus use 5000 evaluations each on evolutionary search and gradient search. This halfway switch is chosen in order to allow the evolutionary search to progress sufficiently in discrete space, while also giving the gradient search enough time to advance and adjust step sizes as needed. Note that although the evaluation budgets are equal, the number of sampled points is not. During evolutionary search each evaluation equates to a sampled point, while during HVI gradient search ten evaluations are used per sampled point.

Each algorithm considers a population size $\mu = 25$. This value is chosen to ensure a high likelihood of having a well covered PFA. Moreover, it is not so large that it would prohibit applying gradient approximation to the whole population. Note that with 25 individuals the initialisation costs 25 evaluations, leaving 9975 for the rest of the process. This means HIGA-MO-SC is not split exactly in two halves of 5000. Since HIGA-MO-SC stops when it has insufficient budget left to generate a new point (in this case 10 evaluations), it ultimately uses five evaluations less than the two other algorithms.

Settings for the SMS-EMOA-SC algorithm are given in Table 1. Parameters MT and MC control the probability to perform a discrete or continuous mutation, and the probability of mutation per continuous decision variable respectively (see Sect. 5.2 for details). A reference point of $(1.1e9, 1.1e9)$ is used as in previous work [2]. The settings for HIGA-MO-SC are available in Table 2,

Table 1. Settings for SMS-EMOA-SC

μ	MT	MC
25	0.4993	0.4381

Table 2. Settings for HIGA-MO-SC

μ	λ	σ	c	α	h
25	25	0.0025	0.7	0.8	0.01

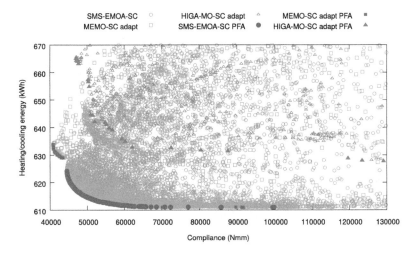

Fig. 2. Scatter plot of the PFA region for a single execution of the SMS-EMOA-SC, adaptive HIGA-MO-SC, and adaptive MEMO-SC approaches

details on their values are available in Sect. 4. Finally, MEMO-SC uses settings from either of the other two algorithms depending on whether it is in the global or local search phase.

7   Results

In Fig. 2 results are shown for a single execution of the SMS-EMOA-SC, adaptive HIGA-MO-SC, and adaptive MEMO-SC algorithms. Both the Pareto front approximations (PFAs), and the points considered during the search (limited to those sufficiently close to the PFAs) are shown. Evidently, SMS-EMOA-SC and MEMO-SC seem to perform similarly well. While, unsurprisingly, HIGA-MO-SC lags behind, unable to navigate the discrete landscape. Even so, HIGA-MO-SC is clearly able to navigate the continuous landscape within the discrete subspaces it is confined to upon initialisation.

Figures 3 and 4 display the behavioural difference in the search strategies of the SMS-EMOA-SC and MEMO-SC approaches during the second half of the optimisation process. MEMO-SC strongly focuses on local improvements to the PFA, while SMS-EMOA-SC continues to explore as well as exploit. Another interesting observation is that for this specific execution MEMO-SC seems to find two partially overlapping discrete subspaces that both contribute to the PFA. This results in a PFA consisting of two parts, one similar to what is found by SMS-EMOA-SC in Fig. 3, and an extra part in the upper-left corner of Fig. 4. Note that the differences in discrete subspaces that are discovered are an artifact of comparing single executions. Given a second execution, the discovered discrete subspaces might be reversed.

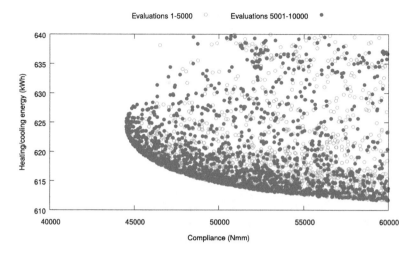

Fig. 3. Scatter plot of the PFA region for a single execution of SMS-EMOA-SC

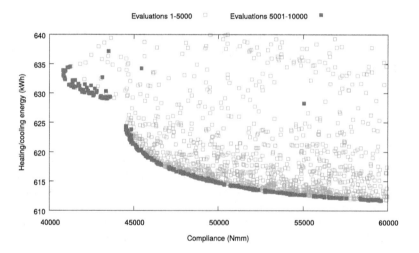

Fig. 4. Scatter plot of the PFA region for a single execution of adaptive MEMO-SC

A visual comparison of the results over multiple repetitions is done using median attainment curves [14]. Figure 5 shows the high level overview of the results, including all of the approaches. Moreover, the results are split in a first and a second half, to indicate how much the algorithms improved during the second half. From this figure it is clear that, as expected, the pure HVI gradient methods are not competitive. Even so, it is also evident that these methods work, and effectively improve their Pareto front approximations (PFA). It also becomes clear form this figure that the use of step size adaptation has a significant effect on the optimisation progress.

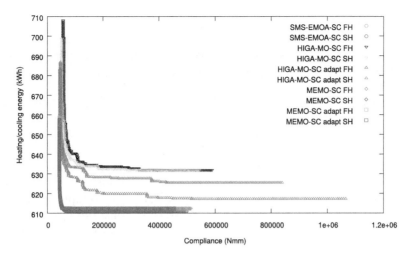

Fig. 5. Median attainment curves per algorithm (35 repetitions each), first halves (FH) and second halves (SH)

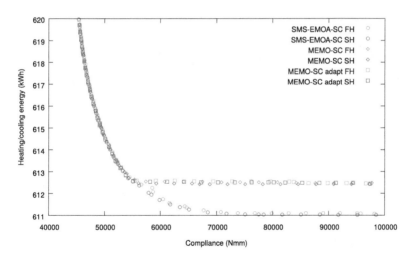

Fig. 6. Median attainment curves per algorithm (35 repetitions each), first halves (FH) and second halves (SH); zoomed in on the knee-point area

When zoomed in on the knee-point area of the median attainment curves in Fig. 6, it can be seen that there is not much difference between the adaptive MEMO-SC, and the regular MEMO-SC algorithms. While SMS-EMOA-SC appears to be able to find better solutions in the heating and cooling energy objective, even during the first half of the search. This is a striking result, given that these three algorithms behave exactly the same during the first half of the search process. Note that despite their equivalent behaviour, given their separately generated random seeds, they can still find different results due to chance.

To understand what is happening Fig. 7 shows the nondominated solutions of every repetition for all considered algorithms. In this figure multiple different PFAs, that are frequently found by all of the competitive approaches, can clearly be identified. These evidently represent the PFAs for different discrete subspaces. Looking back at Fig. 6, it appears that despite using 35 repetitions, the number of times each algorithm ends up in each discrete subspace differs sufficiently to end up with differing median attainment curves. After all, the median attainment curve may be different even if one of the algorithms ends up in (for instance) the optimal discrete subspace only a single time more than the other algorithms.

Based on the points found in Fig. 7 it is also possible to visualise the trade-off between the two objectives. In Fig. 8 three example solutions are shown. One for each objective, and one from the knee point area. For compliance it seems that long, evenly distributed walls with short floor spans are optimal for the distribution of strain across the structural elements. On the other hand, optimal

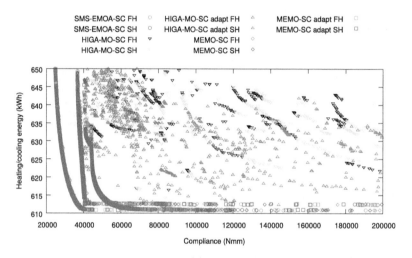

Fig. 7. Nondominated solutions from each of the 35 repetitions per algorithm, first halves (FH) and second halves (SH); zoomed in

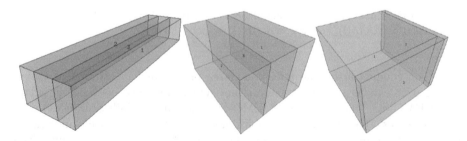

Fig. 8. Example solutions: optimal compliance (left), a knee-point solution (centre), and optimal energy efficiency (right)

energy efficiency is found by using a cubic shape and some spaces as padding to the outside, in order to provide insulation.

Based on the nondominated solutions found over all repetitions of all the approaches (Fig. 9) the objective values are normalised. From these nondominated solutions it is found that for compliance a range of $[0, 500000]$ can be considered, while in energy use a range of $[610, 660]$ is sufficient. All objective values are normalised from those ranges to a $[0, 1]$ range.

Given the normalised objective values, statistics over the hypervolume indicator (HVI) can be computed, with reference point $(1,1)$. Table 3 shows these results per algorithm for the first half. Considering that SMS-EMOA-SC and both MEMO-SC approaches are equivalent in the first half, it is not surprising to see their very similar performance here. Although SMS-EMOA-SC performs slightly better overall, this is purely based on chance.

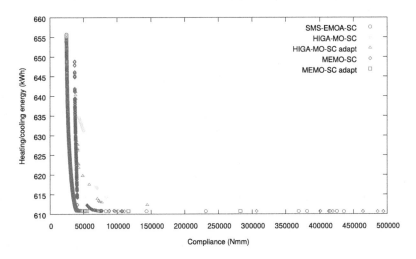

Fig. 9. Nondominated solutions over all 35 repetitions per algorithm, second halves only

Table 3. Statistics of the normalised HVI per algorithm after the first half, over 35 repetitions, best values in bold

Algorithm	Min	Max	Mean	Median	Std. dev.
SMS-EMOA-SC	0.86052	**0.92815**	**0.88674**	**0.88899**	0.01764
HIGA-MO-SC	0.25969	0.75421	0.45662	0.45798	0.11305
HIGA-MO-SC adapt	0.40257	0.76252	0.57560	0.57260	0.09315
MEMO-SC	**0.86112**	0.90129	0.88068	0.87780	**0.01220**
MEMO-SC adapt	0.85973	0.92788	0.88049	0.87633	0.01892

Further, it should be noted that these results largely depend on which discrete subspace an algorithm ends up in. For instance, consider two partially overlapping Pareto front approximations PFA_1 and PFA_2, and two equivalent algorithms A_1, A_2, executed for 10 repetitions each. Now, by chance A_1 could end up in PFA_1 8 out of 10 times, while A_2 does the reverse. Seemingly A_1 would then be better in one objective, and A_2 in the other, although they are actually the same algorithm. In other words, even for a simple case a reasonably large number of repetitions is required. Alternatively each discrete subspace could be analysed separately, but this is contrary to the goal of finding high quality discrete subspaces in the first place. Moreover, recall that the problem at hand does not merely consider two discrete subspaces, but many – often hard to reach – subspaces.

Table 4. Statistics of the normalised HVI per algorithm for the second half, over 35 repetitions, best values in bold

Algorithm	Min	Max	Mean	Median	Std. dev.
SMS-EMOA-SC	**0.86337**	**0.92910**	**0.89087**	**0.89048**	0.01602
HIGA-MO-SC	0.31499	0.77975	0.49822	0.48678	0.10857
HIGA-MO-SC adapt	0.49186	0.85363	0.69118	0.71397	0.08543
MEMO-SC	0.86146	0.90300	0.88163	0.87950	**0.01197**
MEMO-SC adapt	0.86193	0.92797	0.88168	0.87793	0.01843

Table 4 contains the results after completion of the second half of the search process. Once more, SMS-EMOA-SC seems to outperform all other approaches. Moreover, when comparing the results in Table 4 to those in Table 3 it can be observed that SMS-EMOA-SC shows the greatest improvement during the second phase of the search. Although this could be taken as surprising, when taken together with Figs. 6 and 7, it can be postulated that SMS-EMOA-SC simply found the best PFA more often. Moreover, SMS-EMOA-SC is still able to make discrete moves in the second phase, and may therefore find a new, better, front while the MEMO-SC approaches cannot. Regardless of the reason SMS-EMOA-SC appears to be better during the local search phase, the performance of the HIGA-MO-SC approach is also striking. It significantly improves in all metrics during the second phase, and is clearly a viable method for this problem, as long as it is given a good discrete subspace to work in.

All in all, it appears that both the SMS-EMOA-SC and MEMO-SC approaches are able to converge to good Pareto front approximations. Since this is already true after the first half of the search process, there is simply little to improve for either the HVI gradient in MEMO-SC, or the evolutionary approach in SMS-EMOA-SC during the second half. Further, it is observed that the quality of the found PFA depends more on the discrete, than on the continuous decision variables.

8 Conclusion and Discussion

In this work a building spatial design problem for two objectives has been considered. For this problem the shape of a building had to be optimised for structural performance and energy performance. Provided an existing mixed-integer representation, three algorithms have been applied to this optimisation problem. An algorithm based on the hypervolume indicator (HVI) gradient (HIGA-MO-SC), an adapted version of SMS-EMOA (SMS-EMOA-SC), and a memetic algorithm combining the two methods (MEMO-SC).

Results showed that the HVI gradient method by itself could not compete with the evolutionary and memetic approaches. Which, considering the mixed-integer nature of the problem, was not surprising. It has also been shown that the evolutionary approach performed slightly better during the local search phase than the memetic algorithm. However, this may be the result of larger global moves, rather than of its implied local search abilities.

Although the algorithm does not improve significantly in most cases, the HVI gradient remains useful in providing a guarantee of local convergence in the continuous subspace. The non-deterministic evolutionary algorithm cannot provide such guarantees. However, based on the results, in many cases it has a good practical performance in adjusting the continuous variables. Further, the effort spent on solving the integer problem appears to have a more significant impact on the overall result.

Improvement of the MEMO-SC algorithm may be possible by focusing on moving nondominated points, rather than all points. This could reduce the number of used evaluations on points that might never reach the Pareto front, because they are either simply too far away, or worse, stuck in a discrete subspace that is completely dominated by another. Challenges herein are found in how it is ensured that there are sufficiently many points on the Pareto front, and subsequently, how to ensure they remain on the Pareto front during the search.

Other hybridisation strategies could also be explored for the MEMO-SC algorithm. In this work a relay-hybrid, where first evolutionary search is applied and then HVI gradient ascent, has been considered. An alternate-hybrid strategy, which continuously alternates between the two approaches, may produce different results. Additionally, a comparison between HIGA-MO-SC and SMS-EMOA-SC on a single discrete subspace (such that only continuous variables are considered by both of them) could give interesting insights.

One limitation of the HVI gradient is that it cannot navigate mixed-integer space. To overcome this, metamodels can be considered. For instance, metamodels are used for the mixed-integer case in [19]. Use of the HVI gradient is possible in this case by taking the HVI gradient of the metamodel, rather than of the actual objective functions. Otherwise, it may be that methods integrating the HVI gradient are not well suited to problems with far more integer than continuous variables, as is considered here. To investigate this, a comparison to a problem with a small number of integer variables would be interesting as future work.

If the MEMO-SC algorithm as it was presented in this work is improved by the suggestions above or by other means, such that it provides advantages beyond the evolutionary approach, new directions open up. Then, in the future, combining an improved MEMO-SC algorithm and the cooperation between super-structure and free representations presented in [5] can lead to an optimisation strategy covering all levels of the building spatial design problem. Specifically, it would allow exploration with the free representation using co-evolutionary design simulation, followed by global search with the evolutionary algorithm and local search with the HVI gradient method when using the superstructure representation.

Despite the existence of a plethora of different measures to compare the quality of Pareto front approximations (PFAs), determining which PFA is better, or how two PFAs differ remains challenging. In particular this holds if, as in this work, there is a desire for statistical significance and comparisons are done over multiple repetitions per algorithmic approach. Moreover, mixed-integer landscapes further complicate the process, where different repetitions may end up in distinct, but possibly overlapping, discrete subspaces. Evidently, much work remains in the area of multi-objective quality measures.

Acknowledgements. This work is part of the TTW-Open Technology Programme with project number 13596, which is (partly) financed by the Netherlands Organisation for Scientific Research (NWO). The authors express their gratitude to the reviewers for their valuable comments.

References

1. Ahnert, K., Mulansky, M.: Odeint - solving ordinary differential equations in C++. AIP Conference Proceedings vol. 1389, no. 1, pp. 1586–1589 (2011). https://doi.org/10.1063/1.3637934
2. van der Blom, K., Boonstra, S., Hofmeyer, H., Bäck, T., Emmerich, M.T.M.: Configuring advanced evolutionary algorithms for multicriteria building spatial design optimisation. In: IEEE Congress on Evolutionary Computation (CEC), pp. 1803–1810. IEEE (2017). https://doi.org/10.1109/CEC.2017.7969520
3. van der Blom, K., Boonstra, S., Hofmeyer, H., Emmerich, M.T.M.: Multicriteria building spatial design with mixed integer evolutionary algorithms. In: Handl, J., Hart, E., Lewis, P.R., López-Ibáñez, M., Ochoa, G., Paechter, B. (eds.): Parallel Problem Solving from Nature – PPSN XIV, Lecture Notes in Computer Science, vol. 9921, pp. 453–462. Springer, Cham (2016). https://doi.org/10.1007/978-3-319-45823-6_42
4. van der Blom, K., Boonstra, S., Hofmeyer, H., Emmerich, M.T.M.: A super-structure based optimisation approach for building spatial designs. In: Papadrakakis, M., Papadopoulos, V., Stefanou, G., Plevris, V. (eds.): VII European Congress on Computational Methods in Applied Sciences and Engineering – ECCOMAS VII, vol. 2, pp. 3409–3422. National Technical University of Athens (2016). https://doi.org/10.7712/100016.2044.10063

5. Boonstra, S., van der Blom, K., Hofmeyer, H., Emmerich, M.T.: Combined super-structured and super-structure free optimisation of building spatial designs. In: Koch, C., Tizani, W., Ninić, J. (eds.): 24nd International Workshop of the European Group for Intelligent Computing in Engineering, pp. 23–34. University of Nottingham (2017)
6. Boonstra, S., van der Blom, K., Hofmeyer, H., Emmerich, M.T., van Schijndel, J., de Wilde, P.: Toolbox for super-structured and super-structure free multidisciplinary building spatial design optimisation. Adv. Eng. Inf. **36**, 86–100 (2018). https://doi.org/10.1016/j.aei.2018.01.003
7. Deb, K., Pratap, A., Agarwal, S., Meyarivan, T.: A fast and elitist multiobjective genetic algorithm: Nsga-ii. IEEE Trans. Evol. Comput. **6**(2), 182–197 (2002). https://doi.org/10.1109/4235.996017
8. Dellnitz, M., Schütze, O., Hestermeyer, T.: Covering pareto sets by multilevel subdivision techniques. J. Optim. Theory Appl. **124**(1), 113–136 (2005). https://doi.org/10.1007/s10957-004-6468-7
9. Ehrgott, M.: Multicriteria Optimization, vol. 491. Springer, Heidelberg (2005). https://doi.org/10.1007/3-540-27659-9
10. Emmerich, M., Beume, N., Naujoks, B.: An emo algorithm using the hypervolume measure as selection criterion. In: Coello Coello, C.A., Hernández Aguirre, A., Zitzler, E. (eds.): Evolutionary Multi-Criterion Optimization, pp. 62–76. Springer, Heidelberg (2005). https://doi.org/10.1007/978-3-540-31880-4_5
11. Emmerich, M., Deutz, A.: Time complexity and zeros of the hypervolume indicator gradient field. In: Schuetze, O., Coello Coello, C.A., Tantar, A.A., Tantar, E., Bouvry, P., Moral, P.D., Legrand, P. (eds.): EVOLVE - A Bridge Between Probability, Set Oriented Numerics, and Evolutionary Computation III, pp. 169–193. Springer, Heidelberg (2014). https://doi.org/10.1007/978-3-319-01460-9_8
12. Emmerich, M., Deutz, A., Beume, N.: Gradient-based/evolutionary relay hybrid for computing pareto front approximations maximizing the s-metric. In: Bartz-Beielstein, T., Blesa Aguilera, M.J., Blum, C., Naujoks, B., Roli, A., Rudolph, G., Sampels, M. (eds.): Hybrid Metaheuristics, pp. 140–156. Springer, Heidelberg (2007). https://doi.org/10.1007/978-3-540-75514-2_11
13. Fliege, J., Svaiter, B.F.: Steepest descent methods for multicriteria optimization. Math. Methods Oper. Res. **51**(3), 479–494 (2000). https://doi.org/10.1007/s001860000043
14. Fonseca, C.M., Grunert da Fonseca, V., Paquete, L.: Exploring the performance of stochastic multiobjective optimisers with the second-order attainment function. In: Coello Coello, C.A., Hernández Aguirre, A., Zitzler, E. (eds.): Evolutionary Multi-Criterion Optimization, pp. 250–264. Springer, Heidelberg (2005). https://doi.org/10.1007/978-3-540-31880-4_18
15. Guennebaud, G., Jacob, B., et al.: Eigen v3: A C++ linear algebra library (2010). http://eigen.tuxfamily.org. Accessed 7 May 2018
16. Koningklijk Nederlands Metereologisch Instituut: Measured weather data in the Netherlands (2018). http://www.knmi.nl/nederland-nu/klimatologie/daggegevens. Accessed 7 May 2018
17. Kramer, R., van Schijndel, J., Schellen, H.: Simplified thermal and hygric building models: a literature review. Front. Archit. Res. **1**(4), 318–325 (2012). https://doi.org/10.1016/j.foar.2012.09.001
18. Lara, A., Sanchez, G., Coello, C.A.C., Schütze, O.: Hcs: A new local search strategy for memetic multiobjective evolutionary algorithms. IEEE Trans. Evol. Comput. **14**(1), 112–132 (2010). https://doi.org/10.1109/TEVC.2009.2024143

19. Li, R., Emmerich, M.T.M., Eggermont, J., Bovenkamp, E.G.P., Bäck, T., Dijkstra, J., Reiber, J.H.C.: Metamodel-assisted mixed integer evolution strategies and their application to intravascular ultrasound image analysis. In: IEEE Congress on Evolutionary Computation (IEEE World Congress on Computational Intelligence), pp. 2764–2771 (2008). https://doi.org/10.1109/CEC.2008.4631169
20. Martín, A., Schütze, O.: Pareto tracer: a predictorcorrector method for multiobjective optimization problems. Eng. Optim. **50**(3), 516–536 (2018). https://doi.org/10.1080/0305215X.2017.1327579
21. Moscato, P.: On evolution, search, optimization, genetic algorithms and martial arts: Towards memetic algorithms. Caltech concurrent computation program 158-79, Technical Report, pp. 1–68 (1989)
22. Schäffler, S., Schultz, R., Weinzierl, K.: Stochastic method for the solution of unconstrained vector optimization problems. J. Optim. Theory Appl. **114**(1), 209–222 (2002). https://doi.org/10.1023/A:1015472306888
23. Schütze, O., Coello, C.A.C., Mostaghim, S., Talbi, E.G., Dellnitz, M.: Hybridizing evolutionary strategies with continuation methods for solving multiobjective problems. Eng. Optim. **40**(5), 383–402 (2008). https://doi.org/10.1080/03052150701821328
24. Schütze, O., Hernández, V.A.S., Trautmann, H., Rudolph, G.: The hypervolume based directed search method for multi-objective optimization problems. J. Heuristics **22**(3), 273–300 (2016). https://doi.org/10.1007/s10732-016-9310-0
25. Sosa Hernández, V.A., Schütze, O., Emmerich, M.: Hypervolume maximization via set based newton's method. In: Tantar, A.A., Tantar, E., Sun, J.Q., Zhang, W., Ding, Q., Schütze, O., Emmerich, M., Legrand, P., Del Moral, P., Coello Coello, C.A. (eds.): EVOLVE - A Bridge between Probability, Set Oriented Numerics, and Evolutionary Computation V, pp. 15–28. Springer, Cham (2014). https://doi.org/10.1007/978-3-319-07494-8_2
26. Wang, H., Deutz, A., Bäck, T., Emmerich, M.: Hypervolume indicator gradient ascent multi-objective optimization. In: Trautmann, H., Rudolph, G., Klamroth, K., Schütze, O., Wiecek, M., Jin, Y., Grimme, C. (eds.) Evolutionary Multi-Criterion Optimization, pp. 654–669. Springer, Cham (2017). https://doi.org/10.1007/978-3-319-54157-0_44
27. Wang, H., Ren, Y., Deutz, A., Emmerich, M.: On steering dominated points in hypervolume indicator gradient ascent for bi-objective optimization. In: Schütze, O., Trujillo, L., Legrand, P., Maldonado, Y. (eds.): NEO 2015: Results of the Numerical and Evolutionary Optimization Workshop NEO 2015 held at September 23–25 2015 in Tijuana, Mexico, pp. 175–203. Springer, Cham (2017). https://doi.org/10.1007/978-3-319-44003-3_8
28. Zitzler, E., Thiele, L.: Multiobjective optimization using evolutionary algorithms – a comparative case study. In: Eiben, A.E., Bäck, T., Schoenauer, M., Schwefel, H.P. (eds.) Parallel Problem Solving from Nature – PPSN V, pp. 292–301. Springer, Heidelberg (1998). https://doi.org/10.1007/BFb0056872
29. Zitzler, E., Thiele, L., Laumanns, M., Fonseca, C.M., Grunert da Fonseca, V.: Performance assessment of multiobjective optimizers: an analysis and review. IEEE Trans. Evol. Comput. **7**(2), 117–132 (2003). https://doi.org/10.1109/TEVC.2003.810758

Evolutionary and Genetic Computation

Fitting Multiple Ellipses with PEARL and a Multi-objective Genetic Algorithm

Heriberto Cruz Hernández$^{(\boxtimes)}$ and Luis Gerardo de la Fraga

Computer Science Department, Cinvestav, Av. IPN 2508, 07360 Mexico City, Mexico
hcruz@computacion.cs.cinvestav.mx, fraga@cs.cinvestav.mx

Abstract. In this Chapter, we address the problem of identifying and fitting more than one ellipse simultaneously, from a set of data points in the plane. This problem is an active research area with many applications in engineering and biology. Numerous studies attempted to solve this problem by detecting, fitting, and extracting the ellipses in a one-by-one approach from the set of data points. Although the one-by-one approach is effective and useful for many applications, recent studies have show that this approach is ill posed which led to the proposal of novel methods such as PEARL. PEARL is a multi-model fitting algorithm which minimizes an energy function. The PEARL algorithm requires to be initialized with random solutions. In this work we show that the performance of the PEARL algorithm, to solve the multi-ellipse fitting problem, can be improved by initializing it in a smarter way with solutions taken from a multi-objective genetic algorithm. Numerical results show that our approach can solve challenging data points instances, with high amount of outliers, and also with overlapping and nested ellipses.

Keywords: PEARL · Multi-ellipse fitting
Multi-objective optimization · Genetic algorithm
Robust optimization · Computer Vision

1 Introduction

The recognition and fitting of geometrical artifacts described by their mathematical model is an active problem in Computer Vision (CV). Many applications in CV require to solve the problem, and many algorithms to fit lines, circles and ellipses have been developed (Fitzgibbon et al. (1999); Liang et al. (2015); Fornaciari et al. (2014); Duda and Hart (1972); Fischler and Bolles (1981)). In this paper we deal with the multi-model ellipse fitting, which is a problem with many applications in biology (cells counting, detection and segmentation (Bai et al. (2009); Liao et al. (2016)), Computer Vision (heads tracking, camera calibration (Kim et al. (2005))), pattern recognition (shape analysis (Panagiotakis and Argyros (2016); Ma and Ho (2014); Johansson et al. (2013))), medicine (eye tracking and ultrasonic medical imaging (Charoenpong et al. (2015); Cheng et al. (2016); Rueda et al. (2013))), autonomous robotics (unmanned aerial vehicles (Masuzaki et al. (2015); Jung et al. (2014); Ma and Ho (2014))).

© Springer International Publishing AG, part of Springer Nature 2019
L. Trujillo et al. (Eds.): NEO 2017, SCI 785, pp. 89–107, 2019.
https://doi.org/10.1007/978-3-319-96104-0_4

The multi-ellipse fitting problem has been widely studied and many approaches have been proposed to solve it. Hough Transform (Duda and Hart (1972)) is one of the most widely used techniques, which is a general and deterministic method that explores the entire parameter space (5 dimensions for ellipses) to find those model instances with more data points passing through them. Although the original version of the Hough Transform is deterministic, in literature there also exist probabilistic versions (Mukhopadhyay and Chaudhuri (2015)).

Traditionally, except for Hough Transform, the existing methods focus on solving the problem considering a single ellipse in data points. To deal with many ellipse instances, various authors Kanatani (2001); Vincent and Laganiere (2001); Cruz- Díaz et al. (2011) propose to divide the observations point set in order to fit multiple ellipse instances, one-by-one.

Recently, Isack and Boykov (2012); Delong et al. (2012) showed that fitting multiple geometrical artifacts, assuming repetitively a single model, is an ill posed approach. The problem is ill posed since, on one hand, the number of present models is unknown, and on the other hand, depending on the data set, the single model approach can try to consider as inliers points from different model instances. This effect would lead to incorrect model fitting, and even more, if multiple executions are performed, the subsequent executions will receive as input data sets degenerated with incomplete points for the model instances.

In this manner, Isack and Boykov (2012); Delong et al. (2012) propose to solve the multi-model fitting problem in a very different form that the one-by-one approach: as an optimization problem that reduces all the costs –geometrical error, neighborhood, and number of models– at the same time to compute the final solution This PEARL (Delong et al. 2012)) approach automatically determines the number of present model instances in the observations, fits the found models and additionally classifies the outliers (observations that does not belong to any found model).

A curious observation here: PEARL is not an acronym. It is only the name that authors Isack and Boykov (2012) give to their energy-based multi-model fitting algorithm.

The form how PEARL deals with the number of models presented in the data set is to initialize the algorithm with many models. These models are generated randomly, and these random models uniformly sample the parameters space in order to allow the algorithm to guide the initialization to a local optimum. As long as the input data points complexity (number of models, level of noise, and outliers) increases, PEARL requires more random modes at the initialization step.

In this paper we propose to accelerate the PEARL approach to solve the multi-ellipse fitting problem by using a multi-objective genetic algorithm (MOGA) to initialize PEARL. The idea is to use the MOGA to generate high quality solutions, instead of the random ones, to allow PEARL solve very challenging ellipse data sets instances.

The paper is structured as follows. In Sect. 2 a fast review of the background related to this paper is presented. In Sect. 3 the related work is reviewed. In Sect. 4 our approach is presented from a general to specific point of view. In Sect. 5 the scenario for all the performed experiments is described, and also the obtained results are presented. In Sect. 6 some discussion about this work is given, and finally in Sect. 7 our conclusions and path for possible future work are presented.

2 Background

An ellipse in its parametric form is defined as:

$$\begin{bmatrix} x(t) \\ y(t) \end{bmatrix} = \begin{bmatrix} \cos\theta & -\sin\theta \\ \sin\theta & \cos\theta \end{bmatrix} \begin{bmatrix} a_1\cos t \\ b_1\sin t \end{bmatrix} + \begin{bmatrix} c_x \\ c_y \end{bmatrix}, \tag{1}$$

where the five ellipse parameters can be written in vector $l = [a_1, b_1, c_x, c_y, \theta]^T$ with a_1 as the semi major ellipse axis, b_1 the semi minor ellipse axis, $[c_x, c_y]^T$ the ellipse center, and θ the rotation of the ellipse with respect to axis x, as it is visualized in Fig. 1.

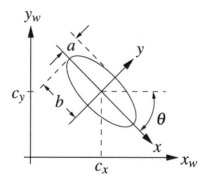

Fig. 1. An ellipse and its five parameters: Semi-axes a and b are referred to a local coordinate system; ellipse center $[c_x, c_y]^T$ and rotation angle θ are referred to a world coordinate system

The single ellipse fitting problem is defined as the task to find the parameters vector b from a set of at least five data points $\mathscr{P} = \{p_1, p_2, \ldots, p_k\}, p_i = [x_i, y_i]^T$, in presence of noise and outliers.

The multi-model ellipse fitting problem consist in finding the set of parameter vectors $\{l_1, l_2, l_3, \ldots, l_q\}$ for a unknown number (q) of ellipse models in presence of noise and outliers. Notice that any points belonging to an ellipse in the data set are outliers for the other ellipses. Therefore, we have several ellipses to detect, and this is the challenge of the problem. The problem to detect the number of data sets on the related k-means problem is NP-hard (Mahajan et al. (2012)) even with points in a plane.

We are going now to define several terms that are used in this work.

Noise is a displacement added to the original ellipse point locations. This noise come from image acquisition and real values to pixels conversion. We consider this noise has normal distribution with zero mean and variance σ. A Gaussian distribution is the most common used because it is the model considered implicitly in least square problems.

Outliers are observation that lies outside the overall pattern of a distribution (Rousseeuw and Leroy (2003)). If one have a single model and outliers, it is clear that inliers are the observations (or points) at a certain distance (selected given a threshold) near to the geometrical model; in other case those points can be consider outliers. Isack and Boykov (2012) mention that the definition of outlier in a multi-model fitting problem is *somewhat philosophical*: it is always possible to find models which have more inliers that outliers and are completely wrong. In this work we consider the outliers as Isack and Boykov (2012): these are points that cover the total area of the image and follows a bi-variate uniform distribution.

We compute the outliers percentage as:

$$\%\text{Outliers} = \frac{|\text{number of outliers}|}{|\text{total ground truth ellipses data points}|} \times 100$$

Multi-objective optimizer is a numerical tool suitable to solve those problems that consider more than one objective function, which are often in conflict. This kind of problems are very common since, in real world, in general there exist more than one aspect to optimize.

A continuous **multi-objective optimization problem** (MOOP), as we will consider in this work, can be expressed mathematically as

$$\min_{\mathbf{x} \in Q} \mathbf{f}(\mathbf{x})$$
$$\text{s.t. } g(\mathbf{x}) \leq 0 \tag{2}$$
$$h(\mathbf{x}) = 0,$$

where $\mathbf{f} : \mathbb{R}^n \rightarrow \mathbb{R}^k$, $\mathbf{f}(\mathbf{x}) = [f_1(\mathbf{x}), \ldots, f_k(\mathbf{x})]^\mathrm{T}$ is defined by the objective functions f_i, $i = 1, \ldots, k$, and $g(\mathbf{x}) = [g_1(\mathbf{x}), \ldots, g_m(\mathbf{x})]^\mathrm{T}$ and $h(\mathbf{x}) = [h_1(\mathbf{x}), \ldots, h_p(\mathbf{x})]^\mathrm{T}$ are the inequality and equality constraints, respectively. $Q := \{\mathbf{x} \in \mathbb{R}^n : g(\mathbf{x}) \leq 0 \text{ and } h(\mathbf{x}) = 0\}$ is called the domain of $\mathbf{f}$. The optimality of a MOOP is defined by the concept of *dominance*: a vector $\mathbf{y} \in Q$ is *dominated* by a vector $\mathbf{x} \in Q$ ($\mathbf{x} \prec \mathbf{y}$) with respect to (2) if $f_i(\mathbf{x}) \leq f_i(\mathbf{y})$, $i = 1, \ldots, k$, and there exists an index $j \in \{1, \ldots, k\}$ with $f_j(\mathbf{x}) < f_j(\mathbf{y})$. Else $\mathbf{y}$ is non-dominated by $\mathbf{x}$. A point $\mathbf{x} \in Q$ is called *(Pareto) optimal* or a *Pareto point* if there is no $\mathbf{y} \in Q$ that dominates $\mathbf{x}$. The set of all Pareto optimal solutions is called the *Pareto set* (denoted by $\mathscr{P}$, and its image $F(\mathscr{P})$ the *Pareto front*. One important characteristic of MOOPs is that the Pareto sets and fronts from $(k-1)$-dimensional objects under certain (mild) smoothness assumptions on the model.

Least squares ellipse fitting algorithm. Given the general equation of a conic:

$$ax^2 + 2bxy + cy^2 + 2dx + 2fy + g = 0, \tag{3}$$

the *algebraic distance* from a point $\boldsymbol{x} = [x, y]^{\mathrm{T}}$ to the ellipse $\boldsymbol{v} = [a, 2b, c, 2d, 2f, g]^{\mathrm{T}}$ is defined as:

$$d_a(\boldsymbol{x}, \boldsymbol{v}) = [x^2, xy, y^2, x, y, 1]\boldsymbol{v}. \tag{4}$$

The general conic equation (3) is an ellipse if $b^2 - 4ac < 0$. The fastest algorithm to fit an ellipse with at least five points is to minimize the sum of squared algebraic distances subject to the constraint $b^2 - 4ac < 0$. This problem leads to a generalized eigen-decomposition problem with matrices of size 6×6 (Fitzgibbon et al. (1999)), or equivalently to solve three cubic equations (de la Fraga and Cruz-Diaz (2011)).

The parameters in vector $\boldsymbol{v}$ in (4) can be converted to the five ellipse parameters in (1) (de la Fraga and Cruz-Diaz (2011)).

3   Related Work

In this section we present PEARL, which is the base algorithm of our approach, and also we briefly present RANSAC, which is the most representative approach for the single ellipse fitting problem.

3.1   PEARL

PEARL Isack and Boykov (2012) considers the multi-model fitting problem as the task of finding a labeling $\mathbf{L}$, i.e., assigning a label l_i from a set of finite label models $\mathscr{L}_{\mathbf{L}} = \{l_1, l_2, l_3, \ldots, l_n\}$ to each data point $\boldsymbol{p}_i \in \mathscr{P}$, to minimize the energy function $E(\mathbf{L})$ in Eq. (5).

$$E(\mathbf{L}) = \underbrace{\sum_{\boldsymbol{p}} ||\boldsymbol{p} - L_{\boldsymbol{p}}||}_{\text{geometrical error}} + \underbrace{\lambda \cdot \sum_{(\boldsymbol{p},\boldsymbol{q}) \in \mathscr{N}} w_{pq} \cdot \delta(L_{\boldsymbol{p}} \neq L_{\boldsymbol{q}})}_{\text{smooth prior}} + \underbrace{\beta \cdot |\mathscr{L}_{\mathbf{L}}|}_{\text{label cost}}, \tag{5}$$

where L_p in the geometrical error is a label (model) assigned to a point $\boldsymbol{p}$, $\mathscr{N}$ in the smooth prior is a set of p, q indexes pairs that are neighbors in $\mathscr{P}$, $\delta(\cdot)$ takes a value 1 if the condition inside parenthesis holds, 0 otherwise. w_{pq} weights penalties for neighborhood discontinuity, and $|\mathscr{L}_{\mathbf{L}}|$ is the number of models with non-empty support, i.e., the number of models with L_p assigned to at least the minimal data points to fit the model (five for ellipse). The terms λ and β are weighing terms for smoothness and the label cost terms respectively.

The PEARL algorithm is shown in Algorithm 1. It proposes to use a graph cut optimization, specifically the so-called α−expansion Delong et al. (2012) to minimize Eq. (5). With each α−expansion execution, those labels with not

---

Algorithm 1. PEARL algorithm

Input: Data points ($\mathscr{P} = \{\boldsymbol{p}_1, \boldsymbol{p}_2, \ldots, \boldsymbol{p}_n\}$), initialization set size: (γ), regularization parameters (λ, w_{ii}, β), outlier model value v.

Output: Labeling $\mathbf{L}_{i+1}$, and models $\mathscr{L}_\mathbf{L} = \{l_0 \, l_1, \ldots, l_n\}$

1: Propose:

    a) set $i = 0$, randomly sample data points to obtain γ initial models $\mathscr{L}_0$. Add the outliers model $l_0, |\boldsymbol{p} - \boldsymbol{l}_0| = v$.

    b) Optionally sample more or merge/split current models in $\mathbf{L}_i$

2: Expand:

    a) Perform α-expansion to minimize (5).

    b) If no changes are performed, stop the algorithm.

3: Re-estimate labels:

    a) Fit models with their current spatial support to update $\mathbf{L}_i$, i.e., solve $\mathbf{L}_{i+1} = \arg \min_l \sum_{p \in P(\mathbf{L}_i)} \|p - \boldsymbol{l}\|$

    b) Increment i and go to step 2) or optionally to 1(b).

4: **return** L_{i+1}

---

null spatial support are re-estimated, i.e. the model is fitted using the current associated data points.

Unlike other methods, PEARL solves the multi-model fitting as well as the outlier classification simultaneously. PEARL introduces an outlier model l_0, which has a fixed geometrical error value, i.e. $\|\boldsymbol{p} - L_0\| = v$. Those points $\boldsymbol{p}_i$ associated with weak models are automatically added by $\alpha-$expansion to the outliers model l_0 or added to other models with non-empty spatial support.

3.2 RANSAC

This is one of the most popular algorithms for fitting geometrical models in presence of noise and outliers. The method randomly selects a subset of points from $\mathscr{P}$ (the minimum to fit the model) and counts the number of inliers (those with distance less than a threshold). This selection and counting process is repeated a certain number of times, and the final solution is the one with the highest number of inliers. This method is the most representative of the traditional approach, since it fits the model expecting to find only one instance per execution, and it can handle high amounts of noise and outliers.

4   Proposed Approach

Our proposal consists in using evolved ellipses instead of random ones to initialize the PEARL method. The idea is presented in Algorithm 2. We perform the MOGA to obtain candidate ellipses from the Pareto set $\mathscr{S}$. We perform the

MOGA multiple times, each time we select from S the ellipses that will be used by PEARL as initialization (to build the set $\mathscr{L}_0$) but also we select some of these ellipses in S to keep a record of the already explored regions by the MOGA (to build the set $\mathscr{L}_{\text{explored}}$).

The aim of $\mathscr{L}_{\text{explored}}$ is to allow the MOGA to explore different search space regions in each execution, this process is detailed in Sect. 4.1. The MOGA is executed repeatedly until the PEARL initialization set is bigger enough ($|\mathscr{L}_0| \geq \gamma$) to start the PEARL algorithm. In subsequent subsections we present details of our approach as the multi-objective problem formulation, the record of already explored regions, and also the selection steps.

Algorithm 2. Proposed approach overview

Input: point set ($\mathscr{P}$), initialization set size: (γ), MOGA configuration parameters ($\mathscr{C}_{\text{MOGA}}$)
Output: found models $\mathscr{L}_{\text{final}} = \{l_1, l_2, \ldots, l_n\}$
1: **repeat**
2:    Perform the MOGA:
       $\mathscr{S}=\text{MOGA}(\mathscr{P},\mathscr{C}_{\text{MOGA}},\mathscr{L}_{\text{explored}})$
3:    Select models from S to be used by PEARL:
       $\mathscr{L}_0 = \mathscr{L}_0 + \text{select}(\mathscr{S})$
4:    Select representative models from S to be saved as already explored regions:
       $\mathscr{L}_{\text{explored}} = \mathscr{L}_{\text{explored}} + \text{select}(\mathscr{S})$
5: **until** $|\mathscr{L}_0| \geq s$
6: $\mathscr{L}_{\text{final}} = \text{PEARL}(\mathscr{P}, \mathscr{L}_0)$
7: **return** $\mathscr{L}_{\text{final}}$

4.1   Multi-objective Problem Formulation

We use the so-called Non-dominated Sorting Genetic Algorithm II (NSGA-II) Deb et al. (2002), which is known in literature to perform well for two objective functions. We define MOGA decision variables as the vector $\boldsymbol{m} = [i_1, i_2]^{\mathrm{T}}$, where i_1 and i_2 are indexes of data points $\boldsymbol{p}_{i_1}$ and $\boldsymbol{p}_{i_2}$. The two points defined by $\boldsymbol{m}$ are the start to perform a local search algorithm to fit an ellipse from the points in the neighborhood. We denote this process as $E(\boldsymbol{m})$ to obtain $\boldsymbol{l_m}$. This ellipse fitting algorithm from two points is shown in Algorithm 3.

With $\boldsymbol{l_m}$ we minimize two objective functions f_1 and f_2. f_1 defines the ellipse area $f_1(\boldsymbol{l_m}) = ab\pi$, and f_2 the negative cardinality of the $\boldsymbol{l_m}$ neighborhood, i.e., $f_2(\boldsymbol{l_m}) = -n(\boldsymbol{l_m})$. All f_1 functions for all the ellipses during the search are multiplied by π, thus f_1 can be simplified as $f_1(\boldsymbol{l_m}) = ab$. $n(\boldsymbol{l_m})$ defines a neighborhood around the ellipse $\boldsymbol{l_m}$, this is all point inside a certain distance to $\boldsymbol{l_m}$ are considered part of the ellipse.

Although other possibilities of objectives functions can be chosen, the used here give us the best results with lesser number of evaluations of those functions. We believe here that the multi-objectivization of ellipse fitting problem increases the exploration characteristic of the heuristic (Cruz Hernández and de la Fraga (2018)).

Algorithm 3. Ellipse fitting from two points through local search

---

Input: data points ($\mathscr{P}$), two data point indexes ($\boldsymbol{m} = [i_1, i_2]^{\mathrm{T}}$), radius ($r$).
Output: Ellipse parameters $\boldsymbol{l_m}$
1: Select points $\boldsymbol{p}_{i_1}$ and $\boldsymbol{p}_{i_1}$ from $\mathscr{P}$.
2: Use the indexes in $\boldsymbol{m}$ to define two circles with center $\boldsymbol{p}_{i_1}$ and $\boldsymbol{p}_{i_1}$ respectively, and radius r.
3: Use the points inside the circles inside obtained in step 2 to fit an ellipse l using the algebraic method.
4: Use the points in l neighborhood, $n(l)$, to fit a second ellipse l_m .
5: **return** $\boldsymbol{l_m}$

One example of this approach is shown in Fig. 2. In the Fig. 2(a) it is shown the Pareto front of one result to find a single ellipse using the data set in Fig. 2(b). In this last Fig. it is also shown the Pareto set and the ellipse that solves the problem. More details about this approach can be found in (Cruz Hernández and de la Fraga (Cruz Hernández and de la Fraga [2018])).

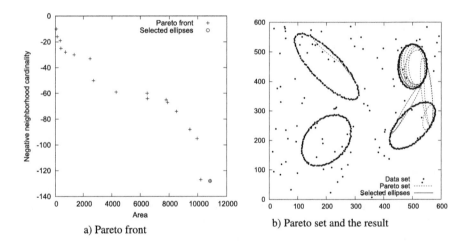

Fig. 2. Pareto front in (a) and set in (b) to find one ellipse using the multi-objective approach. Data set has a 20% of outliers

MOGA Search Record
Traditionally, the MOGA starts with the generation of random solutions that complain with the specified box constraints for decision variables. Since we perform MOGA repeatedly, we desire the algorithm to explore different search space regions with each execution. For this purpose we propose to build a solutions archive $\mathscr{L}_{\text{explored}}$ which contains representative instances of the already explored regions. The idea is to init the MOGA with individuals far from the elements in the archive. The algorithm to generate solutions for MOGA is the following: (1) generate a random solution inside the box constrains $\boldsymbol{m}$ and compute its

associated ellipse l, with the algorithm in Algorithm 3, (2) find the closest solution in the archive to the ellipse l and compute distance d, 3) if $d > T_{\text{MOGA}}$, then accept the solution m for the MOGA initialization step, else, go to step 1. The iteration process can be repeated up to a MAX_TRIES number, in case MAX_TRIES is reached the solution m is accepted. The distance between archive elements and l can be up to a specified number of ellipse parameters, we can use the five ellipse parameters to compute d. Since not all parameters have the same range, d can be computed as a Mahalanobis distance. In this paper we compute d using only the ellipse center, i.e, c_x and c_y from l.

4.2   Selection of Ellipses

Since the MOGA output is a Pareto set (ideally solutions distributed in all the Pareto front) some of the obtained solutions/ellipses are more useful than other for our approach. For each execution of the MOGA we perform two kinds of selection: selection for initialization and selection for the MOGA archive.
Selection for Initialization
The aim of this selection is to choose from $\mathscr{S}$ those ellipses that are going to be used as PEARL initialization. The idea is to select high quality ellipses that are prone to be near a local optimum. Ellipses with a high number of points in their neighborhood are good candidates that contribute to the PEARL initialization set. We add to the PEARL initialization set only those solutions from $\mathscr{S}$ with neighborhood cardinality higher than a threshold.
Selection for MOGA Archive
We select the k ellipses from $\mathscr{S}$ with highest completeness value. The MOGA queries $\mathscr{L}_{\text{explored}}$ in the initialization to init its population with individuals different from those in the archive.

We define completeness as:

$$\text{completeness}(l) = \text{perimeterPolygon}(l)/ab\pi,$$

where perimeterPolygon(l) is the perimeter of the polygon formed by the orthogonal projection of the points in the ellipse neighborhood, $n(l)$ to the ellipse l. Completeness tends to one when the number of points tends to infinite and the points in $n(l)$ are well distributed over the ellipse perimeter.

An example of the ellipses selection is shown in Fig. 3. The input set has a 20% of outliers. As it is a case of disjoint ellipses, 50 ellipses are used to initialize PEARL algorithm (see Table 3). Six iterations of the MOGA are used to obtain those 50 ellipses. In Fig. 3 are shown all the Pareto front and the selected solutions (the obtained ellipses) after 2, 3, and 6 iterations of the MOGA.

5   Experiments

To validate our approach (MOGA+PEARL) we compared against the original PEARL for multi-ellipse fitting. We used data points conformed by four ellipses

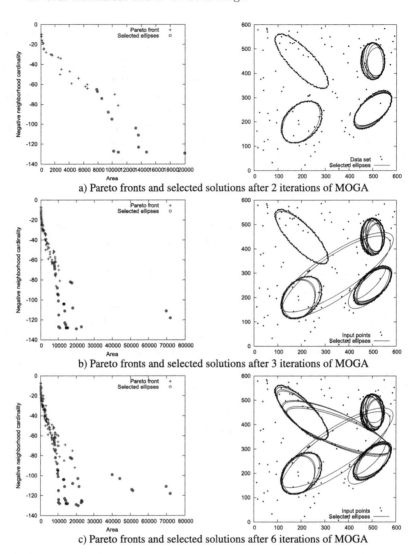

a) Pareto fronts and selected solutions after 2 iterations of MOGA

b) Pareto fronts and selected solutions after 3 iterations of MOGA

c) Pareto fronts and selected solutions after 6 iterations of MOGA

Fig. 3. An example of the ellipses generated with MOGA. These ellipses are the initialization set used by PEARL. It is shown all the Pareto fronts and the selected solutions after 2, 3 and 6 executions of the MOGA

in three different arrangement cases and with distinct levels of outliers. In subsequent paragraphs we detail aspects about: data sets, PEARL considerations, MOGA parameters, and the results for our experiment.

Test Cases

We used the same three cases considered in Cruz-Díaz et al. Cruz-Díaz et al. [2011]: disjoint, nested and overlapped. Each case is composed by four ellipses (126 points per ellipse) and different amount of outliers. For each case we varied outliers percentage from 20% to 100% in steps of 20%, i.e., a total of 15 tests.

Instances with 60% of outliers for the three cases are shown in Figs. 4, 5 and 6. Data points are given in a 600×600 resolution and outliers are uniformly distributed.

PEARL Settings

As neighborhood, each point $p \in \mathscr{P}$ is associated to its k nearest neighborhood, for all our experiments we fixed a value $k = 5$. For computation of the smooth cost term in Eq. (5) we consider a weight based in the Euclidean distance: $w_{pq} = 1/\sqrt{(p_p - p_q) \cdot (p_p - p_q)}$ and $\lambda = 150$, for the label cost term we consider $\beta = 700$, as outliers model value we set $|p - l_0| = -2\ln(1.0/600^2)$. This value is proposed by using the probability density function of the uniform distribution of outliers, it is scaled by a factor of two for tuning the energies distribution and thus the final result. We compute all weights parameters value a priori, and we compared both approaches with the same values.

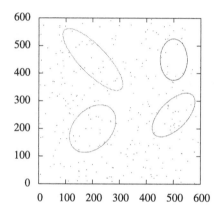

Fig. 4. Disjoint case with 60% of outliers.

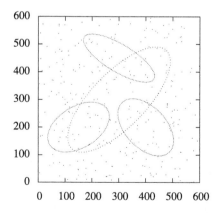

Fig. 5. Overlapped case with 60% of outliers.

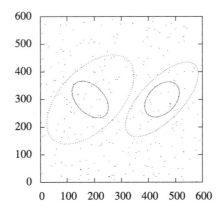

Fig. 6. Nested case with 60% of outliers.

To compute the data energy term in Eq. (5) we assume that pixels on the ellipses are affected by noise with normal distribution with mean zero and standard deviation $\sigma = 1$, such that the probability of a point at a distance d of the ellipse l, $d(\boldsymbol{p}_i, l)$, is given by:

$$P(d, \sigma) = \frac{1}{\sqrt{2\pi\sigma^2}} \exp\left(-\frac{d^2}{2\sigma^2}\right).$$

The probability is converted to energy by computing the negative of $\ln[P(d, \sigma)]$, such that:

$$D = \sum_{\boldsymbol{p}} -\ln[P(|\boldsymbol{p} - L_{\boldsymbol{p}}|, \sigma)],$$

which allows to obtain maximum log-likehood estimations.

MOGA/NSGA-II Settings

For all our experiments we maintained the settings in Table 1. Depending on the case and the outliers percentage we varied the number of generations; this information is shown in Table 2.

Table 1. Design parameters for NSGA-II

NSGA-II parameter	Value
Population size	20 individuals
Crossover probability	0.8
Mutation probability	0.2
Distribution index for crossover	15
Mutation index for mutation	20

Table 2. Number of generations used in our experiments.

%Outliers	Disjoint	Overlapped	Nested
20%	10	25	2
40%	15	40	2
60%	25	80	2
80%	35	125	2
100%	80	150	2

For the selection of ellipses from $\mathscr{S}$ to $\mathscr{L}_{\text{explore}}$ we set a threshold $T_{\text{init}} = 40$ for all the experiments and we set manually the PEARL initialization set size v as shown in Table 3.

Table 3. PEARL initialization size for each experiment (number of solutions generated by the MOGA).

%Outliers	Disjoint	Overlapped	Nested
20%	50	30	70
40%	50	30	70
60%	70	30	70
80%	80	30	70
100%	70	30	70

5.1   Comparison

In order to make MOGA+PEARL comparable with the basic version of PEARL approach, we execute both methods with the same number of initial solutions $|\mathscr{L}_0|$. The idea is to compare PEARL's performance with $\mathscr{L}_0$ generated randomly against it generated using MOGA.

First, we run the MOGA+PEARL counting the number of MOGA executions required and the correctly found ellipses. Then we run the PEARL approach using the product of the MOGA executions by the MOGA's population size as the size for $\mathscr{L}_0$ and we count the correctly found ellipses. We perform this sequence for each data set case, outliers percentage and execution number, and we compare the statistics for the correctly found ellipses of both approaches.

5.2   Results

For each of the 15 data sets we performed 30 independent executions of the two algorithms, MOGA+PEARL and PEARL. Since we know a priori the number of points conforming the ground truth ellipses (126 data points) we consider an ellipse correctly detected if the number of neighborhood points is at lest 85% (107 data points).

Table 4. Results for the disjoint case

%Outliers	PEARL+MOGA			PEARL		
	Min	Max	Mean	Min	Max	Mean
20	2	4	3.67	0	4	1.40
40	2	4	3.93	0	3	1.00
60	3	4	3.77	0	3	1.13
80	2	4	3.63	0	2	0.60
100	2	4	3.60	0	3	0.63

Table 5. Results for the nested case

%Outliers	PEARL+MOGA			PEARL		
	Min	Max	Mean	Min	Max	Mean
20	3	4	3.80	1	4	3.23
40	2	4	3.57	0	4	2.63
60	2	4	3.23	0	4	1.57
80	2	4	3.13	0	4	1.20
100	1	4	3.17	0	3	1.03

Table 6. Results for the overlapped case

%Outliers	PEARL+MOGA			PEARL		
	Min	Max	Mean	Min	Max	Mean
20	3	4	3.90	0	3	1.60
40	3	4	3.80	0	3	0.97
60	3	4	3.87	0	3	0.63
80	3	4	3.87	0	2	0.83
100	3	4	3.60	0	2	0.53

The statistics of this experiment are shown in Tables 4, 5 and 6, i.e., the minimum, maximum, and the mean of the correctly found ellipses in the 30 execution.

In the Tables 4, 5 and 6 we observe that (PEARL+MOGA) obtains a better performance that PEARL. We observe that mean of the correctly found ellipses is higher for all data sets. This behavior is consistent for the minimum and the maximum, (PEARL+MOGA) obtains the four ellipses as max in all data sets, while PEARL obtains the four ellipses for the nested case, except when the 100% of outliers. For min, MOGA+PEARL obtains more than 2 ellipses in 13 of the 15 data sets, contrasting PEARL obtains one ellipse for only one data set.

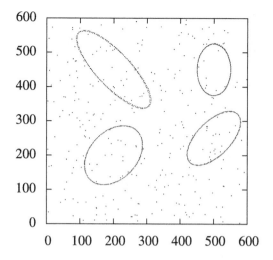

Fig. 7. Final result of MOGA+PEARL for the disjoint case with 100% of outliers.

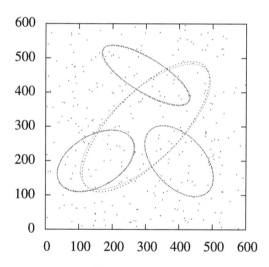

Fig. 8. Final result of MOGA+PEARL for the overlapped case with 100% of outliers.

From other point of view, we observe that the nested case is harder to solve by PEARL+MOGA, from the 15 cases, we obtain the minimum mean for the nested case with 80% of outliers, which is very close to the 100% of the same case. Some instances of the final output are shown in Figs. 7, 8 and 9, we show the three cases with 60% of outliers.

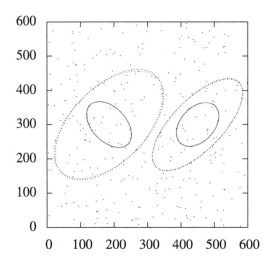

Fig. 9. Final result of MOGA+PEARL for the nested case with 100% of outliers.

6   Discussion

From the results in Sect. 5.2 we observe an improvement of the MOGA+PEARL over the PEARL approach in terms of the correctly found ellipses. We attribute this improvement to the robustness to local optima of the MOGA algorithm. The crossover operator improves ellipses over the generations and the mutation operator allows to avoid local optima. These two characteristics allow the MOGA+PEARL approach to obtain multiple ellipses with high amount of points in their neighborhood and also with small geometric error. The solutions obtained by the MOGA algorithm can not be considered alone as the final solution. PEARL solves various aspects that MOGA alone don't. These aspects include the estimation of the minimal number of model instances that explain the data points, reduction of similar or repeated models, local optimization for ellipses, and data points segmentation.

Analyzing MOGA+PEARL, we observe some aspects to mention. Depending on MOGA and PEARL own parameters as the number of generations and the $\mathscr{L}_0$ size, one can tune the grade that PEARL or MOGA work to compute the final solution. From Tables 3 and 2 we observe that the nested case requires more from PEARL than from the MOGA. We attribute this behavior to the proposed MOGA exploration strategy. To guide the MOGA to explore different search space regions we generate solutions far from archive; depending on the ellipses arrangement it can happen that the use of the archive obstruct the solution process. Since for archive we compute distance from ellipse centers, and because in the nested data sets we have concentric ellipses, it results better to use more random solutions, i.e., less generations for MOGA and more initial solutions for PEARL, although (from Tables 5) we observe that it is still better to use the hybrid approach over PEARL alone. The proposed archive building requires MOGA executions to be executed sequentially. This aspect is an aspect to improve since multiple MOGA could be executed in parallel.

7 Conclusions

We presented an hybrid approach to solve the multi-ellipse fitting problem based in initializing the PEARL method with solutions from a multi-objective genetic algorithm. We were able to solve challenging data sets, up to 100% of outliers with the proposed approach and we were able to improve the performance of the original PEARL algorithm with our approach. As main results we present:

- A combined approach that exploits the global search of the MOGA and the local search of the PEARL method.
- A method capable to solve the multiple ellipse fitting problem in presence of high percentage of outliers.

As future work we plan to study better techniques for MOGA exploration strategies, parallel executions for MOGA, and also the study of our approach with different levels of ellipse normal noise.

References

Bai, X., Sun, C., Zhou, F.: Splitting touching cells based on concave points and ellipse fitting. Pattern Recognit. **42**(11), 2434–2446 (2009). https://doi.org/10.1016/j.patcog.2009.04.003

Charoenpong, T., Jantima, T., Chianrabupra, C., Mahasitthiwat, V.: A new method to estimate rotation angle of a 3D eye model from single camera. In: 2015 International Conference on Intelligent Informatics and Biomedical Sciences (ICIIBMS), pp. 398–402 (2015). https://doi.org/10.1109/ICIIBMS.2015.7439474

Cheng, CW., Ou, WL., Fan, CP.: Fast ellipse fitting based pupil tracking design for human-computer interaction applications. In: 2016 IEEE International Conference on Consumer Electronics (ICCE), pp. 445–446 (2016). https://doi.org/10.1109/ICCE.2016.7430685

Cruz-Díaz, C., de la Fraga, LG., Schütze, O.: Fitness function evaluation for the detection of multiple ellipses using a genetic algorithm. In: 2011 8th International Conference on Electrical Engineering Computing Science and Automatic Control (CCE), pp. 1–6 (2011). https://doi.org/10.1109/ICEEE.2011.6106652

Cruz Hernández, H., de la Fraga, LG.: A multi-objective robust ellipse fitting algorithm. In: NEO 2016 Results of the Numerical and Evolutionary Optimization Workshop NEO 2016 and the NEO Cities 2016. Studies in Computational Intelligence, vol. 731, pp. 141–158. Springer, Cham (2018)

Deb, K., Pratap, A., Agarwal, S., Meyarivan, T.: A fast and elitist multiobjective genetic algorithm: NSGA-II. IEEE Trans. Evol. Comput. **6**(2), 182–197 (2002). https://doi.org/10.1109/4235.996017

Delong, A., Osokin, A., Isack, H.N., Boykov, Y.: Fast approximate energy minimization with label costs. Int. J. Comput. Vis. **96**(1), 1–27 (2012). https://doi.org/10.1007/s11263-011-0437-z

Duda, R.O., Hart, P.E.: Use of the Hough transformation to detect lines and curves in pictures. Commun. ACM **15**(1), 11–15 (1972). https://doi.org/10.1145/361237. 361242

Fischler, M.A., Bolles, R.C.: Random sample consensus: a paradigm for model fitting with applications to image analysis and automated cartography. Commun. ACM **24**(6), 381–395 (1981). https://doi.org/10.1145/358669.358692

Fitzgibbon, A., Pilu, M., Fisher, R.B.: Direct least square fitting of ellipses. IEEE Trans. Pattern Anal. Mach. Intell. **21**(5), 476–480 (1999). https://doi.org/10.1109/34.765658

Fornaciari, M., Prati, A., Cucchiara, R.: A fast and effective ellipse detector for embedded vision applications. Pattern Recognit. **47**(11), 3693–3708 (2014). https://doi.org/10.1016/j.patcog.2014.05.012

de la Fraga, LG., Cruz-Diaz, C.: Fitting an ellipse is equivalent to find the roots of a cubic equation. In: 2011 8th International Conference on Electrical Engineering, Computer Science and Automatic Control, pp. 743–746 (2011)

Isack, H., Boykov, Y.: Energy-based geometric multi-model fitting. Int. J. Comput. Vis. **97**(2), 123–147 (2012). https://doi.org/10.1007/s11263-011-0474-7

Johansson, E., Johansson, D., Skog, J., Fredriksson, M.: Automated knot detection for high speed computed tomography on Pinus sylvestris L. and Picea abies (L.) Karst. using ellipse fitting in concentric surfaces. Comput. Electron. Agric. **96**, 238–245 (2013). https://doi.org/10.1016/j.compag.2013.06.003

Jung, Y., Lee, D., Bang, H.: Study on ellipse fitting problem for vision-based autonomous landing of an UAV. In: 2014 14th International Conference on Control, Automation and Systems (ICCAS), pp. 1631–1634 (2014). https://doi.org/10.1109/ICCAS.2014.6987819

Kanatani, K.: Motion segmentation by subspace separation and model selection. In: Proceedings of the Eighth IEEE International Conference on Computer Vision, ICCV 2001, vol. 2, pp. 586–591 (2001). https://doi.org/10.1109/ICCV.2001.937679

Kim, J.S., Gurdjos, P., Kweon, I.: Geometric and algebraic constraints of projected concentric circles and their applications to camera calibration. IEEE Trans. Pattern Anal. Mach. Intell. **27**(4), 637–642 (2005)

Liang, J., Wang, Y., Zeng, X.: Robust ellipse fitting via half-quadratic and semidefinite relaxation optimization. IEEE Trans. Image Process. **24**(11), 4276–4286 (2015). https://doi.org/10.1109/TIP.2015.2460466

Liao, M., Zhao, Y.Q., Li, X.H., Dai, P.S., Xu, X.W., Zhang, J.K., Zou, B.J.: Automatic segmentation for cell images based on bottleneck detection and ellipse fitting. Neurocomputing **173**, 615–622 (2016). https://doi.org/10.1016/j.neucom.2015.08.006

Ma, Z., Ho, K.C.: Asymptotically efficient estimators for the fittings of coupled circles and ellipses. Digit. Signal Process. Rev. J. **25**(1), 28–40 (2014). https://doi.org/10.1016/j.dsp.2013.10.022

Mahajan, M., Nimbhorkar, P., Varadarajan, K.: The planar k-means problem is NP-hard. Theor. Comput. Sci. **442**, 13–21 (2012). https://doi.org/10.1016/j.tcs.2010.05.034

Masuzaki, T., Sugaya, Y., Kanatani, K.: Floor-wall boundary estimation by ellipse fitting. In: 2015 IEEE 7th International Conference on Cybernetics and Intelligent Systems (CIS) and IEEE Conference on Robotics, Automation and Mechatronics (RAM), pp. 30–35 (2015). https://doi.org/10.1109/ICCIS.2015.7274592

Mukhopadhyay, P., Chaudhuri, B.B.: A survey of Hough transform. Pattern Recognit. **48**(3), 993–1010 (2015). https://doi.org/10.1016/j.patcog.2014.08.027

Panagiotakis, C., Argyros, A.: Parameter-free modelling of 2D shapes with ellipses. Pattern Recognit. **53**, 259–275 (2016). https://doi.org/10.1016/j.patcog.2015.11.004

Rousseeuw, P., Leroy, A.: Robust Regretion and Outlier Detection. Wiley, New York (2003)

Rueda, S., Knight, CL., Papageorghiou, AT., Noble, JA.: Oriented feature-based coupled ellipse fitting for soft tissue quantification in ultrasound images. In: 2013 IEEE 10th International Symposium on Biomedical Imaging, pp. 1014–1017 (2013). https://doi.org/10.1109/ISBI.2013.6556649

Vincent, E., Laganiere R.: Detecting planar homographies in an image pair. In: Proceedings of the 2nd International Symposium on Image and Signal Processing and Analysis ISPA 2001, In Conjunction with 23rd International Conference on Information Technology Interfaces, pp. 182–187. IEEE (2001). https://doi.org/10.1109/ISPA.2001.938625

Analyzing Evolutionary Art Audience Interaction by Means of a Kinect Based Non-intrusive Method

Francisco Fernández de Vega[1]([⊠]), Mario García-Valdez[2]([⊠]), J. J. Merelo[4],
Georgina Aguilar[2], Cayetano Cruz[1], and Patricia Hernández[3]

[1] Universidad de Extremadura, Badajoz, Spain
fcofdez@unex.es
[2] Instituto Tecnológico de Tijuana, Tijuana, Mexico
mario@tectijuana.edu.mx
[3] University of Seville, Seville, Spain
patriciapahr@gmail.com
[4] Computer Architecture and Technology, University of Granada,
Granada, Spain
jjmerelo@gmail.com

Abstract. This paper analyzes the perception by the audience of Evolutionary Works of Art which were produced by means of the *unplugged evolutionary algorithm*. The long term goal is to study if genetic operations applied by artists in an evolutionary art work are consistently understood by the audience visiting the art exhibit. Yet, we need to involve the audience in the experiment, so that enough data can be retrieved and compared with the way the artists work. Thus, by means of a series of experiments that took place in an art exhibit environment, we analyze audience behavior using two different approaches, the more traditional survey based approach and a new non-intrusive methodology which relies on a depth camera. We show how the latter can largely increase the amount of data collected. Furthermore, it allows to measure both the time spent by a given person in front of an art work, but also to collect additional features about his face, where he is looking and basic gesture recognition. These features can be used to predict their personal attitude and feelings when facing a given art work. Although we describe here preliminary results, they allow us to conclude the pertinence of the approach.

Keywords: Evolutionary Algorithms · Art · Creativity

1 Introduction

Evolutionary Algorithms (EAs), as well as its interactive version, the Interactive EAs (IEA), have been successfully applied to the solution of optimization problems, as well as to promote creativity when producing art works [10].

© Springer International Publishing AG, part of Springer Nature 2019
L. Trujillo et al. (Eds.): NEO 2017, SCI 785, pp. 108–123, 2019.
https://doi.org/10.1007/978-3-319-96104-0_5

Human beings have played the role of fitness evaluators within the IEA, applying aesthetic selection when creative processes are faced. But still new possibilities for human involvement have been proposed with the *unplugged evolutionary algorithm*: it provides new means for the user to run the algorithm in a different fashion; main EA operations are now performed by users, including selection, crossover, mutation, evaluation, etc., while computers are in charge of more administrative tasks: storing and managing populations [13].

Since its proposal, Unplugged Evolutionary Algorithms (UEAs) have been successfully applied to generate collective art works that have gained public attention: the latest experiment has produced a collective art work that has been selected as one of the winners at the international *Show Your World 2017* art competition[1]. But before that, other art works produced by means of the UEA were successfully shown to the public in different cities around the world: Cancn, Amsterdam, Madrid, Vancouver, etc., being one of them awarded with the ACM Gecco evolutionary art design and creativity award in 2013 [5].

Results obtained encouraged us to continue analyzing other components of the algorithm, but we also understood that not only artists but also the audience should be included in the loop, to check whether the evolutionary approach and operations artists apply are coherently interpreted by the audience. Although an initial experiment was performed involving the audience, using a series of questions to be answered, we quickly understood, as described below, that new non-intrusive approaches would be needed to properly evaluate audience experience without disturbing their participation on art exhibits [3]. Actually, previously applied surveys typically produced fatigue in the audience, when a large number of questions must be answered by carefully examining the art works shown. New alternatives are required that allows to obtain information from the audience, avoiding fatigue and allowing them to more effectively enjoy the experience.

This paper presents a proposal for attaining such a goal: being able to extract information from the audience while they are simply looking at the artworks in the exhibit. By using off-the-shelf human-computer interaction devices we are able to extract useful information. Not only that, the first experiments show how a correlation may be found between the data extracted from the audience and some of the genetic operations applied by artists. The methodology thus provides a better understanding of audience perception, and relevant information that may allow in the future to improve the UEA. In addition, the data extracted allows a first comparison between audience perception and artists' operations. This comparison may help to understand if creativity developed by artists are properly grasped by the audience.

The rest of the paper is organized as follows: Sect. 2 presents the literature review. Section 3 proposes the methodology aimed at analyzing public reaction. Section 4 describes the experiments performed while Sect. 5 presents results. Finally, we draw our conclusions in Sect. 6.

---

[1] http://www.reartiste.com/juried-exhibition-show-your-world-2017/.

2   Humans in the EA Loop

Several decades ago, researchers noticed the interest of allowing users to interact with the EA, so that some part of the algorithms were run by users. This way, the IEA was successfully applied to problems requiring aesthetic evaluations. Moreover, new means to allow users play different roles within the EA were envisioned over the years, thus producing new versions of the EA particularly adapted to analyze creative processes developed by artists, and the Unplugged Evolutionary Algorithm (UEA) was born [14].

The UEA is a version of the IEA that is "run" by artists when producing a collective art work. Instead of using the computer to run the EA and allow users -artists- to interact, computers are removed, and the human beings are thus in charge of applying every step of the algorithm, not just evaluation, to produce results: evaluation, selection and crossover. Somehow, the UEA provided artists with a methodology for artistic creation, while computer scientists gain access to human creativity analyzed from the point of view of EAs.

The basic idea behind UEAs is to allow artists to apply the EA as the methodology along the art creation process. Therefore, all operations, including evaluation, selection, crossover, mutation, etc., are performed by human beings, artists, that simultaneously complete a form describing the reasons behind every decision they take, and operation they apply. Thus, once the experiment has been completed, we have a bunch of information to be analyzed and contextualized under the EA perspective, that could help to better understand artists creativity.

Yet, in every artistic process, artists are not the only ones involved. The work is always produced to be confronted with the public, and thus the audience response is very important for the artists behind the work. Therefore in any computer mediated art process, the reaction of the audience should be studied as carefully as the algorithm is designed. Thus, we could summarize the main steps of the methodology provided by UEAs as follows: (i) Unplugging the EA to be executed by artists; (ii) artists produce a collective work while applying the algorithm; (iii) operations are then analyzed, and the creative process evaluated; (iv) Audience perception is analyzed and compared with data collected from the previous analysis.

Although in previous works we conducted a number of surveys to check whether the audience properly understood what the artists did [7], fatigue of the audience was seen as a drawback of the method. Actually, the problem of fatigue had already been studied in the context of IEA [4]. In both cases, a long interaction with the medium produce boredom: in the IEA the need for the user to evaluate a large number of *individuals* for many generations; in the UEA when all of the works collectively produced by the artists team -*individuals*- must be analyzed by the audience to answer a series of questions about the genetic operations for every work produced and also figure out the genealogy of the whole work. Given that 50 or 60 works were produced in previous experiments [13,14] a complete series of questions would include five to ten per work, which produce more than 500 questions. We had instead to focus in just some of the works, thus avoiding to repeat every question 50 or 60 times. Thus the surveys include about thirty questions. Even in this case, the audience felt tired and bored, given that for every question a detailed analysis of the picture was required. Unfortunately, when all of the surveys are completely filled, which is the best situation, partial information is only retrieved, given that many questions had to be removed from the ideal survey for practical reasons, as described above.

Therefore, an alternative is required if we want to capture as much information as possible from the audience, and simultaneously want them to enjoy the experience of visiting an art exhibit: a different approach must be applied, so that information from the audience can be extracted without disturbing the *art experience*.

We can find in the literature interactive installations where sensors has been already used [11]. For instance, sensor based interactive installations were set up as a learning auxiliary tool for children with autism [2]; similarly, remote operation of machinery and drone robot controls have been implemented using the same kind of sensors [1]. Examples can be also found in medical applications, such as those related to rehabilitation of patients of Parkinson's disease [9], sport related injuries [1] and burn rehabilitation [8], all of them using sensors to analyze patients behavior. We can thus conclude that the flexibility of this kind of devices could be also useful for the goal we pursue.

We thus propose in the third step of our methodology to use a non-intrusive human-computer interaction system, which should be capable of obtaining similar information than that acquired before by means of surveys. Although this goal may be ambitious, we present below our preliminary work towards that goal, which allows us to be optimistic with the idea. The proposed non-intrusive interaction system has other applications, for instance, the aesthetic evaluation of computer-generated art, graphic design or texts [6]. But in UEAs, the non-intrusive property is mandatory, because the whole point of them is to take out as much as possible the computer from the loop.

3   Non-intrusive Human-Computer Interaction System

In order to measure the amount of engagement users have with each painting, a natural user interface device is used for facial tracking. In this project in particular the *Kinect sensor Version 2.0 for Xbox One* was chosen, but other similar sensors could be used; for instance the *RealSense D400* to be available at the beginning of 2018. The Kinect sensor, uses an infrared camera and is capable of measuring depth even in dim light. The software developer kit offers libraries for face and whole body tracking but a disadvantage of the solution is that it has proprietary software. At the time of writing the unit has been discontinued and is no longer been manufactured [15]. Yet, other alternatives are available from different manufacturers, so that the experiments described below could be reproduced using a different sensor.

The sensor we use can track up to 6 persons at once, and for each face it provides a basic set of information: where the face is, where it is looking, basic expressive information (if the person is smiling), if the face has glasses, and if

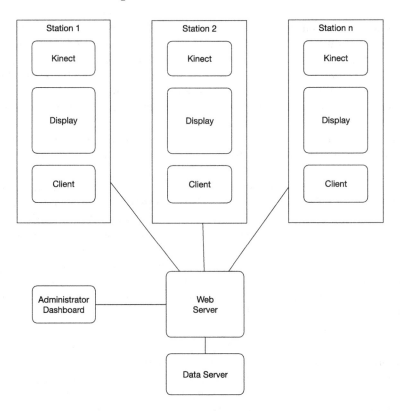

Fig. 1. Components of the non-intrusive human-computer interaction system.

the eyes are open. With this information the total amount of interaction can be measured and correlated with engagement. User engagement in this case is defined as the amount of time users spend staring at a painting. This is a simplification, because user engagement in general is a deeper behavior, that occurs when a person is fully concentrated or absorbed in the action that he or she is doing.

The hardware components of the installation are presented in Fig. 1. For each painting or display presenting a group of images there is a sensor station installed. Each station consists of a compact computer, needed to run the Kinect application, and a Kinect sensor on the top or bottom of the display. For displaying the art a monitor or LCD projector appropriate to the size of the venue or a physical canvas is installed together with the station. Another computer is needed to act as the server, collecting information from all the stations and serving the image files to be presented in each station. A detailed description of each component is presented next.

3.1   Kinect Client-Server Architecture

When installing a station, the Kinect client application has to be configured simply by setting the Kinect manufacture identifier and the server IP. The Kinect

ID will identify all readings sent from the current station, this is important because there could be many sensors reading and sending at the same time. The sensor was set to work at a frame rate of one second, firing a Frame Event only if a face was recognized in that frame. For each Frame Event there could be up to six Face Frames. For each Face Frame the client application sends a `FrameEvent` object to the server with the following attributes:

Fig. 2. Event Frame data, showing face boxes and expressive information for two subjects. We may notice that in the image captured by the web camera, the positions of the subjects is inverted.

frame_id: This is the frame identifier, several Frame Events could have the same frame_id because there will be a frame event for each face detected.

kinect_id: As said before, this is the identifier of the station.

date_time: This is a time stamp of the event, this is an important data because it will tie the currently displayed image with the event.

happy: This is information is included in the API's `FreameEvent`, it could have the values: **yes**, **no** and **unknown**.

engaged: This attribute indicated if the face is looking at the screen. It has one of the values: **yes**, **no** and **unknown**.

looking_away: Indicates if the user is looking to away from the screen. It has one of the values: **yes** and **no**.

face_index: This attribute indicates the index of the Recognized Face is a value from 0 to 5. Once a face is tracked is assigned a value and is kept until losing track. If the same person is out of frame for a few seconds a new face_id could be generated.

face_box: Screen coordinates of a box drawn for the face. See Fig. 2.

Data returned by the stations are stored by the node.js server in a central database, in these experiments both the redis in-memory key-value store and

MySQL database system were used. In order to synchronize the reading with the media being displayed at the station, an additional time stamp is added by the server at the time of insertion.

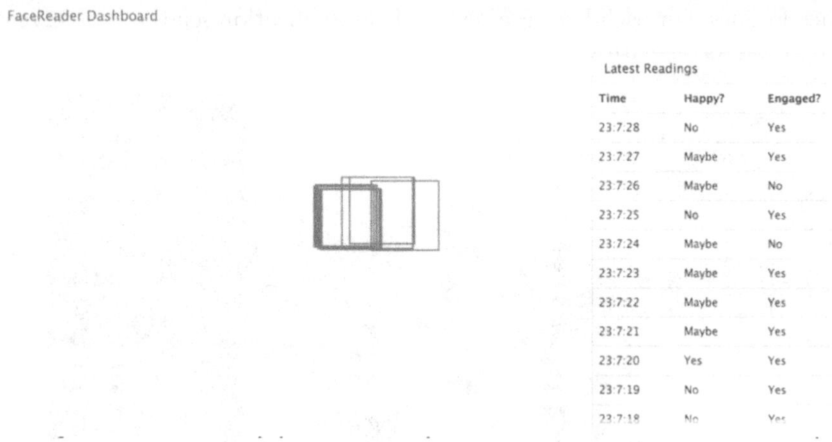

FaceReader Dashboard

	Latest Readings		
	Time	Happy?	Engaged?
	23:7:28	No	Yes
	23:7:27	Maybe	Yes
	23:7:26	Maybe	No
	23:7:25	No	Yes
	23:7:24	Maybe	No
	23:7:23	Maybe	Yes
	23:7:22	Maybe	Yes
	23:7:21	Maybe	Yes
	23:7:20	Yes	Yes
	23:7:19	No	Yes
	23:7:18	No	Yes

Fig. 3. Web-based dashboard application showing readings from a remote station

3.2   Web Client Application

In each station, a JavaScript client application is constantly polling the server requesting a file to be displayed. The request also includes the Kinect identifier of the station, if there is a new image available the new image is displayed. The Web application, can display images or videos, play audio, or run scripts written in the Processing language. There is also a dashboard type application for the system administrator. In that application, the current activity of the system is displayed. There is a screen displaying almost in real-time, face-boxes and other Face Frame information from each station. The graphical user interface is presented in Fig. 3. On the left side there is an area where the information about the last frames is presented. Each box represents the coordinates of a face as reported by the Kinect sensor, in this case a there are a few overlapped boxes indicating the slight movement of the user. All this information can also be used later to analyze the behavior of users in front of a painting over time. On the right, again the information about the last frames is presented, but indicating the facial expressions.

The web server is responsible for coordinating all the stations. A REST API is implemented to deliver media to each station, and also to receive frame events. The server communicates with the back-end databases and file system. Databases contain the Event Frame records and in the file system, images and videos are stored.

Post processing is required to join Frame Event information with the corresponding media presented in the station at the same time. Data is aggregated at different granularities and stored in a MySQL database. With this data, further analysis can be conducted or a fitness estimation could be calculated for each media.

4   Experiments

The idea behind the methodology is to confirm that non-intrusive methods for analyzing audience perception can obtain useful information, comparable to those extracted with previous approaches based on surveys, but now avoiding the user fatigue produced by those surveys.

Therefore, we present below a series of experiments aiming at comparing behavior of the audience when surveys are provided, and also when non-intrusive methods are applied. The idea is to conduct a number of experiments that can be compared to results already available for audience perception on the XY, ans also XYZ experiment developed under the UEA methodology[2].

We will first present a summary of each of the methods, and then will proceed by comparing results obtained with the first one, and published in [7], with those that we may obtain using the second one.

4.1   Analyzing Audience Perception by Means of Surveys

As already described, a systematic analysis of the audience comments on the results shown are very useful to see if creative processes developed by artists are properly understood by the viewer. Thus, the first approach for such an analysis was based on surveys conducted among the audience.

We planned a number of art exhibits, and surveys were designed to be collected with useful information for analyzing audience perception. Unfortunately, as we will show below in the Results section, surveys had to be elaborated again and even in this case the success was not properly attained. Two cities from Spain (Merida and Jerez de la Frontera) an another one from Argentina (Buenos Aires) were selected for the experiments.

4.2   Capturing Data from Audience by Means of Non-intrusive Methods

Alternatively, a different setting was designed to test the non-intrusive methodology described before. As an experiment, four paintings were exposed in pairs, each pair for one day. Two stations were deployed on the campus of the Tijuana Institute of Technology, in a busy hallway next to the cafeteria, displaying one painting each. The number of students on each day was similar as there was no special event or activity that could increase traffic. Two days before the first experiment a dummy experiment was conducted in order to eliminate the extra attention generated by the new stations in the hallway. On each station a high

---

[2] http://xyz-project.herokuapp.com.

resolution web camera was also installed, to record the interaction for validation purposes. The four paintings (two pairs) selected for the experiments are: The Kiss by Gustav Klimt (1908) Fig. 4 and Ophelia by John Everett Millais (1852) Fig. 5 for the first pair to be shown, and The Birth of Venus by Sandro Botticelli (1486) Fig. 6 and Impossible Love by Marc Brunet (2006-2016) Fig. 7 as the second pair.

The idea is the to analyze the interaction of users for each of the pairs shown.

Fig. 4. The Kiss by Gustav Klimt (1908)

Fig. 5. Ophelia by John Everett Millais (1852)

5   Results

This section presents and compare results obtained with both methods previously described: (i) survey based approaches; (ii) non-intrusive methods for audience analysis.

5.1   Analyzing Audience Perception by Means of Surveys

As described above, three art exhibits were organized with collective works XY and XYZ, and 50 surveys were finally collected, which included questions about relationships among parents-children, elitism, figurativeness, feelings translated (love, fragility, mystery, etc), initial population, etc. Although surveys allow to

Fig. 6. The Birth of Venus by Sandro Botticelli (1486)

Fig. 7. Impossible Love by Marc Brunet ()

Fig. 8. Students completing the XY survey

ask anything required, the process of collecting surveys, as we show below, was not easy: more than one thousand people visited the art exhibits but only 50 surveys were correctly filled.

The first art exhibit where surveys were distributed among the audience took place in Jerez de la Frontera, a mid-size southern spanish city. Art students attended the exhibition and were asked to collaborate (see Fig. 8). Unfortunately, as described in [7] "The result of the analysis could not be performed because there were incomplete surveys. The reason was that students didn't understood properly some of the questions posed. Another reason is the fatigue due to the number of questions included". Thus, this first failed attempt to use surveys as the main source of information led us to re-elaborate the surveys again, and a new questionnaire was formulated, with smaller number of questions more clearly specified.

Therefore, a second attempt to collect information from the audience took place in Palermo, Argentina, during the 10th Ibero-American Design Meeting 2015, at the University of Palermo. All viewers were professional designers. Yet, few people from the audience completed the survey (just 8 surveys completed). Again, the problem was that the audience typically attended an art exhibit to enjoy, and usually refuse surveys because they disturb the otherwise rewarding experience.

Finally, we also showed the work to Industrial Design Engineering and Product Development students, at the University of Extremadura, Spain (42 surveys). They were our own students, and this was the reason for the success: completing the survey was a compulsory assignment for the subject. Therefore, only when completing the survey was a compulsory activity for a subject we were able to collect enough information.

Although interesting information was obtained, which finally allowed an analysis of some of the operations applied by artists (see [7]), our main goal was not achieved, as summarized in Table 1. The main conclusion is that audience refuse to complete surveys when attending art exhibits, so a different mechanism is required to collect useful information for analyzing public's reaction to the artwork on display.

Although we have obtained further information regarding how feelings are understood, or mechanism for building the initial generation, we will focus on those shown above for the experiments involving survey-free non-intrusive methods.

Table 1. Surveys collected.

Type	City	Surveys collected	Audience
Elective	Jerez (Spain)	0	Hundreds
Elective	Palermo (Argentina)	8	Hundreds
Mandatory	Merida (Spain)	42	42

Table 2. Readings from each station.

Painting	Expression	Yes	No	Maybe	Unknown	Total
The Kiss	Happy	107	266	188	211	772
	Engaged	183	466	123	0	
Ophelia	Happy	179	488	294	348	1269
	Engaged	332	681	256	0	
The Birth of Venus	Happy	188	907	313	1759	3167
	Engaged	374	2545	248	0	
Impossible Love	Happy	337	1218	446	3260	5261
	Engaged	405	4474	382	0	

5.2 Capturing Data from Audience by Means of Non-intrusive Methods

In Table 2, the total number of frames captured by each sensor is presented for each painting. As mentioned in Sect. 3.1 only those frames with at least one face recognized by the sensor are recorded at a frame rate of two seconds. Each record has the following attributes: painting, happy, engaged. Both *happy* and *engaged* attributes can have one of the following values: *yes*, *no*, *maybe* or *unknown*. In the table, the column named *total* has the total number of frames captured for each painting. In each row, a breakdown for each of the expressions is presented. For instance, The Kiss had a total of 772 records or frames with faces in them. If we consider the *happy* attribute, out of those 772 records 107 had *yes*, 266 *no*, 188 *maybe* and 211 *unknown*. Even regardless of the expression, the total number of records indicate persons facing the painting, but not necessarily looking directly at it or smiling. The table is ordered by the total number of records, we can see that the paintings of the first day The Kiss and Ophelia had considerably less attention than those of the second day. Again, the traffic on the hallway was the same on both days and they where presented the same amount of time. More people stared at The Impossible Love painting with 5261 records but only a small number of people stared directly at it. Summarizing, even when the exhibit was located in an open area and the audience was not directly interested in visiting an art exhibit, the methodology has shown how a larger amount of data can be captured.

Yet, we were not only interested in the amount of data, but also in whether data obtained could be trusted. Therefore an additional survey based experiment was designed so that we can establish a correlation between data obtained by means of surveys and those extracted by sensors.

Therefore, a *Likert* survey was designed and answered by 102 students in an additional experiment at the same university. A small gallery with the previous paintings was presented to students, after receiving a brief introduction to genetic algorithms they were asked to assign fitness values to each painting according to their tastes. The five options where: Like very much, Like, Neutral, Dislike, Dislike very much. These answers can be seen also as a five star rating system, from one to five stars. The result of the survey is shown in Table 3.

Table 3. Survey results.

Painting	NA	Like much	Like	Neutral	Dislike	Dislike much
The Kiss	0	18	41	29	14	0
Ophelia	1	23	47	19	11	1
The Birth of Venus	0	32	41	24	5	0
Impossible Love	0	55	23	17	4	3

As we may notice, Ophelia was preferred from the first pair of paintings over the Kiss, while Impossible love was the preferred one from the second pair of paintings. We can compare these results with those shown in Table 2: Ophelia and Impossible love were both the works displaying higher values of happiness and engagement for each of the pairs shown. Therefore, we see that data extracted with the kinnect device are coherent with data that surveys provide.

But we can also compare this results with artists preferences when developing the evolutionary art work, which allows us to analyse whether best valued art works are the same for the audience and the artists. If we check *elitism* information for both experiments XY (Ophelia and Birth of Venus) and XYZ (The

Kiss and Impossible Love), we see that Ophelia was second most selected works for XY [12], while Impossible Love was also second most selected one by artists in XYZ experiment [7]. Although there was no exact coincidence in the most valued work of art, in both cases the chosen works were in the group of the most selected. Thus we can conclude that although not a perfect coincidence in the preferred work, there is a correlation between artist's preferences and audience ones. To the best of our knowledge, this is the first time such an information is automatically captured from audience when evolutionary art is displayed.

Although still preliminary, we can thus conclude, that the results show that sensor-based non-intrusive methods allows to capture useful information when analyzing audience perception: (i) the audience is not disturbed; (ii) the amount of data increases significantly; (iii) information obtained is coherent with surveys performed, and (iv) data obtained allows to understand and confront audience perception with artists creativity. Future and more extensive experiments be useful to confirm the appropriateness of the approach presented. Another aspect that was not considered in these experiments is the effect of cultural background or context in the preferences of both audience and artists. No special attention was put on capturing additional data from the audience other than the basic survey.

6   Conclusions

This paper describes a new non-intrusive approach to the analysis of audience perception when attending art exhibits. To the best of our knowledge, this is the first time a non-intrusive device allows to analyse audience perception in the context of evolutionary art.

Although survey-based methods were used in previous works to extract useful information from the audience, the problem of user fatigue for completing surveys keep us from obtaining enough data to perform a wider analysis and comparison with the evolutionary process that artists had developed. Yet, this information should always be considered as is the case for more conventional art work: audience, critics, museums and galleries have always a role in the art world.

This paper shows for the first time how a kinnect device allows to capture a much larger amount of data otherwise impossible with surveys, given that audience generally rejects them and simply prefers to enjoy the art exhibits.

Thus a new sensor based methodology has been developed that allow us to capture audience perception avoiding public nuisance with polls.

A couple of experiments has been employed to capture several thousand interaction data, with useful information about the interest, engagement and feelings. Moreover, data has been confronted with another survey based experiment, and the correlation between sensor based information and answer provided by users is direct, which allow us to conclude that similar information -although in much larger amounts- can be captured without disturbing user experience.

Finally, the extracted data have been confronted with the way artists work in the evolutionary context, and have provided useful information to see the relationship between the operations of elitism and the favourite works of the public: we have seen that the most favourite works of the artists - which become the "elite" works of the next generations - are increasingly correlated with the works preferred by the public.

These encouraging results show the interest of the approach, and will allow us to design new art exhibitions that provide a massive capture and analysis of the public's reaction, which will hopefully be useful for designing better evolutionary art algorithms in the future.

Acknowledgments. The authors would like to thank Spanish Ministry of Economy, Industry and Competitiveness and European Regional Development Fund (FEDER) under projects TIN2014-56494- C4-4-P (Ephemec) and TIN2017-85727-C4-4-P (Deep-Bio); Junta de Extremadura FEDER, projects GR15068, GRU10029 and IB16035 Regional Government of Extremadura, Consejería of Economy and Infrastructure, FEDER.

References

1. Berra, E., Cuautle, J.R.: Interfaz natural para el control de drone mediante kinect. J. Cienc. Ing. **5**(1), 53–65 (2013)
2. Boutsika, E.: Kinect in education: a proposal for children with autism. Procedia Comput. Sci. **27**, 123–129 (2014)
3. Chou, S.Y., Lin, S.W.: Museum visitor routing problem with the balancing of concurrent visitors. In: Complex Systems Concurrent Engineering pp. 345–353 (2007)
4. Frade, M., Fernández de Vega, F., Cotta, C.: Evolution of artificial terrains for video games based on accessibility. In: European Conference on the Applications of Evolutionary Computation, pp. 90–99. Springer (2010)
5. Loiacono, D.: Gecco-2013 competitions. ACM SIGEVOlution **6**(2), 27–28 (2014)
6. Madera, Q., Castillo, O., García-Valdez, M., Mancilla, A.: A method based on interactive evolutionary computation and fuzzy logic for increasing the effectiveness of advertising campaigns. Inf. Sci. **414**, 175–186 (2017)
7. Moreno, L.N., Fernández de Vega, F., Rondán, P.H., García, C.J.C., Fernández, J.V.A.: Analysing creative models based on unplugged evolutionary algorithms. In: 2016 IEEE Congress on Evolutionary Computation (CEC), pp. 4570–4577. IEEE (2016)
8. Parry, I., Carbullido, C., Kawada, J., Bagley, A., Sen, S., Greenhalgh, D., Palmieri, T.: Keeping up with video game technology: objective analysis of xbox kinect and playstation 3 move for use in burn rehabilitation. Burns **40**(5), 852–859 (2014)
9. Pompeu, J.E., Torriani-Pasin, C., Doná, F., Ganança, F.F., da Silva, K.G., Ferraz, H.B.: Effect of kinect games on postural control of patients with parkinson's disease. In: Proceedings of the 3rd 2015 Workshop on ICTs for improving Patients Rehabilitation Research Techniques, pp. 54–57. ACM (2015)
10. Takagi, H.: Interactive evolutionary computation: fusion of the capabilities of ec optimization and human evaluation. Proc. IEEE **89**(9), 1275–1296 (2001)
11. Tieben, R., Sturm, J., Bekker, T., Schouten, B.: Playful persuasion: designing for ambient playful interactions in public spaces. J. Ambient Intell. Smart Environ. **6**(4), 341–357 (2014)

12. Fernández de Vega, F., Cruz, C., Hernández, P., Navarro, L., Gallego, T., Espada, L.: Arte Evolutivo y Computaci'on: Di'alogos de un experimento. Universidad de Extremadura (2015)
13. Fernández de Vega, F., Cruz, C., Navarro, L., Hernández, P., Gallego, T., Espada, L.: Unplugging evolutionary algorithms: an experiment on human-algorithmic creativity. Genet. Program. Evolvable Mach. **15**(4), 379–402 (2014)
14. Fernández de Vega, F., Navarro, L., Cruz, C., Chavez, F., Espada, L., Hernandez, P., Gallego, T.: Unplugging evolutionary algorithms: on the sources of novelty and creativity. In: 2013 IEEE Congress on Evolutionary Computation (CEC), pp. 2856–2863. IEEE (2013)
15. Wilson, M.: Exclusive: Microsoft has stopped manufacturing the kinect. https://www.fastcodesign.com/90147868/exclusive-microsoft-has-stopped-manufacturing-the-kinect. Accessed 11 Dec 2017

Optimal Control

Applying Control Theory to Optimize the Inventory Holding Costs in Supply Chains

Pablo M. Ayllon-Lorenzo, Selene L. Cardenas-Maciel,
and Nohe R. Cazarez-Castro$^{(\boxtimes)}$

Instituto Tecnológico de Tijuana, Tecnológico Nacional de México, Calz. del
Tecnológico S/N, Fracc. Tomás Aquino, 22414 Tijuana, Baja California, Mexico
`pablo.ayllon17@tectijuana.edu.mx`, {`lilettecardenas,nohe`}`@ieee.org`

Abstract. Based on the design of a proportional integral controller, the
intention is to eliminate the existing error between the level of planned
inventory and the actual level in the different parts of the supply chains
of goods and services. It is considered necessary to increase the performance of this indicator to optimize the inventory holding costs by
improving the profitability of the companies of the multiple chains. A
PI controller was developed according with the material balance equation, adjusted to the dynamic models of the supply chains. Likewise, the
network elements were classified into producers and non-producers and
saturation functions were assigned to each group. Subsequently, the system was simulated incorporating two types of deterministic demands: the
first one constant in time and the second one with a variation by season.
A linear relationship between the supply volumes was also established
for those segments of the chain that have more than one supplier at the
same time. It can be argued as a result of this simulation that the control
system designed for the supply networks solves the problems of regulation and monitoring. As a consequence of this effect, the satisfaction of
the demand of the end customer of the chain is guaranteed, maintaining
in an optimal state the levels of inventory as long as the models used in
the planning process are adequate.

Keywords: Supply chains · Control theory inventory · Holding cost
Optimize

1 Introduction

Supply chain management has a direct impact on the financial performance
of the companies that comprise it. In such a competitive environment, it is
extremely important to achieve an efficient flow to reduce the inventory holding
and transportation costs.

In principle, a supply chain is composed of a set of sequentially connected
production or distribution units, giving rise to information and material flow

© Springer International Publishing AG, part of Springer Nature 2019
L. Trujillo et al. (Eds.): NEO 2017, SCI 785, pp. 127–137, 2019.
https://doi.org/10.1007/978-3-319-96104-0_6

between its members (see Fig. 1). At the beginning of the chain the market demand acts as the driver of the dynamic processes evolving in the system, while at the end the producer of raw materials provides the material input. Typically, the market demand for a product exhibits different types of variations [1]:

1. small stochastic oscillations around a mean value (random changes of the demand),
2. rapid change with large amplitude (caused by promotions),
3. slow variations with large amplitude (due to seasonal, economical tendencies).

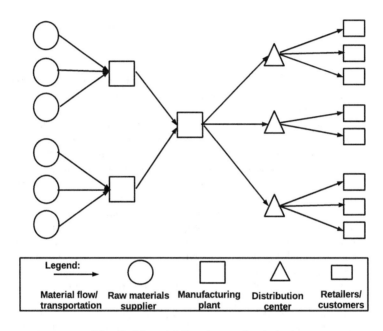

Fig. 1. Material flow in supply chains.

Formulating effective strategy requires a good understanding of what drives cost and service in a supply chain. This note introduces an inventory model that will enable you to quantifying the often subtle impact of both operational and structural changes in a supply chain. In addition - and perhaps more importantly - the model can help sharpen your intuition about what factors affect the operating performance of a supply chain [2].

Inventory is the stock of any item held in an organization. The aim is, naturally, to have the right amount, in the right place, at the right time and the right cost. Inventory management sets out to achieve just that. It is the process of directing and administering the holding, moving and converting of raw materials through value-adding processes to deliver finished products to the customer. The efficient and effective management of inventory (or stock) is important to almost every organization.

In simple terms, inventory management [3]:

1. is the set of policies and controls that monitor levels of inventory;
2. determines what levels of each product should be maintained;
3. identifies when stock should be replenished; and
4. decides how large orders should be.

The main objective of inventory management in the supply chain is to prevent the bullwhip effect. The causes and remedies of this effect are exposed in [4,5].

In recent years, classic and modern control theory have been developed to keep inventory in the chain within the desired parameters [6–8]. To this is added the analysis of multiple mathematical models to describe the behavior of the flow of the merchandise through the supply chains [9,10]. In the same way, several stability strategies have been designed, guaranteeing the functionality of these controllers [11].

In this context, this research presents a PI controller to optimize the inventory holding cost in supply chains based on the mathematical modelling [12] to design the control system. To solve the problem, a methodological sequence is used, starting by model selection of material flow in producer and non producers nodes, through the design of general system architecture, until the simulation results with the help of MATLAB Simulink Toolbox.

The rest of the work is organized as follows. The plant structure for the supply chain is developed in Sect. 2 (Mathematical modeling). Section 3 explains in detail the problem and formulation of control law. On the other hand, Sect. 3.2 explains the control law and saturation parameters. Finally, the main results and final comments are shown in Sects. 5 and 6 respectively.

2   Mathematical Modeling

According to the model [12], the inventory rate of change in all supply chain nodes is determined by the material balance equation

$$\frac{\mathrm{d}N_i}{\mathrm{d}t} = \lambda_i(t) - \lambda_{i+1}(t) \tag{1}$$

where N_i is the inventory level of node i, λ_i is the entry rate of node i, λ_{i+1} is the demand rate of node i, all this parameters $\in \mathbb{R}^+$ and in this article they are given in physical units; and following the sequence shown in Fig. 2.

For the analysis simplification, nodes are classified as producers nodes to those chain parts that add some modification on the final product (manufacturing plants), while the rest of elements will be considered non-producers nodes.

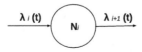

Fig. 2. Material flow sequence.

The producer nodes include other variables that determine their behavior, and is modeled as

$$\frac{\mathrm{d}\lambda_i}{\mathrm{d}t} = \frac{1}{T}\left[W_i\left(t\right) - \lambda_i\left(t\right)\right] \tag{2}$$

where T is the time to adapt to change of node i given in time units, W_i is the production policy of node i given in physical units.

3 Problem Statement and Formulation of the Control Objective

In this section, the problem statement and the control objective are presented. It is important to note that in the considered approach, the optimization of the inventory holding cost of a supply chain depends on having an optimal predicted final client demand (reference signal) and on having a controller capable to solve the regulation or tracking problem. In the rest of this chapter it is considered that the optimal predicted final client demand is a solved problem, and the main effort is applied to design the controller.

3.1 Optimization Problem

The main objective of this work is to guarantee the optimal inventory cost in the links of the chains. For this it is necessary to evaluate the structure of the total costs according with [13] and Fig. 3.

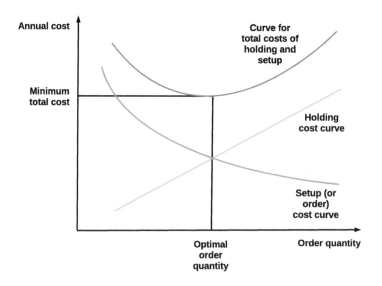

Fig. 3. Inventory costs structure.

The variable to be optimized will be the inventory holding cost and it is given by [14]:

$$Ihc_i = \frac{N_i}{2} * T_s * P * I \tag{3}$$

where $Ihc_i \in \mathbb{R}$ is the inventory holding cost of node i given in monetary units, $T_s \in \mathbb{R}$ is the storage time given in time units, $P \in \mathbb{R}$ is the unit price of material given in monetary units, $I \in \mathbb{R}$ is the storage rate expressed as a percentage of the unit price, it is a dimensionless number.

An optimization problem must be solved whose objective function is to minimize the Ihc, so the quantity of merchandise to be stored must be reduced since both variables are directly proportional.

3.2 Controller Design

The controller to use in this work is based on the study done in [15]. The Non-producer nodes are modeled as

$$\lambda_i = \begin{cases} \lambda_{max} \ if & u(t) \geq \lambda_{max} \\ u(t) \ if & 0 < u(t) < \lambda_{max} \\ 0 \ if & u(t) \leq 0 \end{cases} \tag{4}$$

with control law

$$u(t) = -k_p(N_i - N_{planned}) - k_i \int (N_i - N_{planned})dt + \lambda_{i+1} \tag{5}$$

where λ_{max} is the maximum capacity of entry rate of node i given in physical units, $N_{planned}$ is the level of inventory desired given in physical units. Note: For the purposes of this paper, this variable will guarantee maintaining the optimal inventory holding cost in the chain, $u(t)$ is the signal that belongs to the PI controller associated with the saturated control, k_p and k_i are the proportional and integral gains, λ_{max}, $N_{planned}$, k_p, $k_i \in \mathbb{R}^+$ and $u(t) \in \mathbb{R}$.

The system is in closed loop when $0 < u(t) < \lambda_{max}$, and replacing (1), (4) and (5) is obtained

$$\dot{N}_i = -k_p(N_i - N_{planned}) - k_i \int (N_i - N_{planned})dt. \tag{6}$$

The producer nodes are modeled as

$$W_i = \begin{cases} \lambda_{max} \ if & u(t) \geq \lambda_{max} \\ u(t) \ if & 0 < u(t) < \lambda_{max} \\ 0 \ if & u(t) \leq 0 \end{cases} \tag{7}$$

with control law

$$u(t) = -k_p(N_i - N_{planned}) - k_i \int (N_i - N_{planned})dt + \lambda_{i+1} \tag{8}$$

The system is in closed loop when $0 < u(t) < \lambda_{max}$ and replacing (1), (2) and (7) is obtained

$$\dot{\lambda}_i = \frac{1}{T}(-k_p(N_i - N_{planned}) - k_i \int (N_i - N_{planned})dt - \lambda_i + \lambda_{i+1}) \quad (9)$$

The control objective is clearly defined by

$$\lim_{t \to \infty} |N_i - N_{planned}| = 0 \quad (10)$$

4   Simulation Design

To carry out the simulation, a synthetic case composed of four nodes was designed, the first two and the last are non-producers, while the central node is producer with a time to adapt to de change of 15 days. As part of the design it is considered that nodes 1 and 2 are distributed to 50% the supply of the next node. The supply chain architecture is shown in Fig. 4, which mathematical model is

$$\frac{dN_1}{dt} = \lambda_1(t) - \frac{\lambda_3(t)}{2}$$
$$\frac{dN_2}{dt} = \lambda_2(t) - \frac{\lambda_3(t)}{2}$$
$$\frac{dN_3}{dt} = \lambda_3(t) - \lambda_3(t) \quad (11)$$
$$\frac{d\lambda_3(t)}{dt} = \frac{1}{T}[W_3(t) - \lambda_3(t)]$$
$$\frac{dN_4}{dt} = \lambda_4(t) - \lambda_{planned}(t)$$

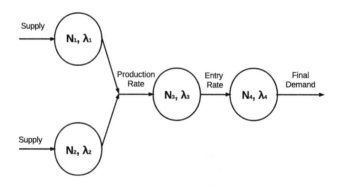

Fig. 4. Supply chain architecture.

It is important to note that each node of Fig. 4 is an element like the described in Fig. 2.

To evaluate the controller efficiency, two cases were designed. In the first, the demand is constant over time, while in the second it experiences quarterly variations, which is known in logistic terms as a stationary demand. In both cases the demand is deterministic. In the same way it is considered as the initial condition for the supply chain that its inventory is zero.

The rest of the simulation parameters are shown in Tables 1, 2 and 3, where the PI gains are tuned using the MatLab PID tuning algorithm.

Table 1. Cost elements

Nodes	T - Storage time [days]	P - Unit price [dollars]	I - Storage rate percentage of the unit price
1	15	15.6	15%
2	15	15.6	15%
3	25	21.05	22.3%
4	15	24.17	15%

Table 2. PI controller gains - Case 1

Nodes	Proportional gain	Integral gain
1	8.66	50.00
2	8.66	50.00
3	5.48	0.16
4	8.66	50.00

Table 3. PI controller gains - Case 2

Nodes	Proportional gain	Integral gain
1	8.66	50.00
2	8.66	50.00
3	9.87	0.5
4	8.66	50.00

5 Results

Using Matlab Simulink Toolbox, the supply chain was simulated. The results are divided according to each one of the designed cases.

- Case 1 (constant demand).

The Figs. 5, 6 and 7 show the behavior of the costs in the nodes of the chain when the demand of the final customer is known and constant. For each case the upper figure depicts the tracking signals (blue = reference and red = inventory holding cost) and lower depicts errors.

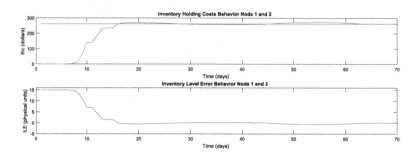

Fig. 5. Inventory error and Ihc behavior, node 1 and 2.

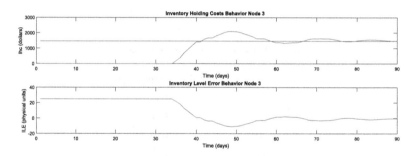

Fig. 6. Inventory error and Ihc behavior, node 3.

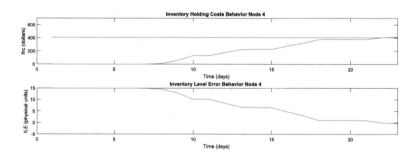

Fig. 7. Inventory error and Ihc behavior, node 4.

The controller is able to solve regulatory problems by guaranteeing a transitory state of less than 60 days for the non-producer nodes while it manages to stabilize the costs in the producer node before 130 days.

- Case 2 (demand with variation by season).

The Figs. 8, 9 and 10 show the behavior of the costs in the nodes of the chain when the demand of the final customer is known and has a variation by season. For each case the upper figure depicts the tracking signals (blue = reference and red = inventory holding cost) and lower depicts errors.

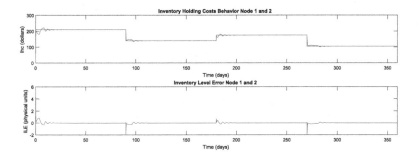

Fig. 8. Inventory error and Ihc behavior, node 1 and 2.

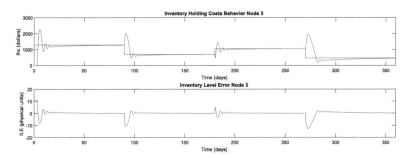

Fig. 9. Inventory Error and Ihc behavior, node 3.

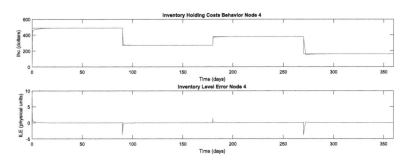

Fig. 10. Inventory Error and Ihc behavior, node 4.

This controller can solve trajectories tracking problems by guaranteeing a transitory state of less than 15 days for the non-producer nodes while it manages to stabilize the costs in the producer node before 100 days.

6   Final Comments

The proposed controller is able to solve the problems of regulation and trajectories tracking for supply chains non-producers and producers nodes. Particularly, for producers nodes it works effectively as long as the time to adapt to the change is less than 10 days. This design guarantees the optimization of inventory maintenance costs in the supply chains as long as the demand of the end customers is deterministic and does not suffer drastic variations.

The current results have direct application on continuous supply chains like chemical continuous reaction processes, which considers a wide type of scenarios including continuous and distributed systems.

Acknowledgements. This research was partially funded by project number 6351.17-P and "Estabilización orbital de sistemas mecánicos - Parte II: Estudio de aspectos de estabilidad" from "Tecnológico Nacional de México".

References

1. Csík, A., Földesi, P.: A bullwhip type of instability induced by time varing target inventory in production chains. Int. J. Innov. Comput. Inf. Control **8**, 5885–5897 (2012)
2. van Ryzin, G.J.: Analyzing Inventory Cost and Service in Supply Chains. Columbia Business School (2001)
3. Relph, G., Milner, C.: Inventory Management. Advanced methods for managing inventory within business systems, The Institute Operations Management (2015)
4. Grabara, J.K., Starostka-Patyk, M.: The bullwhip effect in supply chain. Adv. Logis. Syst. **3** (2012)
5. Balasubramanian, S., Whitman, L., Ramachandran, K., Sheelavant, R.: Causes and Remedies of Bullwhip Effect in Supply Chain. Department of Industrial and Manufacturing Engineering Wichita State University, Wichita, Kansas (2000)
6. Rodriguez-Angeles, A., Morales-Diaz, A., Blanco, A.M., Sanchez, A.: Modeling and decentralized control of inventories of linear dynamic supply chains. Elsevier, IFAC Publications (2005)
7. White, A.S., Chooi Yan, Y., Akpunar, E., Kai Hock, K.: Control strategies for inventory management. Engineering Systems Group. Middlesex University (2010)
8. Ivanov, D., Dolgui, A., Sokolov, B.: On applicability of optimal control theory to adaptive supply chain planning and scheduling. Professorship for International Supply Chain Management (2014)
9. Kastsian, D., Mönnigmann, M.: Optimization of a vendor managed inventory supply chain with guaranteed stability and robustness. Automatic Control and Systems Theory, Germany (2011)
10. Antić, S., Kostić, K., Dordević, L.: Spreadsheet model of inventory control based on modern control theory. Innovative Management and Business Performance, Germany (2012)
11. Rodriguez-Angeles, A., Molrales Diaz, A., Sanchez, A.: Dynamic Analysis and Control of Supply Chain Systems. Supply Chain, The Way to Flat Organisation, p. 404 (2009)

12. Nagatani, T., Helbing, D., Rodriguez-Angeles, A.: Stability analysis and strategies for linear supply chains. Dresden University of Technology, Institute for Economics and Traffic (2004)
13. Meng, Y.: The effect of inventory on Supply Chain. School of Technology and Design. Linneo University, Suecia (2006)
14. Zapata, J.A.: Cortes Fundamentos de la gestion de inventarios. Institucion Universitaria Esumer, Colombia (2014)
15. Trejo-Rodriguez, D.A., Rodriguez-Angeles, A.: Control acotado para la regulacion de inventario en cadenas de suministro. Memorias del Congreso Nacional de Control Automtico, Queretaro (2016)

On the Selection of Tuning Parameters in Predictive Controllers Based on NSGA-II

R. C. Gutiérrez-Urquídez[1], G. Valencia-Palomo[1(✉)], O. M. Rodríguez-Elías[1], F. R. López-Estrada[2], and J. A. Orrante-Sakanassi[3]

[1] Tecnológico Nacional de México, Instituto Tecnológico de Hermosillo,
Av. Tecnológico y Periférico Poniente S/N,Hermosillo, Mexico
{ro_gutierrez,gvalencia,omrodriguez}@ith.mx
[2] Tecnológico Nacional de México, Instituto Tecnológico de Tuxtla Gutiérrez,
Carretera Panamericana Km. 1080, C.P. 29050, 599 Tuxtla Gutiérrez, Mexico
frlopez@ittg.edu.mx
[3] CONACYT-Tecnológico Nacional de México, Instituto Tecnológico de Hermosillo,
Av. Tecnológico y Periférico Poniente S/N,Hermosillo, Mexico
jaos@ieee.org

Abstract. In the design of linear (model) predictive controllers (MPC), tuning plays a very important role. However, there is a problem not yet fully resolved: how to determine the best strategy for the selection of the optimal tuning parameters in order to obtain good performance with a large feasibility region, but maintaining a low computational load of the control algorithm? Because these objectives determine the proper functioning of the controller and are committed to each other, adjusting the controller parameters becomes a difficult task. The main contribution of this paper is to revise a method that uses the Nondominated Sorting Genetic Algorithm II (NSGA-II) for the parameter selection of a predictive control algorithm that has been parameterized with Laguerre functions (LOMPC) in order to explore the efficiency and provide statistical significance of the algorithm. Numerical simulations show that NSGA-II is a useful tool to obtain consistently good solutions for the selection of MPC tuning parameters.

Keywords: Predictive control · NSGA-II · Tuning
Multi-objective optimization

1 Introduction

Model based predictive control (MPC) is the general name for different computer control algorithms that use past information of inputs and outputs and a mathematical model of the plant to optimize the predicted future behavior. Although MPC has its background well established [6,34], and is widely used [26,37], there are still some theoretical and practical problems to be solved.

© Springer International Publishing AG, part of Springer Nature 2019
L. Trujillo et al. (Eds.): NEO 2017, SCI 785, pp. 138–157, 2019.
https://doi.org/10.1007/978-3-319-96104-0_7

For instance, one key conflict is between feasibility and performance: if a MPC controller is well tuned to provide high performance, feasibility will have (in general) a very small region, unless one uses a large number of decision variables (degrees of freedom or d.o.f.), which implies increasing the computational load of the algorithm. On the other hand, if one aims to get better feasibility with a fixed d.o.f., the result will be a poorly tuned controller with relatively poor performance [36]. Several authors have proposed different strategies to implement the MPC algorithm with the purpose of finding a suitable balance on the different performance indexes, such as multiparametric solutions [3], interpolations of different control laws [2], among others. Recently, some works have been proposed for more efficient predictive control algorithms by parametrization of the d.o.f. of the optimization problem. An alternative is to use orthogonal functions either with Laguerre/Kautz polynomials or through Generalised orthonormal functions [24,25,44], since they have proven to be very effective for improving the volume of the feasible region with a limited number of d.o.f. with almost no performance loss. These MPC algorithms differ in the way of parameterizing the d.o.f., but introduce one or more tuning parameters that have to be selected by the user. In the case of MPC parametrized by Laguerre polynomials, there are two parameters that must be considered in the tuning of the controller: the number of d.o.f. (n_c) and the Laguerre time-constant, a. Traditionally, conventional MPC controllers have been tuned by trial and error simulations or using thumb rules for the selection of some parameters [20]; other authors try to solve this problem by iterative methods [5] or metaheuristic techniques [29,43]. However, there is still no systematic way to define the controller tuning parameters, this is mainly due to the compromise between the objectives to be optimized; it makes difficult to efficiently handle the trade-off between feasibility, computational burden and controller performance. In the particular case of controllers with parameterized d.o.f., the original papers [24,25,44,45] fail to give guidelines in this selection, since in most cases, the selection of tuning parameters is still performed based on trial and error simulations. Therefore, the main debate is to establish the best strategy for the selection of controller tuning parameters that allow to efficiently handle the trade-off between feasibility, computational load and controller performance. On the other hand, the use of evolutionary algorithms is becoming more accepted in the control community, since they offer a flexible representation of the decision variables, which facilitates the evaluation of the controller performance. Some applications of evolutionary algorithms in predictive control formulations include [1,19,27,31] but for different purposes, e.g. economic optimization in plantwide control, adjusting the closed-loop response specifications, using the evolutionary optimization online. An alternative for a systematic selection of tuning parameters for a parameterized predictive controllers using an evolutionary algorithm was presented in [21]. In this paper, the method proposed in [21] is revised and more experiments were conducted in order to determine its efficiency. For each set of solutions obtained with this method, the feasibility, performance and computational cost values of a Laguerre Optimal MPC (LOMPC) controller are calculated and compared with the corresponding values of an Optimal MPC (OMPC) controller for two similar examples used in various articles. This work further explores the method in order to support the results with respect to the statistical significance. This paper is organized as follows: Sect. 2 gives the basic background about predictive control, the optimal predictive control (OMPC) and the Laguerre optimal predictive control (LOMPC). Section 3 includes multi-objective optimization concepts and characteristics of the multi-objective evolutionary algorithm NSGA-II. Section 4 develops the proposed tuning algorithm for LOMPC with NSGA-II. Section 5 shows the results of numerical examples that demonstrate the effectiveness of the proposed method. And finally, the conclusions are presented in Sect. 6.

2   Model Predictive Control and Laguerre Functions

2.1   Model Predictive Control (MPC)

Model based predictive control is the general name for different computer control algorithms with a similar structure that use past information of the inputs and outputs and a mathematical model of the plant to optimize the predicted future behavior. Many other names exist for MPC, e.g. IDCOM (Identification Command), DMC (Dynamic Matrix Control), GPC (Generalized Predictive Control), QDMC (Quadratic Dynamic Matrix Control), etc. [26]; and all of them have in common the following components:

- **The process model**. The mathematical model is used to generate system predictions. The plant model must be able to fully represent the dynamics of the process to allow predictions to be adequately calculated. A discrete-time state-space model is assumed, which has the following form:

$$\mathbf{x}_{k+1} = \mathbf{A}\mathbf{x}_k + \mathbf{B}\mathbf{u}_k; \quad \mathbf{y}_k = \mathbf{C}\mathbf{x}_k + \mathbf{D}\mathbf{u}_k; \tag{1}$$

with $\mathbf{x}_k \in \mathbb{R}^n$, $\mathbf{y}_k \in \mathbb{R}^l$, $\mathbf{u}_k \in \mathbb{R}^m$ which are the state vectors, the measured output and the plant input respectively. This work also adopts an independent model approach with optimal feedback $\mathbf{K}$ [34]. Let $\mathbf{w}_k$ the output of the independent model, hence, the estimated disturbance is $\hat{\mathbf{d}}_k = \mathbf{y}_k - \mathbf{w}_k$ [8]. Disturbance rejection and offset free tracking will be achieved using the offset form of state feedback [32], that is:

$$\mathbf{u}_k - \mathbf{u}_{ss} = -\mathbf{K}(\mathbf{x}_k - \mathbf{x}_{ss}), \tag{2}$$

where $\mathbf{x}$ is the state of the independent model and $\mathbf{x}_{ss}$, are estimated values of the steady-states giving no offset; these depend upon the model parameters and the disturbance estimate [35].

- **The performance index** (objective function or cost function). This is used to quantify the deviation of the measured output with respect to the desired output and the control effort. A way of defining the cost function is by using deviation variables to penalize the deviation of absolute input on the expected steady state input:

$$\mathbf{J}_k = \sum_{i=0}^{n_y}(\mathbf{x}_{k+i} - \mathbf{x}_{ss})^T\mathbf{Q}(\mathbf{x}_{k+i} - \mathbf{x}_{ss}) + \sum_{i=0}^{n_u-1}(\mathbf{u}_{k+i} - \mathbf{u}_{ss})^T\mathbf{R}(\mathbf{u}_{k+i} - \mathbf{u}_{ss}), \quad (3)$$

with $\mathbf{Q}$ and $\mathbf{R}$ positive definite state and input cost weighting matrices, n_y is the prediction horizon and n_u, the control horizon. Because in practice all real processes are subject to constraints, these need to be considered in the cost function. One of the major selling points of MPC is its ability to do online constraint handling in a systematic fashion, retaining to some extent the stability margins and performance of the unconstrained law. Let the system be subject to constraints of the form:

$$\left.\begin{array}{c}\mathbf{u}_{min} \leq \mathbf{u}_k \leq \mathbf{u}_{máx}; \\ \triangle\mathbf{u}_{min} \leq \mathbf{u}_{k+1} - \mathbf{u}_k \leq \triangle\mathbf{u}_{máx}; \\ \mathbf{y}_{min} \leq \mathbf{y}_k \leq \mathbf{y}_{máx}.\end{array}\right\} \forall k. \quad (4)$$

- **The optimization algorithm.** It is needed to minimize the performance index imposed by a set of restrictions. Minimizing the cost function subject to both current and future constraints and obtaining control action, the optimization results in a quadratic program (QP).
- **The receding horizon.** At each sampling time, an optimal finite horizon control problem is solved over a prediction horizon. The sequence of the control signal is calculated by minimizing the cost function, the actual control input to the plant only takes the first sample of the control signal, $\mathbf{u}_k$, while neglecting the rest of the trajectory. At next sampling interval, the procedure is restarted [28].

2.2   Optimal Predictive Control (OMPC)

The key idea of OMPC [41] is to embed into the predictions the unconstrained optimal behaviour and handle constraints by using perturbations about this. The closed-loop paradigm uses perturbations as d.o.f. for the optimal control law without constraints. Disturbance rejection and offset free tracking will be achieved using the offset form of state feedback of (2) [35]. For convenience, d.o.f. can be reformulated in terms of a new variable $\mathbf{c}_k$. Hence, assuming $\mathbf{K}$ is the optimal feedback, the input predictions are defined as follows:

$$\mathbf{u}_{k+i} - \mathbf{u}_{ss} = \begin{cases} -\mathbf{K}(\mathbf{x}_{k+i} - \mathbf{x}_{ss}) + \mathbf{c}_{k+i}; & i \in \{1, 2, ...n_c - 1\} \\ -\mathbf{K}(\mathbf{x}_{k+i} - \mathbf{x}_{ss}); & i \in \{n_c, n_c + 1, ...\} \end{cases}, \quad (5)$$

where the perturbations $\mathbf{c}_k$ are the d.o.f. for optimization; conveniently summarized in vector: $\underline{\mathbf{c}}_k = [\mathbf{c}_k^T, \mathbf{c}_{k+1}^T, ..., \mathbf{c}_{k+n_c-1}^T]^T$. Input predictions (5) and state associated (1) which satisfies constraints (4): $\mathbf{Mx}_k + \mathbf{N}\underline{\mathbf{c}}_k \leq \mathbf{d}(k)$. In practice, if an unconstrained optimal prediction may violate a constraint defined in (4), prediction class more suitable shall be used according (5). The OMPC algorithm can be summarized [35]:

$$\underset{\rightarrow k}{\mathbf{c}}^* = \arg\min_{\underset{\rightarrow k}{\mathbf{c}}} \underset{\rightarrow k}{\mathbf{c}}^T \mathbf{W}_j \underset{\rightarrow k}{\mathbf{c}} \quad s.t. \ \mathbf{M}_j \mathbf{x}_k + \mathbf{N}_j \underset{\rightarrow k}{\mathbf{c}} \leq \mathbf{d}(k). \tag{6}$$

Use $\underset{\rightarrow k}{\mathbf{c}}^*$ to construct the input (5). OMPC algorithm has implied LQR theory and is able to find a global optimum on the objective function. If one chooses a value for $\mathbf{K}$ in (5) to become a optimal Linear-Quadratic-Regulator(LQR) [40], the feasible region depends only on the class of prediction and hence also the number of free movements, that is, n_c.

Definition 2.1. *Maximum Admissible Set (MAS). A common method to achieve recursive feasibility is to find the region of the state space where positively invariant sets ensure the action of an unconstrained control law but satisfy all constraints in the future. This is achieved using the dual-mode paradigm. And the greatest invariant set possible for use as the terminal state set is referred as Maximum Admissible Set (MAS) [30, 35]. For a linear discrete system, observable, pre-stabilized by a gain $\mathbf{K}$ of state feedback, associated with a set of constraints (4), there exists a set, MAS, finite and where the constraints are satisfied for all future time intervals: $MAS = \{\mathbf{x}_k \in \mathbb{R}^n \mid \mathbf{M}\mathbf{x}_k \leq \mathbf{d}\}$.*

Definition 2.2. *Maximal Controllable Admissible Set (MCAS). It is also possible to define a region in $\mathbf{x}$ in which it is possible to find a $\mathbf{c}_k$ such that at the future trajectory satisfying the constraints: $MCAS = \{\mathbf{x}_k \in \mathbb{R}^n \ \exists \ \underset{\rightarrow k}{\mathbf{c}} \in \mathbb{R}^{n_c m}$ s.t. $\mathbf{M}\mathbf{x}_k + \mathbf{N}\underset{\rightarrow k}{\mathbf{c}} \leq \mathbf{d}\}$; and this is named Maximal Controllable Admissible Set (MCAS).*

2.3   Laguerre Polynomials and LOMPC

The fundamental weakness of OMPC algorithms is that the d.o.f. are parameterized as individual values at specific samples and have an impact over just one sample and thus have a limited impact on feasibility. If the initial state is far away from the MAS associated to $\mathbf{c}_k = 0$, the n_c steps will be insufficient to move into the MAS. LOMPC algorithm proposes that d.o.f within the input predictions, i.e., $\mathbf{u}_k$ and $\mathbf{c}_k$, be parametrized in terms of Laguerre polynomials, L_i, in OMPC. The $z-$transform of discrete Laguerre polynomials are defined as follows:

$$\Gamma_i(z) = \sqrt{1 - a^2} \frac{(z^{-1} - a)^{n-1}}{(1 - az^{-1})^n}; \qquad 0 \leq a \leq 1. \tag{7}$$

With the inverse $z-$ transform of $\Gamma_n(z, a)$, denoted by $l_{(k,n)}$, the Laguerre functions set are the vector: $\mathbf{L}_k = \{l_{k,1}, \ l_{k,2}, \ ..., \ l_{k,n}\}^T$. The size of the $\mathbf{A}_L$ matrix, is $n \times n$; and it is a function of the parameters a, $\beta = 1 - a^2$ and initial condition $\mathbf{L}_0$, so that:

$$\mathbf{L}_{k+1} = \underbrace{\begin{bmatrix} a & 0 & 0 & 0 & \cdots \\ \beta & a & 0 & 0 & \cdots \\ -a\beta & \beta & a & 0 & \cdots \\ a^2\beta & -a\beta & \beta & a & \cdots \\ \vdots & \vdots & \vdots & \vdots & \ddots \end{bmatrix}}_{A_L} \mathbf{L}_k, \quad \mathbf{L}_0 = \sqrt{1-a^2}\,[1, -a, a^2, \ldots]^T. \quad (8)$$

The basic concept of OMPC is preserved [35,41], that is the predictions take the form of (5) and the optimality dynamics is included in the predictions. However, a key difference is that the disturbance (terms $\mathbf{c}_k$) is defined by the Laguerre polynomials instead of taking the d.o.f. individually. The relevant link between Laguerre and predicted values of $\underset{\rightarrow}{\mathbf{c}}_k$, are summarized in the following equation:

$$\underset{\rightarrow}{\mathbf{c}}_k = [\mathbf{c}_k^T, \ldots, \mathbf{c}_{k+n_c-1}^T]^T, \quad \underset{\rightarrow}{\mathbf{c}}_k(L) = \underbrace{[\mathbf{L}_0^T, \mathbf{L}_1^T, \ldots]^T}_{H_L} \underset{\rightarrow}{\eta}_k = \mathbf{H}_L \underset{\rightarrow}{\eta}_k. \quad (9)$$

Now, with $\mathbf{L}_{k+1} = \mathbf{A}_L * \mathbf{L}_k$, the decision variable is $\underset{\rightarrow}{\eta}_k$; and substituting predictions in (8), get the LOMPC optimization, which is:

$$\underset{\rightarrow}{\eta}_k^* = \arg\min_{\underset{\rightarrow}{\eta}_k} \underset{\rightarrow}{\eta}_k^T \left\{ \sum_{i=0}^{\infty} \mathbf{A}_L^i \mathbf{L}_0 \mathbf{S} \mathbf{L}_0^T \mathbf{A}_L^{i\,T} \right\} \underset{\rightarrow}{\eta}_k = \underset{\rightarrow}{\eta}_k^T \mathbf{S}_L \underset{\rightarrow}{\eta}_k \quad (10)$$

$$s.t. \quad \mathbf{M}\mathbf{x}_k + \mathbf{N}\mathbf{H}_L \underset{\rightarrow}{\eta}_k \le \mathbf{d}. \quad (11)$$

It has been shown that with the parameterization of d.o.f. get an increase in feasibility region (MCAS) of controller LOMPC regarding equal number of d.o.f. in OMPC [44], but now, the number of parameters that must be tuned now are two: n_c and a, also known as the pole of the Laguerre discrete network or time scale factor.

Remark 2.1. *LOMPC has a guarantee of stability and recursive feasibility in the nominal case. This follows conventional arguments whereby one can show that the cost of (10) is Lyapunov so long as the optimization at sample k can reuse the optimal trajectory computed at sample $k-1$ [36].*

3 Multi-Objective Optimization (MOO) and NSGA-II

Many engineering design problems involve more than one objective function where the optimal decisions need to be taken in the presence of trade-offs between two or more conflicting objectives. These problems are called Multi-Objective Problems (MOP). In mathematical terms, a MOP can be formulated as [42]:

$$\min_{k \in Q}\{\mathbf{F}(\mathbf{k})\} = \min_{k \in Q}[f_1(\mathbf{k}), f_2(\mathbf{k}), \cdots, f_n(\mathbf{k})]. \quad (12)$$

where $f_i : Q \rightarrow \mathbf{R}^1$, $\mathbf{F} : Q \rightarrow \mathbf{R}^k$, f_i are n-objective functions. $\mathbf{k} \in Q$ is a q-dimensional vector of design parameters (decision variables). The domain

$Q \subset \mathbf{R}^q$ is the design space, and can in general be expressed in terms of inequality and equality constraints of the form:

$$Q = k \in \mathbf{R}^q \mid g_i(k) \leq 0, i = 1, 2 \cdots l, \text{ and } h_j(k) = 0, j = 1, 2, \cdots m. \qquad (13)$$

The performance criteria is described mathematically by objective functions that allow evaluating the MOP (12). The constraints (13) indicate allowed values or physical limits of a particular problem. The objective functions and constraints are functions of the decision variables [7]. It is observed that two different and complementary spaces are explicitly considered: one for the decision variables and another for the objective functions. A solution will only be valid if it satisfies the constraints for all the objective functions. The constraints delimit the space of solutions and divide the search space into two regions: (i) feasible region, it is the set of elements of the search space that comply with all restrictions; (ii) unfeasible region, it is formed by the elements of the search space that do not comply with at least one of the restrictions of the MOP. There is a wide variety of techniques for solving MOPs (see e.g. [7,10,15,16,46] and references therein). It is clear that when there are multiple objectives to optimize, there is no single solution, but rather a set of compromise solutions together. We can define the optimal solutions of the MOP by using the concept of *dominance* [33].

Definition 3.1.

- *Let $v, w \in R^k$. The vector v is said to be less than w ($v <_p w$), if $v_i < w_i$ for all $i \in \{1, 2, \cdots, k\}$. The relation $\leq_p$ is defined analogously.*
- *A vector $v \in Q$ is called dominated by another vector $w \in Q$ ($w \prec v$), with respect to the MOP (12) if $F(w <_p v)$ and $F(w) \neq F(v)$, otherwise v is called non-dominated by w.*

If a vector w dominates a vector v, then w can be considered as a *better* solution of the MOP.

Definition 3.2.

- *A point $w \in Q$ is called Pareto Optimal or a Pareto point of MOP (12) if there is no $v \in Q$ which dominates w.*
- *The set of all Pareto optimal solutions is called the Pareto set denoted as:*

$$\mathscr{P}_s := \{w \in Q : w \text{ is Pareto point of (12)}\}. \qquad (14)$$

- *The image $F(\mathscr{P}_s)$ is called the Pareto front ($\mathscr{PF}$).*

Within the techniques to solve MOPs are the Multi-objective Evolutionary Algorithms (MOEAs). Some MOEAs use optimization techniques based on the Pareto concept (14) and different mechanisms to preserve the diversity of the population [9,48]; other MOEAs include indicators that can be adapted to the user's preferences in terms of continuous generalizations of the dominance relation [47]. In general terms, MOEAs are systematic methods for solving search and optimization problems that apply the same methods of the biological evolution, based on selection, reproduction and mutation of the population. These algorithms transform a set of individual mathematical objects using operations that are modeled according to the Darwinian principle of reproduction and survival of the fittest [18]. MOEAs allow flexible representation of decision variables and performance evaluation; also are robust and methodological tool for search and optimization, with the ability to work in environments that include discontinuities, time variance, bad behavior, multi-modality, uncertainty and noise. MOEAs are increasingly accepted in the control community, their applications are mainly in the off-line design and online optimization [17,37]. For example, the MPC scheme based on multiobjective optimization [4] shows that it is possible to calculate an optimal Pareto solution as a function by intervals of the state vector and the weighting vector assigned to the different objectives to get a particular solution. Other authors proposed a predictive control scheme based on multiobjective optimization where the control action is chosen, at each sampling time, among the set of Pareto optimal solutions based on a time-varying decision criterion [4]. Several types of MOEAs exist; each uses different mechanisms of selection, crossover and elitism in particular, but for this paper we use the algorithm NSGA-II, proposed in [12]. NSGA-II is an improved version of NSGA that modifies the mechanism for diversity preserving and incorporates an explicit mechanism for elitism; leave the use of Sharing distance of the NSGA to use Crowding tournament selection operator as a method of diversity preservation. Most MOEAs use a binary coding to represent the values of the parameters, but NSGA-II uses the Simulated Binary Crossing operator (SBX) to directly handle the parameters with real values. SBX simulates the operation of a crossover operator at a point on a binary string [11,13]. SBX uses a probability distribution around two parents ($s_{1,t}$ y $s_{2,t}$) to create two descendant solutions ($s_{1,t+1}$ and $s_{2,t+1}$). For the calculation of the descendant solutions, first, a dispersion factor γ is defined as the ratio of the absolute difference of the values of the children to the values of the parents, that is:

$$\gamma = \left| \frac{s_i^{1,t+1} - s_i^{2,t+1}}{s_i^{1,t} - s_i^{2,t}} \right|. \tag{15}$$

and the probability distribution to create a descendant solution is:

$$\mathscr{P}_d(\gamma) = \begin{cases} 0.5(n+1)\gamma^n & \text{if } \gamma \leq 1; \\ 0.5(n+1)\frac{1}{\gamma^{n+2}} & \text{otherwise;} \end{cases} \tag{16}$$

where n is a positive real number that represents the probability distribution index. Furthermore, this MOEA has a simple strategy for handling restrictions based on binary tournament selection, a necessary advantage in the resolution of complex problems in the real world [12]. All the aforementioned characteristics make the NSGA-II a good option to solve the problem posed in this work. Table 1 shows the NSGA-II settings used in this work; most of these according to the recommendations in [9]. The NSGA-LOMPC algorithm is proposed in the next section.

Table 1. NSGA-II parameters.

Operator/Parameter	Type/Value
Individuals	Chromosomes: 1; (**s**)
	Genes: 2; $(g_1 = a)$ y $(g_2 = n_c)$
	Coding: real-code
Population	Size: 30 individuals (NI)
Generations	100
Selection operator	Type: tournament
	Size of tournament: 2 individuals
Crossover operator	Type: simulated binary crossover (SBX)
	Probability: 0.8
Mutation operator	Type: polynomial distribution
	Probability: 0.5
Elitism	Non-dominated sort
	Crowding distance in phenotype

4   Tuning LOMPC

In the work previously presented, it is shown that the parameterization of d.o.f. has advantages over conventional approaches. However, there is still no systematic way to define the controller tuning parameters, this mainly due to the compromise between the objectives to be optimized. It is clear that the problem to be solved is a MOO, where the search space is fairly large (given the continuous nature of the parameter a), which justifies the use of NSGA-II. The purpose of this work is to further evaluate a method to systematically choose the optimal values of the tuning parameters, a and n_c, of an MPC whose d.o.f. have been parameterized with Laguerre functions; in order to guarantee the best trade-off between feasibility, performance and computational load. Therefore, there are two decision variables a and n_c and three optimizing conditions:

1. Maximize the feasibility region, $\max f_{a,n_c}(\vartheta)$.
2. Minimize the performance loss, $\min f_{a,n_c}(\beta)$.
3. Minimize the computational burden, $\min f_{a,n_c}(\rho)$.

Also the constraints associated with the parameter selection must be added to the multi-objective optimization problem:

$$0 \le a \le 1; \qquad 1 \le n_c \le n_{c,max}; \qquad n_c \quad integer. \qquad (17)$$

4.1   Feasibility Evaluation (ϑ)

In order to estimate the normalized volume, a polytope is defined first $\mathbf{P}_{opt}$ as the global MCAS of OMPC with a large number of degrees of freedom, able to represent the largest feasible region that can be obtained by the controller (usually $n_c \geq 20$) [24]: $\mathbf{P}_{opt} = \{(x, c) \mid \mathbf{M}\mathbf{x}_k + \mathbf{N}\underset{\rightarrow}{\mathbf{c}}_k \leqslant \mathbf{d}\}$. Also it is defined a polytope $\mathbf{P}_{H_L}$ corresponding to proposed parameterization MCAS: $\mathbf{P}_{H_L} = \{(x, \eta) \mid \mathbf{M}_L x + \mathbf{N}_L \underset{\rightarrow}{\eta} \leq \mathbf{d}\}$; where $\mathbf{P}_{H_L}$ is the polytope sliced by the parameterization matrix $\mathbf{H}_L$. The volume of $\mathbf{P}_{opt}$ and $\mathbf{P}_{H_L}$ polytopes, represent the feasible regions or feasible volumes for each type of algorithm. The volume calculation of a high dimensional polytope is a complex task and the computing time for these polytopes can be prohibitive; consequently, this paper approximates the volume by computing the average distance from the origin to the boundary of the associated MCAS (radius). First select a large number of equispaced (by solid angle) or random directions in the state space i.e. $x = [x_1, ..., x_n]$ and then, the distance from the origin to the boundary of MCAS is determined by solving a linear programming (LP) for each direction x_i selected. Greater distances imply bigger feasible region. The objective function for evaluating the normalized feasible volume is then:

$$\vartheta = \frac{vol(\mathbf{P}_{H_L})}{vol(\mathbf{P}_{opt})}. \tag{18}$$

4.2   Performance Evaluation (β)

The performance evaluation it is done by realizing the calculation for the n-points x_i selected, they are represented by the optimized values of the associated cost function, i.e. $J_{opt}(x_i)$ and $J_{H_L}(x_i)$. To ensure fairness in comparison of these values, scaling is used (setting) in one direction x_i given [24]. The objective function for evaluating the normalized performance is:

$$\beta = \frac{1}{n} \sum_{i=1}^{n} \frac{\mathbf{J}_{H_L}(x_i)}{\mathbf{J}_{opt}(x_i)}. \tag{19}$$

4.3   Computational Load Evaluation (ρ)

It is demonstrated that the parameterization of d.o.f. proposed in LOMPC algorithm is able to achieve great feasibility regions while maintaining an acceptable local optimality within a relatively low computational complexity compared to conventional OMPC approaches. However, this reduction in d.o.f. not necessarily results in a reduction of the complexity of optimization and therefore the computational load, since the resulting quadratic programming of reassignment is denser (heavier) than for OMPC [24]. One alternative to determine the minimum number of d.o.f. is to compare the online computational load for LOMPC and OMPC as a function of the number of floating point operations per second (flops) required for each algorithm [24]. For OMPC, the computational complexity is linear with respect to the horizon length and cubic respect to state and the input dimension, so that:

$$\rho_{(OMPC)} = n_c + n_x{}^3 + n_u{}^3 \quad (flops). \tag{20}$$

For LOMPC, their computational load is cubic in number of d.o.f., the state and input dimensions:

$$\rho_{(LOMPC)} = n_c{}^3 + n_x{}^3 + n_u{}^3 \quad (flops). \tag{21}$$

The step-by-step procedure of the proposed algorithm is presented in Table 2.

Table 2. Pseudocode of the proposed algorithm.

$t = 0$	$P_0(a, n_c)$	A random population is created and sorted based on nondomination
	$Q_0(a, n_c)$	Create offspring population of size N
$t + 1 = i$	$R_t = P_t \bigcup Q_t$	Combine parent and offspring population
	$\mathbf{F} = \text{fnd}(R_t)$	$\mathbf{F} = (f_{a,n_c}(\vartheta), f_{a,n_c}(\beta), f_{a,n_c}(\rho))$ all fast nondominated fronts (fnd) of R_t
	$P_{t+1} = \varnothing$ and $i = 1$ until $|P_{t+1}| + |\mathbf{F}_i| \leq N$	Until the parent population is filled
	$\text{cda}(\mathbf{F}_i)$	Calculate crowding distance (cda) in $\mathbf{F}_i$
	$P_{t+1} = P_{t+1} \bigcup \mathbf{F}_i$	Include i-th nondominated front
	$i = i + 1$	Check the next font for inclusion
	$\text{Sort}(\mathbf{F}_i \prec_d)$	Sort in descending order
	$P_{t+1} = P_{t+1} \bigcup \mathbf{F}_i[1 : (N - |P_{t+1}|)]$	Choose the first elements of $N - |P_{t+1}|$
	$Q_{t+1} = \text{mnp}(P_{t+1})$	Make new population (mnp)Q_{t+1}
	$t = t + 1$	Increment the generation counter
	Check end criterion	Stop or back to step $t + 1$
$t = NG$	$[a, n_c]$	Set of all Pareto optimal N solutions for NG generations
	$[f_{a,n_c}(\vartheta), f_{a,n_c}(\beta), f_{a,n_c}(\rho)]$	Set of all Pareto Front

5   Numerical Examples

This section presents two experiments similar to those used in [24, 25, 44] in order to compare the results of them with those obtained with this method. The first experiment is a Single Input/Single Output (SISO) system, represented by a discrete-time state-space model with constraints:

$$\mathbf{x}_{k+1} = \begin{bmatrix} 0.6 & -0.4 \\ 1.0 & 1.4 \end{bmatrix} \mathbf{x}_k + \begin{bmatrix} 0.20 \\ 0.05 \end{bmatrix} \mathbf{u}_k; \quad \mathbf{y}_k = \begin{bmatrix} 1.0 & -2.0 \end{bmatrix} \mathbf{x}_k;$$

$$-1.5 \leq \mathbf{u}_k \leq 0.8; \quad |\Delta\mathbf{u}_k| \leq 0.4, \quad \mathbf{x}_k \leq 5; \quad \mathbf{Q} = \begin{bmatrix} 1 & 0 \\ 0 & 1 \end{bmatrix}; \quad \mathbf{R} = 2.$$

This example utilizes the Matlab Multi-objective Optimization Toolbox that allows the use of a variant of NSGA-II [10]. NSGA-II was run 80 times using a search space generated by varying $n_c \in \{2, 3, 4, 5, 6, 7\}$ and $a \in [0, 1]$. Tables 3, 4 and 5 summarize the LOMPC tuning parameters obtained in three different runs of the algorithm with NSGA-II and the objective functions evaluated with these. For the purpose of comparing these results with an OMPC controller, the values of their corresponding objective functions have been added. Table 6 shows heuristical values for the parameters n_c and a found in the literature. The Pareto Front obtained with the LOMPC data from the Table 3 is shown in Fig. 1. It is clear that the values of the objective functions are located in the proximity of the local optima, which shows that there is a good balance between the three objectives to be optimized. Tables 3, 4 and 5 shows that performance and feasibility improve as n_c is increased for LOMPC respect OMPC, and as expected, the computational load too. However, within the various tuning parameters found with the NSGA-II, the user can select any combination of these and achieve the best trade-off between the above objectives instead a heuristical value without guarantee that it is the best option (see Table 6). Using one of the selected tuning options, i.e. $a = 0.7915, n_c = 3$, Fig. 2 shows the plots of the controller simulation for: (i) an initial state in the MCAS of both, OMPC and LOMPC (top); (ii) an initial state in the MCAS of LOMPC, but outside of the MCAS of OMPC (bottom), where the latter is infeasible. Once again, it is clear that LOMPC with NSGA-II is better than OMPC under similar conditions (d.o.f.) and very similar to the global optimum. Furthermore, with NSGA-II, more options of tuning parameters are obtained than those found in the literature. The second experiment is a system with two inputs and two outputs (MIMO system), represented by a discrete-time state-space model with constraints, given by:

$$\mathbf{x}_{k+1} = \begin{bmatrix} 0.9146 & 0 & 0.0405 \\ 0.1665 & 0.1353 & 0.0058 \\ 0 & 0 & 0.1353 \end{bmatrix} \mathbf{x}_k + \begin{bmatrix} 0.0544 & -0.0757 \\ 0.0053 & 0.1477 \\ 0.8647 & 0 \end{bmatrix} \mathbf{u}_k;$$

$$\mathbf{y}_k = \begin{bmatrix} 1.7993 & 13.2160 & 0 \\ 0.8233 & 0 & 0 \end{bmatrix} \mathbf{x}_k;$$

Table 3. LOMPC tuning parameters (a, n_c) obtained with NSGA-II for Example 1 in experiment 1 with values of objective functions for LOMPC and OMPC.

Tuning parameter		Average radius to MCAS (J_ϑ)			Performance loss (J_β)			Computational cost (J_ρ)		
n_c	a	OMPC	LOMPC	$\triangle J_\vartheta$	OMPC	LOMPC	$\triangle J_\beta$	OMPC	LOMPC	$\triangle J_\rho$
2	0.6206	0.4720	0.6990	+48.0%	1.0526	1.0526	−0.01%	0.3793	0.5862	+54.5%
3	0.6313	0.5560	0.7691	+38.3%	1.0525	1.0527	−0.02%	0.4138	1.2414	+200%
3	0.7915	0.5560	0.7999	+43.8%	1.0525	1.0527	−0.02%	0.4138	1.2414	+200%
4	0.6551	0.6262	0.8481	+35.4%	1.0526	1.0528	−0.02%	0.4483	2.5172	+461.5%
5	0.6304	0.6774	0.8708	+28.6%	1.0528	1.0529	−0.01%	0.4828	4.6207	+1002.4%
5	0.7164	0.6774	0.8932	+31.8%	1.0528	1.0530	−0.02%	0.4828	4.6207	+1002.4%

Table 4. LOMPC tuning parameters (a, n_c) obtained with NSGA-II for Example 1 in experiment 2 with values of objective functions for LOMPC and OMPC.

Tuning parameter		Average radius to MCAS (J_ϑ)			Performance loss (J_β)			Computational cost $(J_\rho$		
n_c	a	OMPC	LOMPC	$\triangle J_\vartheta$	OMPC	LOMPC	$\triangle J_\beta$	OMPC	LOMPC	$\triangle J_\rho$
2	0.2212	0.4720	0.5781	+22.5%	1.0526	1.0527	−0.01%	0.3793	0.5862	+54.5%
2	0.2284	0.4720	0.5805	+34.6%	1.0526	1.0528	−0.02%	0.3793	0.5862	+54.5%
3	0.6080	0.5560	0.7659	+38.3%	1.0525	1.0532	−0.06%	0.4138	1.2414	+200%
3	0.7908	0.5560	0.7685	+38.3%	1.0525	1.0535	−0.09%	0.4138	1.2414	+200%
4	0.2034	0.6262	0.7453	+43.8%	1.0526	1.0530	−0.04%	0.4483	2.5172	+461.5%
4	0.2346	0.6262	0.7528	+35.4%	1.0526	1.0527	−0.01%	0.4483	2.5172	+461.5%
4	0.4862	0.6262	0.8194	+38.3%	1.0526	1.0538	−0.11%	0.4483	2.5172	+461.5%
5	0.4862	0.6774	0.8690	+28.6%	1.0549	1.0531	+0.02%	0.4828	4.6207	+1002.4%
5	0.7167	0.6774	0.8933	+31.8%	1.0549	1.0542	+0.06%	0.4828	4.6207	+1002.4%

Table 5. LOMPC tuning parameters (a, n_c) obtained with NSGA-II for Example 1 in experiment 3 with values of objective functions for LOMPC and OMPC.

Tuning parameter		Average radius to MCAS (J_ϑ)			Performance loss (J_β)			Computational cost (J_ρ)		
n_c	a	OMPC	LOMPC	$\triangle J_\vartheta$	OMPC	LOMPC	$\triangle J_\beta$	OMPC	LOMPC	$\triangle J_\rho$
2	0.6268	0.4720	0.6991	+48.1%	1.0526	1.0526	−0.01%	0.3793	0.5862	+54.5%
3	0.4898	0.5560	0.7487	+34.3%	1.0525	1.0526	−0.02%	0.4138	1.2414	+200%
4	0.7240	0.6262	0.8376	+39.5%	1.0526	1.0530	−0.04%	0.4483	2.5172	+461.5%
5	0.6237	0.6774	0.8028	+18.6%	1.0528	1.0530	−0.01%	0.4828	4.6207	+1002.4%
5	0.7077	0.6774	0.8308	+22.6%	1.0528	1.0531	−0.02%	0.4828	4.6207	+1002.4%
6	0.6783	0.7162	0.8576	+19.8%	1.0570	1.0544	+0.24%	0.5172	7.7586	+1002.4%
7	0.6084	0.7465	0.8877	+18.9%	1.0577	1.0549	+0.26%	0.5517	12.138	+1002.4%

Table 6. Heuristic LOMPC tuning parameters (a, n_c) for Example 1 and values of objective functions.

Tuning parameter		Average radius to MCAS (J_ϑ)	Performance loss (J_β)	Computational cost (J_ρ)	Reference
n_c	a				
2	0.5	0.6322	1.0247	0.5862	[23]
2	0.8	0.6322	1.0247	0.5862	[25,45]
3	0.5	0.7487	1.0691	1.2414	[25]

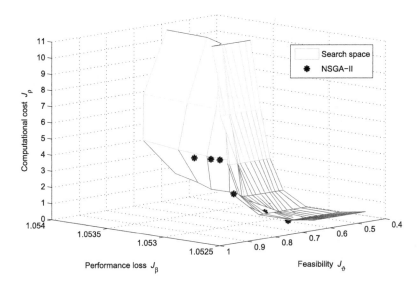

Fig. 1. Pareto front for example 1.

with constraints and weigths:

$$|\mathbf{u}_k| \leq \begin{bmatrix} 1 \\ 2 \end{bmatrix}; \quad |\Delta\mathbf{u}_k| \leq 2; \quad \mathbf{x}_k \leq \begin{bmatrix} 10 \\ 10 \\ 10 \end{bmatrix}; \quad \mathbf{Q} = \begin{bmatrix} 3.91 & 23.77 & 0 \\ 23.7 & 174.6 & 0 \\ 0 & 0 & 0 \end{bmatrix}; \quad \mathbf{R} = \begin{bmatrix} 1 & 0 \\ 0 & 1 \end{bmatrix};$$

As in the previous experiment, NSGA-II is used to determine the parameters n_c and a. For this model, NSGA-II was run 50 times using a search space generated by varying $n_c \in \{2, 3, 4, 5, 6, 7\}$ and $a \in [0, 1]$. Tables 7 and 8 summarizes the tuning parameters obtained in two different runs of the algorithm NSGA-II and the objective functions evaluated for LOMPC and OMPC. These tables shows that performance and feasibility improve as n_c is increased for LOMPC respect OMPC, but the computational load too. Table 9 shows heuristical values for the parameters n_c and a found in the literature. Again, it is observed that with NSGA-II, it is possible to find a larger set of tuning parameters than those suggested in the literature. Figure 3 shows the Pareto front for this MOP. It is easy to see how the values of the functions are located in the vicinity of the local optima, so there is a guarantee of a good trade-off between them.

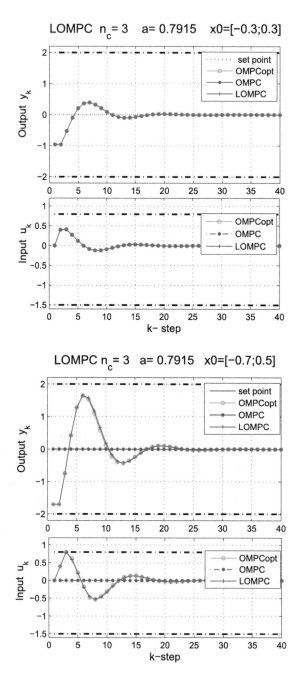

Fig. 2. Controllers OMPC and LOMPC simulation with different initial conditions for Example 1.

Table 7. LOMPC tuning parameters (a, n_c) obtained with NSGA-II for Example 2 in experiment 1 with values of objective functions for LOMPC and OMPC.

Tuning parameter		Average radius to MCAS (J_ϑ)			Performance loss (J_β)			Computational cost (J_ρ)		
n_c	a	OMPC	LOMPC	$\triangle J_\vartheta$	OMPC	LOMPC	$\triangle J_\beta$	OMPC	LOMPC	$\triangle J_\rho$
2	0.6208	0.1048	0.2339	—	Unfeasible	1.0526	—	0.6727	0.7818	—
3	0.7915	0.1418	0.4345	+205%	1.0592	1.0599	-0.06%	0.6909	1.1273	+84.2%
3	0.7916	0.1418	0.4345	+205%	1.0592	1.0598	-0.05%	0.6909	1.1273	+84.2%
4	0.6548	0.1823	0.4833	+165%	1.0613	1.0528	+0.80%	0.7091	1.8000	+153.8%
4	0.6688	0.1823	0.4827	+164%	1.0613	1.0528	+0.80%	0.7091	1.8000	+153.8%
6	0.6783	0.2752	0.6076	+60%	1.0580	1.0535	+0.42%	0.7455	7.7586	+940.7%
7	0.6084	0.3282	0.6078	+85%	1.0569	1.0549	+4.65%	0.7636	12.138	+1480.4%

Table 8. LOMPC tuning parameters (a, n_c) obtained with NSGA-II for Example 2 in experiment 2 with values of objective functions for LOMPC and OMPC.

Tuning parameter		Average radius to MCAS (J_ϑ)			Performance loss (J_β)			Computational cost (J_ρ)		
n_c	a	OMPC	LOMPC	$\triangle J_\vartheta$	OMPC	LOMPC	$\triangle J_\beta$	OMPC	LOMPC	$\triangle J_\rho$
2	0.9800	0.1048	0.3466	—	Unfeasible	1.0486	—	0.6727	0.7818	—
3	0.7614	0.1418	0.3021	+113%	1.0592	1.0555	-0.03%	0.6909	1.1273	+84.2%
3	0.9555	0.1418	0.4488	+216%	1.0592	1.0528	-0.06%	0.6909	1.1273	+84.2%
4	0.8560	0.1823	0.5099	+179%	1.0613	1.0541	+0.60%	0.7091	1.800	+153.8%
5	0.7765	0.2267	0.5190	+128%	1.0592	1.0527	+0.61%	0.7273	2.9091	+299.8%

Table 9. Heuristic LOMPC tuning parameters (a, n_c) for Example 2 and values of objective functions.

Tuning parameter		Average radius to MCAS (J_ϑ)	Performance loss (J_β)	Computational cost (J_ρ)	Reference
n_c	a				
2	0.8	0.3322	1.0260	7818	[25, 45]

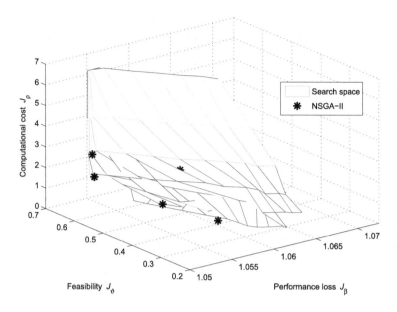

Fig. 3. Pareto Front for Example 2.

6   Conclusions

Correct selection of the tuning parameters of the controller guarantees the best balance between closed-loop performance, feasible region and computational load. In the design of MPC, different strategies to formulate the d.o.f. have been proposed, but, each of these approaches introduces one or more tuning parameters without guidelines for their selection. The main contribution of this work is to explore a systematic tuning for a parameterized predictive control algorithm using a multiobjective optimization evolutionary algorithm. The results of this work suggest that NSGA-II is an useful tool to obtain a set of Pareto non-dominated solutions for the tuning parameters of a LOMPC. The set of solutions found exhibit a proper balance (trade-off) between the three objectives compromised to each other: maximize the feasibility region, minimize the computational burden and minimize the performance loss, even for constrained systems and when the search space is large. The designer only has to choose one and is no longer necessary to explore the whole search space. Also, it can be verified that the tuning combination presented in Tables 3, 4 and 5 (example 1) and Tables 7 and 8 are equivalent or better than the tuning combinations presented in previous papers (Tables 6 and 9), in terms of J_ϑ, J_β and J_ρ. In summary, with the method proposed in this paper, the problem of tuning the controller in a systematized manner is solved, which implies a decrease in the design and test time of the controller, at the same time as it offers a guarantee of good operation. However, in order to demonstrate the effectiveness of this method, it is necessary to perform more tests in systems with more complex dynamics

[14,22,38,39], as well as to implement this method with other optimization techniques, for example, other MOEAs [47,48] or multi-objective cell mapping and subdivision techniques [15,16].

References

1. Aicha, F.B., Bouani, F., Ksouri, M.: A multivariable multiobjective predictive controller. Int. J. Appl. Math. Comput. Sci. **23**(1), 35–45 (2013)
2. Bacic, M., Cannon, M., Lee, Y.I., Kouvaritakis, B.: General interpolation in MPC and its advantages. IEEE Trans. Autom. Control **48**(6), 1092–1096 (2003)
3. Bemporad, A., Filippi, C.: Suboptimal explicit MPC via approximate multiparametric quadratic programming. In: Proceedings of the 40th IEEE Conference on Decision and Control, Orlando, Florida (2001)
4. Bemporad, A., Munoz de la Peña, D.: Multiobjective model predictive control. Automatica **45**, 2823–2830 (2009)
5. Bunin, G., Fraire-Tirado, F., Francois, G., Bonvin, D.: Run-to-run MPC tuning via gradient descent. In: 22nd European Symposium on Computer Aided Process Engineering, London (2012)
6. Camacho, E.F., Bordons, C.: Control predictivo: pasado, presente y futuro. Revista Iberoamericana de Automática e Informática Industrial **1**(3), 5–28 (2004)
7. Coello-Coello, C.A., Lamont, G.B., Van Veldhuizen, D.A.: Evolutionary Algorithms for Solving Multi-Objective Problems (Genetic and Evolutionary Computation). Springer-Verlag New York, Inc., New York (2006)
8. Cutler, C.R., Ramarker, B.C.: Dynamic matrix control - a computed algoritm. In: Proceedings of the 86th National Meeting of the American Institute of Chemical Engineers (AIChe) (1979)
9. Deb, K.: A fast and elitist multiobjetive genetic algorithm NSGA-II. IEEE Trans. Evol. Comput. **6**(2), 182–197 (2002)
10. Deb, K.: Optimization using evolutionary algorithms: an introduction. Technical report KanGAL 2011003, Department of Mechanical Engineering, Kanpur, India, February 2011
11. Deb, K., Agrawal, R.B.: Simulated binary crossover for continuous search space. Complex Syst. **9**, 115–148 (1995)
12. Deb, K., Anand, A., Joshi, D.: A computationally efficient evolutionary algorithm for real-parameter optimization. Evol. Comput. **10**(4), 371–395 (2002)
13. Deb, K., Beyer, H.G.: Self-adaptive genetic algorithms with simulated binary crossover. Technical report CI-61 \ 99, Self-Adaptive Genetic Algorithms with Simulated Binary Crossover. Department of Computer Science University of Dortmund, Dortmund, Germany, March 1999
14. Dellnitz, M., Ober-Blöbaum, S., Post, M., Schütze, O., Thiere, B.: A multiobjective approach to the design of low thrust space trajectories using optimal control. Celest. Mech. Dyn. Astron. **105**(13), 33–59 (2009)
15. Dellnitz, M., Schütze, O., Hestermeyer, T.: Covering Pareto sets by multilevel subdivision techniques. J. Optim. Theory Appl. **124**(1), 113–136 (2005)
16. Fernández, J., Schütze, O., Hernández, C.: Parallel simple cell mapping for multiobjective optimization. Eng. Optim. **48**(11), 1845–1868 (2016)
17. Fleming, P.J., Purshouse, R.C.: Evolutionary algorithms in control systems engineering: a survey. Control Eng. Pract. **10**(11), 1223–1241 (2002)

18. Fogel, D.B.: Evolutionary Computation: Toward a New Philosophy of Machine Intelligence. Series on Computational Intelligence, vol. 3. Wiley-IEEE Press, Hoboken (2006)
19. Gambier, A.: MPC and PID control based on multi-objective optimization. In: American Control Conference, 2008, pp. 4727–4732. IEEE (2008)
20. Garriga, J., Soroush, M.: Model predictive control tuning methods: a review. Ind. Eng. Chem. Res. **49**, 3505–3515 (2010)
21. Gutiérrez-Urquídez, R., Valencia-Palomo, G., Rodrıguez-Elıas, O.M., Trujillo, L.: Systematic selection of tuning parameters for efficient predictive controllers using a multiobjective evolutionary algorithm. Appl. Soft Comput. **31**, 326–338 (2015)
22. Johansson, K.H.: The quadruple-tank process: a multivariable laboratory process with an adjustable zero. IEEE Trans. Control Syst. Technol. **8**(3), 456–465 (2000)
23. Khan, B., Rossiter, J.A.: Robust MPC algorithms using alternative parameterisations. In: UKACC International Conference on Control 2012, Cardiff, UK, pp. 3–5 (2012)
24. Khan, B., Rossiter, J.A.: Alternative parameterisation within predictive control: a systematic selection. Int. J. Control **86**(8), 1397–1409 (2013)
25. Khan, B., Rossiter, J.A., Valencia-Palomo, G.: Exploiting Kautz functions to improve feasibility in MPC. In: Procceedings of the 18th IFAC World Congress (2011)
26. Lee, J.H.: Model predictive control review of the three decades of development. Int. J. Control Autom. Syst. **9**(3), 415–424 (2011)
27. Li, M., Zhou, P., Wang, H., Chai, T.: Nonlinear multiobjective MPC-based optimal operation of a high consistency refining system in papermaking. IEEE Trans. Syst. Man Cybern. Syst. (2017)
28. Maciejowski, J.M.: Predictive Control with Constraints. Prentice Hall, Harlow (2006)
29. Mahmoudi, H., Lesani, M., Arab-Khabouri, D.: Online fuzzy tuning of weighting factor in model predictive control of PMSM. In: 13th Iranian Conference on Fuzzy Systems (IFSC), Qazvin, Iran (2013)
30. Mayne, D.Q., Rawllings, J.B., Rao, C.V., Skokaert, P.O.M.: Constrained model predictive control: stability and optimality. Automatica **36**(6), 789–814 (2000)
31. Merabti, H., Belarbi, K.: Multi-objective predictive control: a solution using meta-heuristics. Int. J. Comput. Sci. Inf. Technol. **6**(6), 147 (2014)
32. Muske, K.R., Badgwell, T.A.: Disturbance modeling for offset-free linear model predictive control. J. Process Control **12**(5), 617–632 (2002)
33. Pareto, V.: Manual of Political Economy. Macmillan, London (1971)
34. Rossiter, J.A.: Model Predictive Control: A Practical Approach. CRC Press, Boca Raton (2005)
35. Rossiter, J.A.: A global approach to feasibility in linear MPC. In: Proceedings of the UKACC ICC (2006)
36. Rossiter, J.A., Wang, L., Valencia-Palomo, G.: Efficient algorithms for trading off feasibility and performance in predictive control. Int. J. Control **83**(4), 789–797 (2010)
37. Ruchika, N.R.: Model predictive control: history and development. Int. J. Eng. Trends Technol. (IJETT) **4**(6), 2600–2602 (2013)
38. Sardahi, Y., Sun, J.Q., Hernández, C., Schütze, O.: Many-objective optimal and robust design of PID controls with a state observer. J. Dyn. Syst. Meas. Control **139**(2), 024502-2–024502-4 (2017)

39. Schütze, O., Vasile, M., Junge, O., Dellnitz, M., Izzo, D.: Designing optimal low thrust gravity assist trajectories using space pruning and a multi-objective approach. Eng. Optim. **41**(2), 155–181 (2009)
40. Scokaert, P.O.M., Rawlingsm, J.B.: Constrained linear quadratic regulation. IEEE Trans. Autom. Control **43**(8), 1163–1169 (1998)
41. Scokaert, P.O.M., Rawlings, J.B.: Feasibility issues in linear model predictive control. Am. Inst. Chem. Eng. (AIChE) J. **45**(8), 1649–1659 (1999)
42. Sun, J.Q., Xiong, F.R.: Cell mapping methods-beyond global analysis of nonlinear dynamic systems. Adv. Mech. **47**(05), 201705 (2017)
43. Suzuki, R., Kawai, F., Ito, H., Nakazawa, C., Fukuyama, Y., Aiyoshi, E.: Automatic tuning of model predictive control using particle swarm optimization. In: IEEE Swarm Intelligence Symposium, HI, Honolulu, p. 2007 (2007)
44. Valencia-Palomo, G., Rossiter, J.A.: PLC implementation of a predictive controller using laguerre functions and multi-parametric solutions. In: Proceedings of the United Kingdom Automatic Control Conference, Coventry, UK (2010)
45. Valencia-Palomo, G., Rossiter, J.A., Jones, C.N., Gondhalekar, R., Khan, B.: Alternative parameterisations for predictive control: how and why. In: American Control Conference, San Francisco, CA, USA (2011)
46. Zitzler, E.: A tutorial on evolutionary multiobjective optimization. Technical report, Swiss Federal Institute of Technology, Computer Engineering and Networks Laboratory, Zurich (2002)
47. Zitzler, E., Künzli, S.: Indicator-based selection in multiobjective search. In: Proceedings of the 8th International Conference on Parallel Problem Solving from Nature (PPSN VIII), Birmingham, UK (2004)
48. Zitzler, E., Laumanns, M., Thiele, L.: Spea2: Improving the strength Pareto evolutionary algorithm. Technical report 103, SPEA2: Improving the Strength Pareto Evolutionary Algorithm. Federal Institute of Technology (ETH), Department of Electrical Engineering Swiss, Zurich (2001)

IDA-PBC Controller Tuning Using Steepest Descent

J. A. Morales[(✉)], M. A. Castro, D. Garcia, C. Higuera, and J. Sandoval

Tecnólogico Nacional de México, Instituto Tecnológico de La Paz, La Paz, Mexico
{jmoralesv,mcastro,diego.garcia,cesarh,jsandoval}@itlp.edu.mx

Abstract. The optimization of controller parameters or gains is a challenge usually approached using empirical methods that consume valuable time, without the certainty that the obtained gains actually produce the desired behaviour of the controlled plant. There are several analytical and numerical methodologies to find the parameters for PID controllers, however currently there is not enough available information regarding the application of optimization methods for nonlinear controllers. The present work describes the application of the maximum descent method to find the gains of IDA-PBC controller for a ball and beam system. The proposed methodology involves implementing a mathematical model to describe the system's dynamics, the design of a objective function to measure how closely the plant follows the desired behaviour, and finally the evaluation of a set of gains obtained by the numerical method. The dynamic model and the optimization algorithm were implemented in C language in order to reduce the computer time compared to the use of frameworks such as MATLAB. Numerical simulations to validate the effectiveness of the proposed methodology are included.

Keywords: Ball and beam system · Controller tuning
Nonlinear control system · Steepest descent · IDA-PBC

1 Introduction

Control systems theory deals with the analysis and design of components and their interaction as a system to produce a desired behaviour [1]. A key configuration in control theory is based on the essential notion of feedback, a closed-loop controller uses feedback to control the states or outputs of a dynamical system. Its name comes from the way that information flows through the system: The system's input is usually a reference point $r(t)$, that represents the desired system output, using a feedback the current system output $y(t)$ is obtained and subtracted from $r(t)$ to calculate an error signal $e(t)$ which is used by the controller to obtain a signal $u(t)$ to modify the plant's behaviour, as shown in Fig. 1.

The theory and practice of control has a wide range of applications on the fields of engineering such as aeronautics, chemistry, mechanical systems, ambient management systems, construction, electrical networks among many other disciplines. The advantages of an efficient control on industrial applications are huge,

© Springer International Publishing AG, part of Springer Nature 2019
L. Trujillo et al. (Eds.): NEO 2017, SCI 785, pp. 158–170, 2019.
https://doi.org/10.1007/978-3-319-96104-0_8

and include quality improvements, reduction of waste and energy consumption, enhanced security levels, and pollution control.

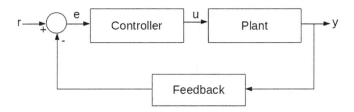

Fig. 1. Closed-loop control system.

The starting point of a control system analysis is the mathematical model of the plant, usually presented as an operator between the system's inputs and outputs, or as a set of partial differential equations. Most mathematical models used in control system are linear and there is a vast and robust knowledge corpus on this specific subject. However, current technological advances have generated a huge variety of new problems and applications where it's nonlinear essence is of importance. For instance, nonlinear dynamics are observed commonly on modern engineering applications such as flight command systems, robotic manipulators, automated highways, plane wing structures, boats, reaction engines, turbo-diesel motors, electrical induction motors and high performance fuel injection systems [2–4]. Such systems cannot be properly described by linear models, that is the main reason for the use of nonlinear models and the development of tools and concepts for nonlinear systems control. Interconnection and Damping Assignment (IDA) an extension of Passivity Based Control (PBC), is a successful methodology for the design of nonlinear control systems [5] that provides certain advantages compared to others. The foremost advantage is that controllers produced using this methodology are designed to achieve asymptotic stability of desired equilibrium states on nonlinear systems, without the need of linearisation or decoupling procedures [6]. Secondly, when using the Hamiltonian structure with a desired energy function, this function qualifies as a Lyapunov candidate for the desired equilibrium. Also, the IDA-PBC method is motivated by the principles of energy and damping injection, and it is demonstrated that the energy shaping and the designs of control systems based on passivity are effective to solve problems that involve underactuated mechanical systems [7]. Ultimately, the IDA-PBC method provides free parameters, which increase the number of possible solutions of the partial differential equations [5]. However a problem with this method is that it does not provide a way to adjust or tune the design parameters and the performance of the controllers that produce, depends on this values.

2   The IDA-PBC Method

A brief review on the IDA-PBC method applied to control a class of underactuated mechanical systems is presented (see the work done by Ortega et al. [5] for further details). The procedure starts from the Hamiltonian description of the system by means of the total energy function (kinetic plus potential energies) given by

$$H(\mathbf{q}, \mathbf{p}) = \frac{1}{2}\mathbf{p}^T M^{-1}(\mathbf{q})\mathbf{p} + V(\mathbf{q}), \tag{1}$$

where $\mathbf{q}$ and $\mathbf{p} \in \mathbb{R}^n$ are the vectors of generalized position and momenta, respectively, $M(\mathbf{q}) = M^T(\mathbf{q}) > 0$ is the inertia matrix and $V(\mathbf{q})$ is the potential energy [8]. If we assume that the system has not natural damping, the equations of motion can be written as

$$\frac{d}{dt}\begin{bmatrix}\mathbf{q}\\\mathbf{p}\end{bmatrix} = \begin{bmatrix}0_{n \times n} & I_{n \times n}\\-I_{n \times n} & 0\end{bmatrix}\begin{bmatrix}\nabla_{\mathbf{q}} H\\\nabla_{\mathbf{p}} H\end{bmatrix} + \begin{bmatrix}0_{n \times m}\\G\end{bmatrix}\mathbf{u}, \tag{2}$$

being $\mathbf{u} \in \mathbb{R}^m$, the vector of control inputs and $G \in \mathbb{R}^{n \times m}$, with $\mathrm{rank}(G) = m$ where $n > m$ for underactuated systems.

The IDA-PBC methods assigns a particular desired structure in closed-loop system, with desired energy function given by

$$H_d(\mathbf{q}, \mathbf{p}) = \frac{1}{2}\mathbf{p}^T M_d(\mathbf{q})^{-1}\mathbf{p} + V_d(\mathbf{q}), \tag{3}$$

where $M_d(\mathbf{q}) = M_d^T(\mathbf{q}) > 0$ and $V_d(\mathbf{q})$ are the desired inertia matrix and the desired potential energy function, respectively.

The desired closed-loop system is

$$\frac{d}{dt}\begin{bmatrix}\mathbf{q}\\\mathbf{p}\end{bmatrix} = \begin{bmatrix}0_{n \times n} & M^{-1}M_d\\-M_d M^{-1} & J_2(q,p) - GK_v G^T\end{bmatrix}\begin{bmatrix}\nabla_q H_d\\\nabla_p H_d\end{bmatrix}, \tag{4}$$

where $J_2 = -J_2^T$ and $K_v = K_v^T$ are free matrices. Now, for this class of systems, the main challenge of the IDA-PBC method consists in solving the following set of partial differential equations, called matching equations

$$G^{\perp}[\nabla_q(\mathbf{p}^T M^{-1}\mathbf{p}) - M_d M^{-1}\nabla_q(\mathbf{p}^T M_d^{-1}\mathbf{p}) + 2J_2 M_d^{-1}\mathbf{p}] = \mathbf{0}, \tag{5}$$

and

$$G^{\perp}[\nabla_q V - M_d M^{-1}\nabla_q V_d] = \mathbf{0}, \tag{6}$$

where $G^{\perp} \in \mathbb{R}^{(n-m) \times n}$, such as $G^{\perp}G = 0_{(n-m) \times m}$ whose solutions M_d and V_d produce the control law given by

$$\mathbf{u} = (G^T G)^{-1}G^T[\nabla_q H - M_d M^{-1}\nabla_q H_d + J_2 M_d^{-1}\mathbf{p}] - K_v G^T M_d^{-1}\mathbf{p}. \tag{7}$$

Furthermore, if M_d is a positive definite in a neighborhood of $\mathbf{q}^*$ and

$$\mathbf{q}^* = argmin(V_d), \tag{8}$$

then $[\mathbf{q}^T \mathbf{p}^T]^T = [\mathbf{q}^{*T} 0^T]^T$ is a stable equilibrium of the closed-loop desired system, with a Lyapunov function H_d. This equilibrium is asymptotically stable if it is locally detectable from the output $G^T \nabla_p H_d$.

3   Ball and Beam System

The ball and beam system is one of the most popular and important benchmark systems for studying control systems. Several classical and modern control methods have been used to stabilize the ball and beam system [9,10]. The sensor placed on one side of the beam detects the ball role along the beam and its position. An actuator drives the beam to a desired angle by applying a torque at the end of the beam. Figure 2 shows the ball and beam system (Quanser Model SRV02 and BB01) which is utilized in this research work. The controller regulates the ball position by moving the beam using the motor and overcoming the disturbances. The ball and beam system is an inherently unstable system. In other words, the ball position can be changed without limit for a fixed input of beam angle. This property has made the ball and beam system a suitable device to test different control strategies. The ball and beam system has 2 Degrees-of-Freedom (DOFs). The ball is assumed to have friction, rotary moment of inertial and coriolis acceleration during motion on the beam. However, some of the dynamic properties are neglected in most of the research work regarding the ball and beam mechanism in order to simplify the dynamic equation of the system [11,12], a linear approximation is not accurate when the angle of the beam is appreciable. Thus a more advanced control technique such as nonlinear control like IDA-PBC should work better (Table 1).

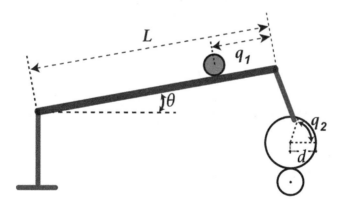

Fig. 2. Ball and beam system.

The behaviour of the system in its Hamiltonian structure (2) can be followed in [13,14] and is described by the following definitions, with the values $a_1 = m_B + \frac{J_B}{R_B^2}$, $a_2 = J_b \frac{d^2}{L^2} + J_{fw}$, $a_3 = m_B \frac{d^2}{L^2}$, $a_4 = m_B g$ and $a_5 = m_b g \frac{L}{2}$, where $\mathbf{q} = [q_1, q_2]^T$ $\mathbf{p} = [p_1, p_2]^T$,

$$M(q_1) = \begin{bmatrix} a_1 & 0 \\ 0 & a_2 + m_B \left(\frac{L}{2} + q_1 \right)^2 \end{bmatrix}, \tag{9}$$

$$V(\mathbf{q}) = a_4 \left(\frac{L}{2} + q_1 \right) + sin(q_2) + a_5 sin(q_2) \tag{10}$$

and

$$G = [0, 1]^T. \tag{11}$$

Table 1. Ball and beam system physical parameters.

Symbol	Description	Units
m_b	Ball mass	0.15 kg
J_b	Inertia of the beam	0.009 kg m^2
J_{fw}	Inertia of the ball	0.002 kg m^2
L	Length of the beam	0.4255 m
d	Coupling union radius	0.0254 m
R_B	Ball radius	0.0127 m
m_B	Ball mass	0.064 kg

4   An IDA-PBC for the Ball and Beam System

The control objective is to bring the ball to a desired position over the beam, which must remain horizontal. Formally, the control objective is established as follows

$$\lim_{t \to \infty} \begin{bmatrix} q_1(t) \\ q_2(t) \end{bmatrix} = \begin{bmatrix} q_{1d} \\ 0 \end{bmatrix}. \tag{12}$$

Muralidharam et al. [13] designed an IDA-PBC control law to a ball and beam system that satisfies the control objective (12), where M_d y V_d are given by

$$M_d = \begin{bmatrix} k_1 a_1 \left(\frac{L}{2} + q_1 \right) & a_1 \left(a_2 + a_3 \left(\frac{L}{2} + q_1 \right)^2 \right) \\ a_1 \left(a_2 + a_3 \left(\frac{L}{2} + q_1 \right)^2 \right) & k_4 \end{bmatrix} \tag{13}$$

and

$$V(\mathbf{q}) = \frac{a_4}{a_1} [1 - cos(q_2)] + \frac{k_p}{2} \left(q_2 - \frac{k_2}{k_1} log \left(\frac{L + 2[q_1 - q_{d1} - q_2]}{L} \right) \right)^2, \tag{14}$$

with $k_1 = \sqrt{2a_1 a_3}$, and k_4 y k_p are free parameters strictly positive.

$$J_d = \begin{bmatrix} 0 & j_2(\mathbf{q}, \mathbf{p}) \\ -j_2(\mathbf{q}, \mathbf{p}) & 0 \end{bmatrix}, \tag{15}$$

$$\alpha = [\alpha_1; \alpha_2]^T, \tag{16}$$

$$W = \begin{bmatrix} 0 & 1 \\ -1 & 0 \end{bmatrix}, \tag{17}$$

$$\alpha_1 = 2a_3\lambda_2 \left(q_1 + \frac{L}{2} \right) (\lambda_4 - \lambda_1) \tag{18}$$

and

$$\alpha_2 = a_3 \left(q_1 + \frac{L}{2} \right) \lambda_4^2. \tag{19}$$

Free matrix J_d is defined by (15) where $j_2 = -\mathbf{p}^T M_d^{-1} \alpha(\mathbf{q}, \mathbf{p})W$, α by (16), W is given by (17), $\alpha_1(\mathbf{q}, \mathbf{p})$ and $\alpha_2(\mathbf{q}, \mathbf{p})$ by (18), (19) respectively.

The functions λ_i are as follows

$$\lambda_1 = k_1 \left(q_1 + \frac{L}{2} \right), \lambda_2 = k_2, \lambda_4 = \frac{k_4}{a_2 + a_3 \left(q_1 + \frac{L}{2} \right)^2}, \tag{20}$$

where $k_1 > 0$ y $k_2 = k_1 \sqrt{a_1/(2a_3)}$.

In [14] q_{d1} is introduced in V_d (14), what extends the results of Muralidharan [13] where only $q_{d1} = 0$ is considered.

Using Eqs. (13) and (14), the control input $\mathbf{u}$ can be obtained by (7) and the system (2) can be simulated.

Once the control law is obtained, remains the problem of assigning suitable values to the free parameters k_4, k_p and k_v in the control law, to ensure the control objective (12) with good performance, this procedure is known as tuning the controller. A problem that can arise with this kind of controller is that a bad tuning may result in a poor performance or even lead to instability.

5   Tuning of the IDA-PBC Controller

There are many tuning methods to determine the gains of PID controllers to obtain good performance and robustness [15]. Some methods with simple formulas use little information of the plant's dynamics resulting in moderate performance and a re-tune process by trial and error depending on those results. More sophisticated tuning method can get rise to considerable improvements in performance, but they are also more computationally demanding and depend on more information of the plant's dynamics [16]. However, there are no methods in the literature for nonlinear controllers, such as the one presented in this work.

We consider the tuning of the control law as an optimization problem. To do this we need the following:

1. Select a performance index to be used as objective function.
2. Select an optimization algorithm.

For the purpose of selecting a performance index we can consider the rise time (the time needed by the control system to reach the desired value after a perturbation), peak overshoot (the highest value reached by the response before reaching the desired value) and settling time.

However, there are other criteria that take into account the transitory regime of the solution, to get a finer tuning and improve the response of the controller. Most of these criteria consider the tracking error as the current value of the position $\mathbf{q}(t)$ minus the desired value $\mathbf{q_d}$, in this particular problem $\mathbf{q_d}, \mathbf{q}(t) \in \mathbb{R}^2$, so the error $\mathbf{e}(t)$ is defined as

$$\mathbf{e}(t) = \begin{bmatrix} q_1(t) - q_{1d} \\ q_2(t) \end{bmatrix}. \tag{21}$$

A performance index commonly applied in control field is the integrated square error or ISE [17], defined as shown below

$$ISE = \int_0^\infty \|\mathbf{e}(t)\|^2 dt. \tag{22}$$

In this performance index, larger errors contribute a lot in the integrate, this tends to give solutions with small overshoots, but unfortunately it is insensitive to small errors. A variation of this index, that was used in this work is the integrate time-square error, given by

$$ISTE = \int_0^\infty \|\mathbf{e}(t)\|^2 t^2 dt, \tag{23}$$

which takes account of errors that remain in time, thus penalizing oscillatory solutions [18].

It was decided to use a numerical method because they are guaranteed to produce at least local optimum when they are used in smooth functions. In contrast, heuristic methods, are designed to produce good answers quickly, but they are not guarantee of finding a true optimum.

The applied numerical method was gradient descent, a first-order iterative algorithm, that consists in taking steps proportional to the negative of the function's gradient at the current point.

The reason to apply this method, in spite of its low order of convergence, is that newtonians or quasinewtonians algorithms would have required the implementation of a numerical estimate of a Hessian matrix or many evaluations of the aptitude functions which was considered expensive and unstable.

Gradient descent is based on the observation that if a multivariable function $F(\mathbf{x})$ is defined and differentiable in a neighborhood of a point a, then $F(\mathbf{x})$ decreases fastest if one goes from a to the direction of the negative gradient of $F(\mathbf{x})$ at a. It follows that, for small enough values of γ, (24) is fulfilled and $F(a_n) \geq F(a_{n+1})$ [19].

$$a_{n+1} = a_n - \gamma_n \nabla F(a_n). \tag{24}$$

Taking this observation into account, the algorithm starts with a guess $\mathbf{x}_0$ for a local minimum of F, and considers the sequence $\mathbf{x}_0, \mathbf{x}_1, \mathbf{x}_2...$ such that the following is fulfilled

$$\mathbf{x}_{n+1} = \mathbf{x}_n - \gamma_n \nabla F(\mathbf{x}_n), \quad n \geq 0, \tag{25}$$

and $F(\mathbf{x}_n)$ in the sequence

$$F(\mathbf{x}_0) \geq F(\mathbf{x}_1) \geq F(\mathbf{x}_2) \geq \ldots \geq F(\mathbf{x}_n) \tag{26}$$

converges to a minimum. Note that the value of step size γ is allowed to change at each iteration.

When the step length γ_k is computed, a trade-off must be solved. It is desirable to choose γ_k to achieve a substantial reduction of F, but at the same time, it is important to limit the time consumed making the choice. The ideal choice would be the global minimizer of the univariable function

$$\phi(\gamma) = F(\mathbf{x}_k + \gamma \nabla F(\mathbf{x}_n)), \tag{27}$$

but in general, it is too expensive to identify this value. A simple condition we could impose on γ_k is that it provides a reduction in $F(\mathbf{x})$ as shown in

$$F(\mathbf{x}_k + \gamma_k \nabla F(\mathbf{x}_k)) < F(\mathbf{x}_k). \tag{28}$$

To find adequate values to γ in each iteration, we used a so-called backtracking approach. In its most basic form, backtracking proceeds as shown in Algorithm 1.

Algorithm 1. Backtracking.

Choose $\gamma_0 > 0$, $\rho \in (0,1)$
$\gamma = \gamma_0$
repeat
  $\gamma = \rho\gamma$
until $F(\mathbf{x}_k + \gamma \nabla F(\mathbf{x}_k)) < F(\mathbf{x}_k)$
return $\gamma_k = \gamma$

Once solved the problem of finding γ in each iteration, we can formulate the gradient descent Algorithm 2.

Algorithm 2. Gradient descent.

Choose $\mathbf{x}_0$, set $\varepsilon > 0, k = 1, \mathbf{x}_k = \mathbf{x}_0$.
while $\nabla F(\mathbf{x}_k) >= \varepsilon$ **do**
  $\mathbf{d}_k = -\nabla F(\mathbf{x}_k)$
  Choose γ using Algorithm 1
  $\mathbf{x}_{k+1} = \mathbf{x}_k + \gamma \mathbf{d}_k$
end while
return $\mathbf{x}_k$

For the particular problem of tuning the ball and beam controller, an objective function of the form (23) was defined as

$$F(k_4, k_p, k_v) = \int_0^T \|\mathbf{e}(t)\|^2 t^2 dt. \tag{29}$$

Since the control law is continuous when $k_4, k_p, k_v > 0$ and M_d, M are invertible, $\mathbf{q}(t)$ is bounded because the control objective (12) is achieved. So we can guess F is a smooth function, thus, can be minimized locally using gradient descent.

However, we can not assure that F has just one optimum point, so a suitable approach consists in generating multiple initial random values for the vector $\mathbf{x}_0 = [k_4, k_p, k_v]$, gradient descent is used to find the local optimum for each of the starting points and the best of them is considered as the solution.

6   Implementation

The evaluation of (29) requires to know the values of $\mathbf{q}$ over a time interval $(0, T)$, so a full simulation of the system's dynamic is needed for each vector $\mathbf{x}_i = [k_4, k_p, k_v]^T$. In order to solve the set of differential equations involved, a fourth order Runge-Kutta (RK) numerical method was implemented, this algorithm guarantees an error of Δt^5 order for each step of size Δt [20].

To compute the next vector $\mathbf{x}_{i+1}$, the gradient of (29) must be estimated, given the complexity of the analytical solution, a divided centred differences method was used, this ensures an numerical error proportional to the square of the step size ϵ used in Algorithm 2.

Since this is a multi-modal problem, it is adequate to perform several executions of the descent using different starting points, a parallel implementation of the algorithm was used to take advantage of the modern processor architecture by evaluation simultaneous starting points.

The program was written in C language and compiled using the GCC 4.8.4 compiler collection, the message passing library OpenMPI 1.8.4 was used to implement process synchronization. The presented results were obtained using a workstation with i7-3770 processor and 4 GB of RAM, running a 64 bits LinuxMint 18.2 installation.

The simulation period limit T was $3\,\mathrm{s}$ with initial resting conditions $q_1(0) = 0.2\,m$, $q_2(0) = -1$ rad and a maximum $1.2N$ in each components of the requested control signal vector was used.

The RK parameters used were $\epsilon = 1e^{-5}$ and $\Delta t = 1e^{-4}$. For the backtracking algorithm, the proposed values in [19], $\gamma_0 = 1$ and $\rho = 0.5$ were used.

The parallelization strategy was to execute eight processes (one per every processor's virtual core) each one applied the gradient descent to ten different random starting position vectors $\mathbf{x}_0$, finally the best result of each process is consolidated on the rank 0 process and displayed as the final result.

7   Results

The obtained results were compared against the performance in simulation of the set of gains empirically obtained presented in [14]. Table 2 shows a comparison of the obtained sets of gains using each method.

Table 2. Comparison of gains.

Tuning method	k_4	k_p	k_v	$F\left(k_4, k_p, k_v\right)$
Empyrical	0.012	40.0	0.795	0.076759
Gradient descent	0.031087	44.509438	0.515262	0.005749

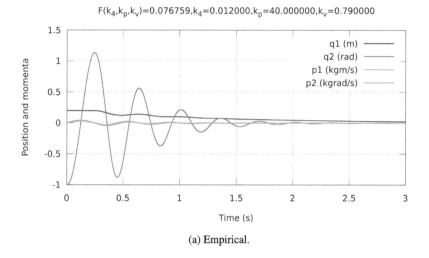

(a) Empirical.

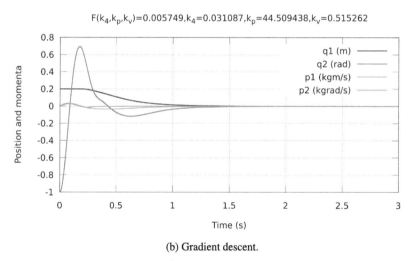

(b) Gradient descent.

Fig. 3. Position and momenta behaviour.

The states and momenta behaviour of the closed loop system using the empirically obtained gains in [14] and the proposed methodology is shown on Fig. 3a and b. Also the control signal required by the controller using each set of gains is depicted on Fig. 4a and b respectively.

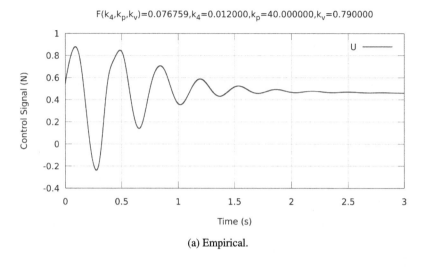

(a) Empirical.

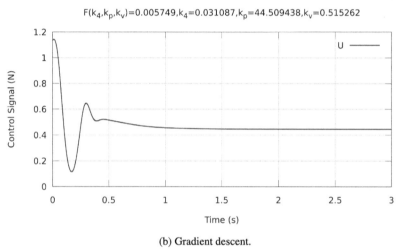

(b) Gradient descent.

Fig. 4. Control signal behaviour.

8   Conclusions and Future Work

The proposed methodology obtains a set of gains that produces a superior behaviour of the simulated system in overshoot reduction and setting time with less oscillations than the gains obtained in [14].

It is also of notice that the range of control signals produced by the tuning process does not exceed the nominal value of the actuator.

The methodology of assuming that the tuning problem is equivalent to the optimization of a soft function and that therefore it is suitable to solve by using numerical methods seems to be valid for this particular ball and beam system.

The implementation of this methodology on different plants for its validation remains as future work, as well as the evaluation of different objective functions and optimization strategies.

References

1. Khalil, H.K.: Nonlinear Systems, 3rd edn. Prentice Hall, Harlow (2002)
2. Zúñiga, J.: Control de un Robot SCARA Basado en Pasividad. Maestría en ciencias en ingeniería electrónica, Centro Nacional de Investigación y Desarrollo Tecnológico, Cuernavaca, Morelos, México (2016)
3. García, D., Sandoval, J., Gutiérrez-Jagüey, J., Bugarin, E.: Control IDA-PBC de un Vehículo Submarino Subactuado. Revista Iberoamericana de Automática e Informática industrial **15**, 36 (2017)
4. González-Cabezas, H.A., Duarte-Mermoud, M.A.: Control de velocidad para motores de inducción usando IDA-PBC. Revista Chilena de Ingeniería **446**, 81–90 (2005)
5. Ortega, R., Spong, M.W., Gómez-Estern, F., Blankenstein, G.: Stabilization of a class of underactuated mechanical systems via interconnection and damping assignment. IEEE Trans. Autom. Control **47**(8), 1218–1233 (2002)
6. Acosta, J.Á., Ortega, R., Astolfi, A., Mahindrakar, A.D.: Interconnection and damping assignment passivity-based control of mechanical systems with underactuation degree one. IEEE Trans. Autom. Control **50**(12), 1936–1955 (2005)
7. Gómez-Estern, F., Van der Schaft, A.: Physical damping in IDA-PBC controlled underactuated mechanical systems. Eur. J. Control **10**(5), 451–468 (2004)
8. De-León-Gómez, V., Santibañez, V., Sandoval, J.: Interconnection and damping assignment passivity-based control for a compass-like biped robot. Int. J. Adv. Rob. Syst. **14**(4), 1–18 (2017). https://doi.org/10.1177/1729881417716593
9. Wang, W.: Control of a ball and beam system. Ph.D. thesis (2007)
10. López, F., Monroy, P., Rairán, J.D.: Control de posición de un sistema bola y viga con actuadores magnéticos. Tecnura **15**, 12–23 (2011)
11. Ortiz, E., Liu, W.Y.: Modelado y control PD-Difuso en tiempo real para el sistema barra-esfera. In: Congreso anual de la AMCA, pp. 1–6 (2004)
12. Meneses Morales, P., Zafra Siancas, H. D.: Diseño e implementación de un módulo educativo para el control de sistema Bola y Varilla. PhD thesis, Pontificia Universidad Católica del Perú (2013)
13. Muralidharan, V., Anantharaman, S., Mahindrakar, A.D.: Asymptotic stabilisation of the ball and beam system: design of energy-based control law and experimental results. Int. J. Control **83**(6), 1193–1198 (2010)
14. Higuera, C., Chargoy, A., Sandoval, J., Bugarin, E., Coria, L.N.: IDA-PBC en Voltaje para la Regulación de un Sistema Barra-Bola Experimental. In: Congreso Nacional de Control Automático, Monterrey, Nuevo León, Mexico, pp. 1–6 (2017)
15. Morales Viscaya, J.A., Rochín Ramírez, A., Castro Liera, M.A., Sandoval Galarza, J.A.: AG y PSO como métodos de sintonía de un PID para el control de velocidad de un motor CD. In: Castro Liera, I., Cortés Larrinaga, M. (eds.) Nuevos avances en robótica y computación, 1st edn. pp. 81–87. ITLP, La Paz (2015)
16. Wu, H., Su, W., Liu, Z.: PID controllers: design and tuning methods. In: Proceedings of the 2014 9th IEEE Conference on Industrial Electronics and Applications, ICIEA 2014 (2014)
17. Ogata, K.: Modern Control Engineering, 5th edn. Pearson, New York (2016)

18. Schultz, W.C., Rideout, V.C.: Control system performance measures: past, present, and future. IRE Trans. Autom. Control **AC–6**, 22–35 (1961)
19. Nocedal, J., Wright, S.J.: Numerical Optimization, 2nd edn. Springer, New York (2006)
20. Nieves Hurtado, A., Domínguez Sánchez, F.C.: Métodos numéricos: aplicados a la ingeniería. 4th edn. Grupo Editorial Patria (2012)

Self-tuning for a SISO-type Fuzzy Control Based on the Relay Feedback Approach

Pablo J. Prieto[1], Nohe R. Cazarez-Castro[1(✉)], Luis T. Aguilar[2], and Selene L. Cardenas-Maciel[1]

[1] Tecnológico Nacional de México - Instituto Tecnológico de Tijuana, Calz. del Tecnológico S/N, Fracc. Tomás Aquino, 22414 Tijuana, Baja California, Mexico
{pablojprieto,nohe,lilettecardenas}@ieee.org
[2] Instituto Politécnico Nacional—CITEDI, Avenida Instituto Politécnico Nacional 1310, Colonia Nueva Tijuana, 22435 Tijuana, Mexico
laguilarb@ipn.mx

Abstract. The chapter describes an alternative of fuzzy-based sliding mode control (FSMC) with a self-adjusting fuzzy system. The introduced auto-tuner is based on the relay control approach where a limit cycle is exhibited at the output of the closed-loop system. The amplitude of oscillations is considered in order to move the centers of membership functions such that the absence of oscillations is assured for the fuzzy controller. Finally, numerical simulations verify the feasibility of the proposed algorithm.

Keywords: Fuzzy sliding mode control
Self-tuning fuzzy inference system · Relay

1 Introduction

The manufacturers widely accept the incorporation of Mamdani-type fuzzy controllers for being a control strategy-based with practical results. Most of fuzzy controllers have two or more inputs and one output [3,13]. However, this type of algorithm increases computational costs due to the implementation of a large number of rules. For this reason, many fuzzy controllers use only the error and the change-of-error as fuzzy input variables [6].

We propose a simple fuzzy controller using a single input fuzzy variable instead of the error and the change-of-error. The single variable so-called sliding surface function encloses the variables of the process. This methodology results attractive because through a single input, which contains the dynamic system, the user can drive the behavior of the plant under study. This concept comes from sliding mode control (SMC) [17] whose objective consists in forcing the dynamic system towards a defined hyperplane in the plane-phase. However, sliding mode controllers tend to exhibit chattering near the hyperplane.

Fuzzy-based sliding mode control (FSMC) has been proposed as an alternative to reduce chattering. In recent decades existing literature was focused on

© Springer International Publishing AG, part of Springer Nature 2019
L. Trujillo et al. (Eds.): NEO 2017, SCI 785, pp. 171–186, 2019.
https://doi.org/10.1007/978-3-319-96104-0_9

introducing 'logic decision' in sliding systems [10]. Proposed options in those researches are based on the substitution of the discontinuous function by a fuzzy inference system. The combination of both algorithms solves two critical problems: the chattering attenuation and the reduction of the number of rules of the fuzzy controller through the sliding surface function [5, 12].

Many control algorithms are based on introducing the fuzzy system into the main parameters of the sliding mode controller [12]. The most common method is based on adapting the main parameters of SMC such as robust gain, sliding surface slope and switching control according to the behavior of the plant. The main advantage of these schemes is that the upper bound of uncertainty and disturbance are not required to be known [1, 4, 11] and also they can adapt parameters of the control law according to pre-specified conditions. However, it is hard to implement the control algorithm because the addition of fuzzy rules increases the computational costs. Adaptive methods are provided in [7] to estimate uncertainties and disturbances using variable structure controllers, however, the uncertainties and disturbances associated could be bounded but parametrically unknown in real-world experiments.

On the other, the inappropriate adjustment of membership functions could lead to vibrations or limit cycles in the control process, which degrades the stability of the dynamic systems. It is important to know that in many cases the trial-and-error method is used to obtain the best performance of the plant. In that sense, adaptive systems can update the parameters of the FSMC. Therefore, the main goal of adaptive systems is to update the variables of the controller according to the measurements of the system and thus the performance of the process is improved [16, Chap. 8].

The adjustment of the adaptive controller considers several stages when the terms of the plant are unknown [15, Chap. 8]. Primarily, a control law is chosen such that it contains the parameters to be adjusted. Subsequently, an adaptation law is implemented such that the controller parameters are updated. Finally, the convergence properties are analyzed as a result of the control system. Adaptive control systems are classified into two major groups [15, Chap. 8]:

1. Model-reference adaptive control (MARC).
2. Auto-tuning regulators.

The main problem for MARC-type control is to synthesize the adaptation mechanism, which ensures the accuracy when the parameters are updating. Furthermore, if the complexity of the plant is high then the adaptation mechanism requires expert knowledge of the dynamics of the plant.

The use of self-tuning is interesting because the control law is automatically updated without prior knowledge of the plant. In this case, we propose a relay feedback approach as the auto-tuner [2] which moves the membership functions of the fuzzy controller to remove chattering. Therefore, self-sustained oscillations are exhibited during the time of experiments and, upon the basis of describing functions (DF) criteria, the parameters of the controller are adjusted by measuring the amplitude of oscillations in order to prevent vibrations. Numerical results verify the proposed technique and its feasibility.

This chapter is outlined as follows: Sect. 2 introduces an overview on the existence of chattering in ideal and real sliding mode controller where describing function method is used to verify its existence. Section 3 refers an overview on the relay feedback approach and its condition of being useful for the PID controller. Section 4 presents the fuzzy-based sliding mode control where DF is used to verify attenuation of chattering. Simulation results to verify the existence and elimination of limit cycles (chattering) via FSMC by means of relay feedback approach are given in Sect. 5. Finally, conclusions are stated in Sect. 6.

2   Preliminaries on the Existence of Chattering in Sliding Mode Control

Consider a nonlinear autonomous system of the form

$$\dot{X} = f(X) + g(X)u + w, \tag{1}$$

where $X = [x \ \dot{x} \ \ddot{x} \cdots x^{(n-1)}]^T \in \mathbb{R}^n$ is the state vector, $u \in \mathbb{R}$ is the control input, $t \in \mathbb{R}$ is the time variable, $f(X)$ and $g(X)$ are piecewise continuous vector functions representing the dynamics of the plant and $w(t)$ represents the matched perturbations where

$$w \le W^+, \tag{2}$$

where $W^+ \in \mathbb{R}_+$ denotes the upper bound of w.

The main idea of the sliding mode control is to force the dynamics of system (1) towards a desired state represented by a pre-defined surface and to maintain it there despite matched disturbances. For first-order sliding mode control, the sliding surface can be defined as [15, Chap. 7].

$$\sigma = CX, \tag{3}$$

where $\sigma \in \mathbb{R}$ denotes the sliding surface function and $C \in \mathbb{R}^{1 \times n}$ where its terms are positive.

On the other hand, the Lyapunov method [8, Chap. 4] is used to conclude the stability properties of the closed-loop system. Let us consider the following Lyapunov candidate function

$$V(\sigma) = \frac{1}{2}\sigma^2, \tag{4}$$

which is positive definite. The time derivative of V along the solution of the closed-loop system is given by

$$\begin{aligned} \dot{V} &= \sigma\dot{\sigma}, \\ &= \sigma C\dot{x}, \\ &= \sigma C(f(x) + g(x)u + w). \end{aligned} \tag{5}$$

It is necessary to design a discontinuous control law $u(t)$ such that $\dot{V} < 0$, which is also known as Utkin's definition that is used to verify the existence of sliding modes [18]. The control law that satisfies this condition is given by

$$u = [Cg(x)]^{-1}\left[-Cf(x) - \rho\,\text{sign}(\sigma)\right], \tag{6}$$

where $\rho \in \mathbb{R}_+$ and

$$\text{sign}(\sigma) = \begin{cases} -1 & \text{if } \sigma < 0 \\ [-1, 1] & \text{if } \sigma = 0. \\ 1 & \text{if } \sigma > 0 \end{cases} \tag{7}$$

The control law (6) is set into the Eq. (5) in such a way that

$$\begin{aligned} \dot{V} &= \sigma C\left(f(x) + g(x)\left[[Cg(x)]^{-1}\left[-Cf(x) - \rho\,\text{sign}(\sigma)\right]\right] + w\right), \\ &= \sigma(-\rho\,\text{sign}(\sigma) + w), \\ &= -\rho|\sigma| + w\sigma, \\ &\leq -\rho|\sigma| + W^+|\sigma|, \\ &\leq -(\rho - W^+)|\sigma|, \end{aligned} \tag{8}$$

therefore, the function $\dot{V}$ will be negative definite if

$$\rho > W^+. \tag{9}$$

The term $-\rho\,\text{sign}(\sigma)$ in the right-hand side of (6) is the discontinuous control, which guarantees the convergence of the trajectories towards the sliding surface despite matched disturbances and uncertainties. However, the chattering appears once the trajectory reaches the sliding surface. In particular, this undesired phenomenon degrades the performance of the dynamic system. For this reason, the chattering is of great concern in the control community.

2.1 Describing Function for Verifying Existence of Chattering

Assume that systems (1) and (3) can be re-written as follows

$$\begin{aligned} \dot{X} &= A_p X + B_p u, \\ \sigma &= CX, \end{aligned} \tag{10}$$

$$A_p = \left.\frac{\partial f}{\partial X}\right|_{X_0, u_0}, B_p = \left.\frac{\partial g}{\partial u}\right|_{X_0, u_0},$$

where $A_p \in \mathbb{R}^{n \times n}$ is assumed Hurwitz, $B_p \in \mathbb{R}^n$ defines the input matrix and X_0 and u_0 are equilibrium points. The sliding mode control law is taken as

$$u(t) = -\rho\,\text{sign}(\sigma). \tag{11}$$

The transfer function, in the Laplace transform, from σ to u can be obtained from

$$G(s) = \frac{\sigma(s)}{U(s)} = C(sI - A_p)^{-1}B_p, \tag{12}$$

where $s = d/dt$ is the Laplace operator. Moreover, the introduction of a parasitic dynamic transfer function with the form

$$D(s) = \frac{d_0}{s^k + d_{k-1}s^{k-1} + \cdots + d_1 s + d_0}, \tag{13}$$

being k the relative degree of the plant. When the un-modeled part has a relative degree of two or above, limit cycles are inevitable even with a perfect switching mechanism. The lag-phase introduction by $D(s)$ is unavoidable for the whole system $DG(j\omega)$, hence, an intersection with the negative real axis occurs [14, 19]. Figure 1 depicts the block diagram which represents Eqs. (11), (12) and (13).

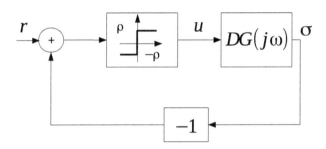

Fig. 1. Block diagram of the closed-loop system using a single relay.

Equation

$$N(A) = \frac{4}{\pi A} \int_0^{\frac{\pi}{2}} u(t) \sin(\omega t) d(\omega t), \tag{14}$$

is the general describing function (DF). Using the above Eq. (14), the DF for the relay (11) is given by

$$N(A)_r = \frac{4\rho}{\pi A}. \tag{15}$$

On the other hand, $DG(j\omega)$ and $-1/N(A)$ can be represented with two curves in a Nyquist diagram where the intersection of both aims to a limit cycle. Formulas to compute the amplitude A and the frequency of oscillations ω are

$$1 + N(A)_r \operatorname{Re}\{DG(j\omega)\} = 0,$$
$$1 + \frac{4\rho}{\pi A} \operatorname{Re}\{DG(j\omega)\} = 0, \tag{16}$$
$$\operatorname{Im}\{DG(j\omega)\} = 0.$$

To suppress chattering in real applications, a smooth version of the discontinuous function in (11) is typically introduced to establish a continuous approximation. This can be achieved if its describing function does not intersects the Nyquist plot of the transfer function $DG(j\omega)$.

3   Brief Background of Relay Feedback Approach

The main idea of relay-based feedback approach is related to the limit cycles exhibited at the time of experiments. Sustained oscillations are generated because the feedback loop is closed by a relay where the control action switches from the minimum to the maximum value periodically [2, Chap. 8]. This method has been focused for auto-tuning PID-type controllers on industrial applications. Figure 2 depicts the whole scheme of auto-tuning PID.

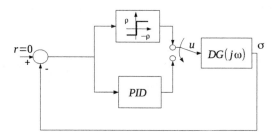

Fig. 2. Scheme of the relay feedback approach [2, Chap. 8].

The limit cycle is characterized by obtaining its oscillation frequency ω_c and amplitude A such that

$$\frac{1}{N(a)_r} = \frac{\pi A}{4\rho} = |DG(j\omega_c)|,$$

$$|DG(j\omega_c)| = \frac{\pi A}{4\rho} = \frac{1}{K_u}, \tag{17}$$

$$T_u = \frac{2\pi}{\omega_c}.$$

The variable K_u and the period T_u are commonly known as *ultimate gain* and *ultimate period* respectively. Moreover, Table 1 depicts the gains K_u and T_u that are obtained for the PID controller. After the gains have been adjusted the controller switches to PID automatically.

4   Fuzzy-Based Sliding Mode Control

Fuzzy systems are proposed to reduce chattering to the sliding mode control systems. This combination represents an alternative as a control algorithm regarding uncertainties and nonlinearities. Moreover, sliding mode systems reduce the number of fuzzy variables and rules [5,12].

Table 1. PID-type controller adjusted by Ziegler & Nichols tuning method.

Control PID	K_c	T_i	T_d
P	$0.5K_u$		
PI	$0.4K_u$	$0.8T_u$	
PID	$0.6K_u$	$0.5T_u$	$0.12T_u$

Therefore, in order to accelerate the convergence towards the sliding surface and to remove the chattering problem, a fuzzy inference system is designed to replace the SMC control law, as shown in Fig. 3.

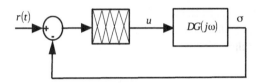

Fig. 3. Block diagram of FSMC control with parasitic dynamics $D(s)$.

In this case $u \in [u_{\min}, u_{\max}] \subseteq \mathbb{R}$ is directly related to σ by means of fuzzy rules, being $u_{\min}$ and $u_{\max}$ the minimum and maximum values of u. The control law

$$u = \psi(\sigma), \tag{18}$$

shows the relationship between the input and the output variables replacing the control law in (6). Here, $\psi(\sigma)$ is the crisp output of the Mamdani-type fuzzy inference system [9]. The fuzzy logic for the static case follows the *if-then* fuzzy rule

$$R_i : \text{If } \sigma \text{ is } M_i \text{ then } u \text{ is } U_i, \tag{19}$$

where M_i is the fuzzy variable at the input. The variable U_i refers to the fired crisp value of the output regarding the value of σ.

For the proposed fuzzy inference system, the whole universe of discourse of σ is separated into several subsets aiming to the values that could take $u(t)$ according to the input state. Membership functions for input and output linguistic variables are written as negative big (**NB**), negative middle (**NM**), zero (**ZR**), positive middle (**PM**), and positive big (**PB**).

Fuzzy rules can be represented as input linguistic variable **Surface σ** and **Output** u as output linguistic variable. The rules matrix is given in Table 2.

The fuzzy sets for the input σ have triangular membership functions which are defined as

$$M_i(\sigma) = \begin{cases} \frac{\sigma - \Phi_{i-1}}{\Phi_i - \Phi_{i-1}} & \text{if } \Phi_{i-1} \leq \sigma < \Phi_i \\ \frac{\sigma - \Phi_{i+1}}{\Phi_i - \Phi_{i+1}} & \text{if } \Phi_i \leq \sigma < \Phi_{i+1}, \\ 0 & \text{elsewhere} \end{cases} \tag{20}$$

Table 2. Fuzzy rules base.

Surface σ	Output u
NB	NB
NM	NM
ZR	ZR
PM	PM
PB	PB

where $\Phi_{-i} = -\Phi_i$. The membership functions are called $M_{-p}, \ldots, M_0, \ldots, M_p$ with $M_0(0) = 0$.

The membership functions for the fuzzy sets of the output u are singleton-type functions U_i such that $U_i = -U_{-i}$ and $U_0 = 0$. Therefore, the control action law for $u(t)$ is formulated as

$$\psi(\sigma) = \sum_i \left\{ \frac{M_i(\sigma)}{\sum\limits_{r=-p}^{p} M_r(\sigma)} \right\} U_i. \tag{21}$$

The fuzzy system holds the following assumptions:

(i) $\psi(\sigma)$ is globally Lipschitz continuous and bounded.
(ii) $\psi(0) = 0$ (steady state).
(iii) The odd conditions $\psi(\sigma) = -\psi(-\sigma)$.
(iv) The fact that only two rules are fired at the same time [9] for a $\Phi_i \le \sigma < \Phi_{i+1}$ then

$$\sum_{r=-p}^{p} M_r(\sigma) = M_i(\sigma) + M_{i+1}(\sigma),$$

$$= \frac{\sigma - \Phi_{i+1}}{\Phi_i - \Phi_{i+1}} + \frac{\sigma - \Phi_i}{\Phi_{i+1} - \Phi_i},$$

$$= 1.$$

The distribution of the membership functions for the input σ and output u variables is depicted in the Fig. 4.

4.1   Describing Function for the Proposed Fuzzy System

Considering the membership functions for σ and $u(t)$ are triangular and singleton fuzzifiers respectively, the control action (21) is calculated for $\Phi_i \le \sigma < \Phi_{i+1}$ as [9]

$$u = \frac{\Delta U_i}{\Delta \Phi_i} \sigma + \frac{1}{\Delta \Phi_i} (\Phi_{i+1} U_i - \Phi_i U_{i+1}), \tag{22}$$

where $\Delta U_i = U_{i+1} - U_i$, $\Delta \Phi_i = \Phi_{i+1} - \Phi_i$.

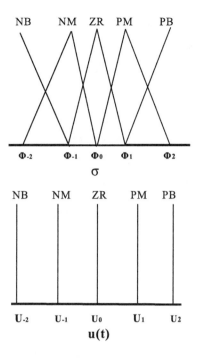

Fig. 4. Membership functions on variable σ as input and $u(t)$ as output.

Finally, according to [9] the describing function of a fuzzy system is

$$N(A)_f = \frac{4}{\pi A} \sum_{i=0}^{n} \left\{ \frac{\Delta U_i A}{2\Delta \Phi_i} \left((\delta_{i+1} - \sin(\delta_{i+1})\cos(\delta_{i+1})) - (\delta_i - \sin(\delta_i)\cos(\delta_i)) \right. \right.$$
$$\left. \left. + \frac{1}{\Delta \Phi_i}(\Phi_i U_{i+1} - \Phi_{i+1} U_i)(\cos(\delta_{i+1}) - \cos(\delta_i)) \right) \right\},$$

(23)

where n satisfies $\Phi_n \le a < \Phi_{n+1}$ and n varies according to A. For $A = 0$

$$N(0)_f = \frac{U_1}{\Phi_1}.$$

is verified. Therefore, the main objective aims to adjust the membership functions such that limit cycles are avoided by satisfying

$$-\frac{1}{N(0)_f} < DG(j\omega_c),$$
$$\frac{\Phi_1}{U_1} > |DG(j\omega_c)|,$$

(24)

where $|DG(j\omega_c)|$ is obtained from the amplitude of oscillation A and Eq. (16). Moreover, the condition (24) guarantees that $DG(j\omega)$ does not encircle $-1/N(A)$

and, in this way, there is no intersection between these functions. Therefore, a constant $\{\varepsilon \in \mathbb{R}_+ | \varepsilon > 1\}$ is introduced such that

$$\frac{\Phi_1}{U_1} = \varepsilon |DG(j\omega_c)|, \tag{25}$$

assuring the non existence of limit cycles. Therefore, the auto-tuning procedure is described by several steps:

- To switch to relay control with the appropriate value of ρ.
- To gather and to save the oscillatory signal of the output σ in a buffer.
- To determine the maximum value of the amplitude A of the limit cycle.
- To determinate $DG(j\omega_c)$ from $N_r(A)$.
- The constants ε, U_1, Φ_1 are stated according to the proposition (24).
- To switch from the relay control to fuzzy controller in T_0 when the parameters of fuzzy rules base have been determined.

5   Numerical Results

Consider the double integrator

$$\dot{x}_1 = x_2,$$
$$\dot{x}_2 = u, \tag{26}$$
$$\sigma = x_1 + x_2,$$

where $x_1(t)$, $x_2(t)$ are the states, $u(t) \in [-5, 5]$ is the control input and $\sigma(t)$ is the output or sliding surface where oscillations occurs. Numerical simulations were carried out with a sampling time $t_s = 0.001$ s using MATLAB/SIMULINK®. The relay control is $u(t) = -5 \operatorname{sign}(\sigma)$ where $\rho = 5$.

Several cases are presented:

(i) Relay control with unmodeled dynamic.
(ii) FSMC where the main criteria about auto-tuning fuzzy system are exposed.

5.1   Relay Control with Unmodeled Dynamic

The transfer function of (26) is given by

$$G(s) = \frac{\sigma(s)}{U(s)} = \frac{s+1}{s^2}. \tag{27}$$

Let us consider a unmodeled dynamic where $k = 2$ according to (13), then, the whole transfer function is given by

$$DG(s) = \frac{s+1}{s^2(0.001s^2 + 0.01s + 1)}. \tag{28}$$

The DF for (28) and $-1/N(A) = -\pi A/4\rho$ is shown in Fig. 5.

The intersection in the real negative axis occurs when $\operatorname{Re}\{DG(j\omega_c)\} = -0.101$ and $\operatorname{Im}\{DG(j\omega_c)\} = 0$. The amplitude and frequency of oscillation are $A = 0.64$ and $\omega_c = 31.46\ rad/s$, which are obtained from (16). Figure 6 depicts the oscillations of σ where the amplitude and frequency of chattering is verified.

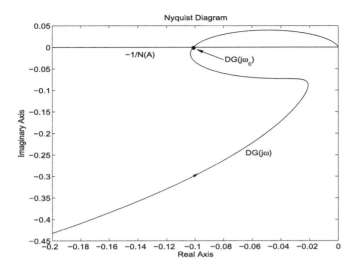

Fig. 5. Nyquist plot of $DG(j\omega)$ and the describing function $-1/N(A)$ for a variable structure system with unmodelled dynamics.

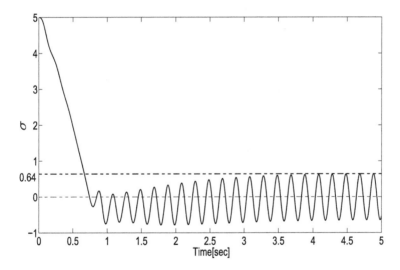

Fig. 6. Sliding surface function behavior under a relay control with unmodeled dynamic.

5.2   Fuzzy-Based Sliding Mode Control

The main objective with the Mamdani-type fuzzy inference system is to avoid limit cycles by adjusting the membership functions according to the criteria proposed in (24), as shown in Fig. 7.

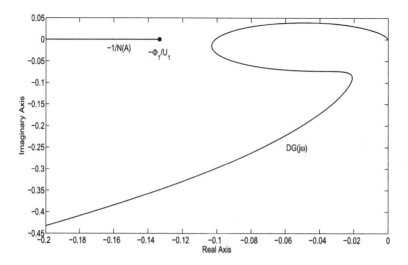

Fig. 7. Nyquist diagram for $DG(j\omega)$ and $-1/N(A)_f$ with the absence of limit cycles.

5.3   Auto-Tuning Fuzzy Inference System

The idea is to exhibit self-sustained oscillations where the auto-tuning mechanism determines the maximum amplitude of oscillation A and $DG(j\omega_c)$ is calculated by means of (16). In this case, the parameter $\varepsilon = 1.7$ assures that, $DG(j\omega)$ does not encircle $N(A)_f$ and the switching time $T_0 = 4$ s. Table 3 shows the centers of the fuzzy system described in the Sect. 4.

Table 3. Fuzzy rules base.

σ_i		U_i	
Φ_{-2}	-12	U_{-2}	-5
Φ_{-1}	-0.1	U_{-1}	$-\hat{U}_1$
Φ_0	0	U_0	0
Φ_1	0.1	U_1	$\hat{U}_1$
Φ_2	12	U_2	5

In this sense, by considering that Φ_2 and U_2 are the maximum values of both σ and $u(t)$, the changes of the membership functions must focus on U_1 and Φ_1. The variable to be changed is U_1, naming it $\hat{U}_1$ and the center of the triangular function Φ_1 takes a fixed value throughout the simulation. Figure 8 shows the whole scheme of the auto-tuning.

Given the amplitude $A = 0.644$, then $DG(j\omega_c) = -0.101$. If it is assigned $\Phi_1 = 0.1$, then, under aforementioned considerations, $U_1 = 0.5796$ (see Table 4).

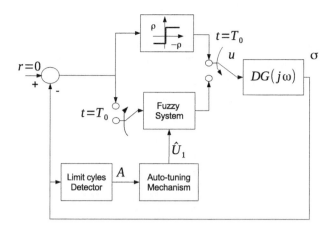

Fig. 8. Block diagram of auto-tuning fuzzy inference system.

Table 4. Fuzzy rules base.

σ_i		U_i	
Φ_{-2}	-12	U_{-2}	-5
Φ_{-1}	-0.1	U_{-1}	-0.5796
Φ_0	0	U_0	0
Φ_1	0.1	U_1	0.5796
Φ_2	12	U_2	5

Figures 9(a) and (b) depict the maximum amplitude of oscillation and the value of U_1 updated. In this case, the auto-tuner switches from the relay to fuzzy control at $T_0 = 4$ s. When U_1 is obtained within the first four seconds, the fuzzy control is turned on.

Figure 9(c) depicts that after T_0 the sliding surface function tends to zero as far as $\sigma = 0$, verifying not only the absence of limit cycles but also the convergence to the origin asymptotically.

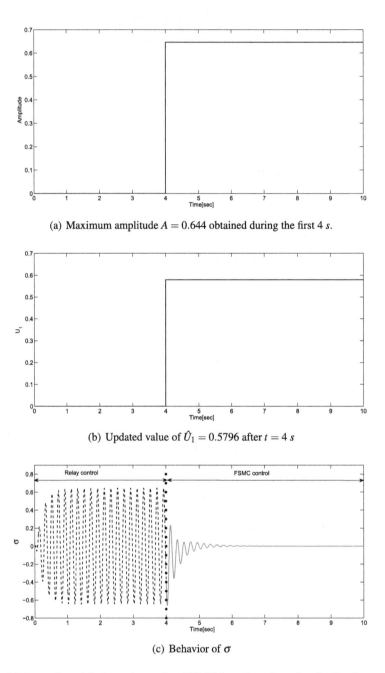

(a) Maximum amplitude $A = 0.644$ obtained during the first 4 s.

(b) Updated value of $\hat{U}_1 = 0.5796$ after $t = 4$ s

(c) Behavior of σ

Fig. 9. Main results with the auto-tuning FSMC based on the relay feedback approach.

6   Conclusion

An auto-tuner was proposed such that the self-sustained oscillations at the output of a relay feedback system are attenuated. Upon the basis of describing function analysis, the membership functions are adjusted in order to suppress chattering through an auto-tuner function. Numerical simulations depict the effectiveness of the proposed self-tuning fuzzy controller where the attenuation of chattering is verified despite the existence of uncertainties.

Acknowledgments. This research was partially funded by project number 6351.17-P and "Estabilización orbital de sistemas mecánicos - Parte II: Estudio de aspectos de estabilidad" from "Tecnológico Nacional de México".

References

1. Amer, A.F., Sallam, E.A., Elawady, W.M.: Adaptive fuzzy sliding mode control using supervisory fuzzy control for 3 DOF planar robot manipulators. Appl. Soft Comput. **11**(8), 4943–4953 (2011)
2. Åström, K.J., Wittenmark, B.: Adaptive Control. Courier Corporation, North Chelmsford (2013)
3. Calcev, G., Gorez, R., De Neyer, M.: Passivity approach to fuzzy control systems. Automatica **34**(3), 339–344 (1998)
4. Cao, L., Sheng, T., Chen, X.: A non-singular terminal adaptive fuzzy sliding-mode controller. In: Proceedings of the IEEE Digital Manufacturing and Automation Conference, pp. 74–80. IEEE (2011)
5. Chang, M.-K.: An adaptive self-organizing fuzzy sliding mode controller for a 2-DOF rehabilitation robot actuated by pneumatic muscle actuators. Control Eng. Pract. **18**(1), 13–22 (2010)
6. Choi, B.-J., Kwak, S.-W., Kim, B.K.: Design and stability analysis of single-input fuzzy logic controller. IEEE Trans. Syst. Man Cybern. Part B Cybern. **30**(2), 303–309 (2000)
7. Islam, S., Robust, P.X.: Robust sliding mode control for robot manipulators. IEEE Trans. Ind. Electron. **58**(6), 2444–2453 (2011)
8. Khalil, H.: Nonlinear Systems. Prentice Hall, Upper Saddle River (2002)
9. Kim, E., Lee, H., Park, M.: Limit-cycle prediction of a fuzzy control system based on describing function method. IEEE Trans. Fuzzy Syst. **8**(1), 11–22 (2000)
10. Kim, S.-W., Lee, J.-J.: Design of a fuzzy controller with fuzzy sliding surface. Fuzzy Sets and Systems **71**(3), 359–367 (1995)
11. Noroozi, N., Roopaei, M., Zolghadri Jahromi, M.: Adaptive fuzzy sliding mode control scheme for uncertain systems. Commun. Nonlinear Sci. Numer. Simul. **14**(11), 3978–3992 (2009)
12. Prieto, P.J., Cazarez-Castro, N.R., Aguilar, L.T., Garcia, D.: Fuzzy slope adaptation for the sliding mode control of a pneumatic parallel platform. Int. J. Fuzzy Syst. **19**, 167–178 (2016)
13. Santibañez, V., Kelly, R., Llama, M.A.: Global asymptotic stability of a tracking sectorial fuzzy controller for robot manipulators. IEEE Trans. Syst. Man Cybern. Part B Cybern. **34**(1), 710–718 (2004)
14. Shtessel, Y., Edwards, C., Fridman, L., Levant, A.: Sliding Mode Control and Observation. Springer, New York (2014)

15. Slotine, J.J., Li, W.: Appl. Nonlinear Control. Prentice Hall, New Jersey (1991)
16. Slotine, J.-J.E., Li, W., et al.: Appl. Nonlinear Control, vol. 199. Prentice Hall, Englewood Cliffs (1991)
17. Utkin, V.: Sliding Modes in Control Optimization. Springer-Verlag, Berlin (1992)
18. Utkin, V., Guldner, J., Shi, J.: Sliding Mode Control in Electro-mechanical Systems. CRC Press, Boca Raton (2009)
19. Xu, J.-X., Lee, T.-H., Pan, Y.-J.: On the sliding mode control for dc servo mechanisms in the presence of unmodeled dynamics. Mechatronics **13**(7), 755–770 (2003)

Optimal Design of Sliding Mode Control Combined with Positive Position Feedback

J. Enríquez-Zárate[1]([⊠]), L. Trujillo[2], C. Hernández[3],
and Claudia N. Sánchez[4]

[1] Facultad de Ingeniería y Ciencias Aplicadas,
Universidad de los Andes, Las Condes, Santiago, Chile
jenriquezza@gmail.com
[2] Tree-Lab, Posgrado en Ciencias de la Ingeniería, Instituto Tecnológico
de Tijuana, Blvd. Industrial y Ave. ITR Tijuana S/N,
Mesa de Otay, 22500 Tijuana, B.C., Mexico
leonardo.trujillo@tectijuana.edu.mx
[3] Centro de Investigación y de Estudios Avanzados del I.P.N. Departamento
de Ingeniería Eléctrica, Sección de Computación, Av. Instituto Politécnico
Nacional no. 2508, C.P. 07360. Col. S.P. Zacatenco, Mexico City, Mexico
chernandez@computacion.cs.cinvestav.mx
[4] Facultad de Ingeniería, Universidad Panamericana, Josemaría Escrivá
de Balaguer 101, Aguascalientes, Aguascalientes 20290, Mexico
cnsanchez@up.edu.mx

Abstract. This work focuses on the application of a discontinuous controller combined with a type of modal control using hybrid optimization techniques to tune the parameters of the controller. The case study is a civil structure with three floors, on which the performance of the control scheme is evaluated by applying an external harmonic force at the ground floor of the structure. The active control is designed to reduce the displacement of the civil structure and the vibrations of the overall system. The Differential Evolution method with the Interior Point Algorithm are used to tune the parameters of the proposed controller, with the goal of maximizing performance relative to hand-tuned parameters. The numerical results are presented comparing the performance of the control in open and closed loop, considering the optimized values of the control parameters.

Keywords: Optimization · Sliding Mode Control · Positive Position Feedback
Differential evolution method

1 Introduction

The application of optimization methods in civil engineering is an interesting alternative, which can be used in several applications as monitoring, and damage prevention, even tuning of control algorithms in structural control applying active control to reduce the displacement of the overall system. Structural control is related to buildings, bridges and power plants, which are commonly perturbed by seismic and wind forces. It provides an alternative to typical schemes of design methods, such as: passive,

© Springer International Publishing AG, part of Springer Nature 2019
L. Trujillo et al. (Eds.): NEO 2017, SCI 785, pp. 187–199, 2019.
https://doi.org/10.1007/978-3-319-96104-0_10

semi-active or hybrid and active control. In particular, the intelligent or active control is governed by the parameters of the controller, to dissipate the energy through such devices as magnetorheological dampers, PZT (Lead Zirconate Titanate) actuators, Alternate Current (AC) or Direct Current (DC) electrical motors among others (Preumont and Seto 2008; Inman 2006).

When it comes to reducing vibration and expanding the lateral integrity of a building due to earthquakes, structural control systems are a good choice. Vibration can be reduced by using an active control system, which sends appropriate control signals. appropriate end elements. However, the control strategy, besides being feasible, must be simple, robust and fault tolerant.

Among the structural control techniques used are those based on intelligent control, like neural networks, interesting for their massive parallel nature and their ability to learn; fuzzy logic that takes advantage of its capacity for nonlinear mapping and robustness; and genetic algorithms that consider the system as a multilevel optimization problem.

One of the techniques used to drive the state path of the system to a pre-specified surface in the state space is Sliding Mode Control (SMC), which allows obtaining an asymptotically stable system (Nguyen et al. 2006; Shtessel et al. 2014).

In this work, we propose a strategy for optimal design, which is implement to obtain the optimal gains of the controllers purposes, which consist of a technique based in the SMC combined with Positive Position Feedback (PPF) applied to reduce the displacement of a civil structure and the vibrations of the overall system.

The rest of this paper is organized as follows. Firstly, the mathematical model of the civil structure and case study are presented. Subsequently, the optimization algorithm is described, as well as the control strategy. Finally, the numerical results obtained are shown and final conclusions are drawn.

2   Motivation

Generally, the schemes of modal control are designed considering mainly two uncertainty parameters, the damping ratio and the gain of controller, ζ, g, respectively. Typically, these variables are tuned applying a methodology, which consists of fixing a value for the damping ratio ζ, and subsequently change the value of the gain g of the controller. This criterion is empirically valid, however is not optimal to determinate the right values of the controller parameters. For this circumstance we propose an alternative method, which consist in the implementation of an optimization algorithm in the specific case of the modal control based on the scheme of PPF controller.

3   Mathematical Model

As stated before, the case study in this work is a civil structure, which consist of three floor's building with concentrated mass on each floor. The building model consist of mechanical elements (mass, springs and dashpots) coupled as shown in the Fig. 1. We assume a generalized case of a multi degree of freedom (MDOF) reducing components in the analysis of the elements of the structure (see Fig. 1). The particular system consists of a building of three story, which is modeled using the Euler-Lagrange formalism obtaining the kinetic and potential energies of the overall structure. In the generalized analysis the building's masses are represented by m_n which consist of a vector denoted as $m_n = \{m_1, m_2, m_3\}$. The damping c_n consist of a vector representing the damping on each floor, which is denoted as $c_n = \{c_1, c_2, c_3\}$ and the stiffness of the columns of the building consist of a vector $k_n = \{k_1, k_2, k_3\}$. The external force representing the perturbation at the base of the civil structure is represented by $\ddot{x}_g$, which is considered as a harmonic force and x_n denotes the displacements of every floor expressed as $x_n = \{x_1, x_2, x_3\}^T$. It's important to comment that in a civil structure, the perturbation forces at the base are caused by the action of phenomena such as earthquakes. Generally, a seismic force is represented in terms of the acceleration, as we consider it in the approach denoted by $\ddot{x}_g \in R$.

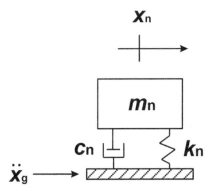

Fig. 1. Reduced schematic diagram of a system of MDOF

The generalized notation of the multi degree of freedom system is expressed as

$$M_n \ddot{x}_n + C_n \dot{x}_n + K_n x_n = -M_n e_n \ddot{x}_n \tag{1}$$

A particular mathematical model of the building of three floors is represented as follow

$$M_3 \ddot{x} + C_3 \dot{x} + K_3 x = -M_3 e_3 \ddot{x}_g \tag{2}$$

where M, C and $K \in R^{3 \times 3}$ and denote the matrices of mass, damping and stiffness respectively. The vector $e_3 = [1, 1, 1]^T \in R^3$ is the influence vector, which relates to the effect of ground motion at the base of the civil structure to each floor. In this work the external force $\ddot{x}_g \in R$ implemented is a harmonic force with an interval of frequency between 0 to 60 Hz with amplitudes $Z \le 0.0127$ m (forces up to 45 N). The masses, damping and stiffness matrices of the simplified civil structures are expressed

$$M_3 = \begin{bmatrix} m_1 & 0 & 0 \\ 0 & m_2 & 0 \\ 0 & 0 & m_3 \end{bmatrix}, \quad C_3 = \begin{bmatrix} c_1 + c_2 & -c_2 & 0 \\ -c_2 & c_2 + c_3 & -c_3 \\ 0 & -c_3 & c_3 \end{bmatrix},$$

$$K_3 = \begin{bmatrix} k_1 + k_2 & -k_2 & 0 \\ -k_2 & k_2 + k_3 & -k_3 \\ 0 & -k_3 & k_3 \end{bmatrix} \tag{3}$$

Particularity, the damping matrix is expressed in terms of the Rayleigh or proportional damping, which depends on the mass and stiffness matrices expressed as $C_3 = a_0 M_3 + b_0 K_3$, where a_0 and b_0 are coefficients related with resonant frequencies and the proportional damping of the civil structure respectively.

3.1   Case Study

The previously described civil structure is monitored using one accelerometer, allocated on the third floor. To reduce the displacement and vibrations on the civil structure an Active Mass Damper (AMD) device is designed to applying a particular scheme of control mounted on the last floor of the civil structure (third floor). For this analysis is necessary to establish the resonant frequencies of the overall civil structure in open-loop, with intention to apply a modal scheme of control like PPF controller. These frequencies are summarized in the Table 1.

Table 1. Civil structure's vibrations modes.

Mode i	Numerical frequency ω_i [Hz]
1	1.7020
2	5.1607
3	8.0274

These resonant frequencies of the civil structure are obtained from the evaluation of the previous mathematical model of the system. For this purpose, its necessary to take into account the physical values of the materials and the dimensions of the columns and masses of every floors. Basically, in this article we consider the values of a building-like structure of three floors, made with alloys of aluminum both the columns and the masses (Enríquez-Zárate and Silva-Navarro 2014).

Generally, in civil engineering the most complicated resonant frequency of the system is referred as the first frequency of the system, in this case $\omega_i = 1.7020$ Hz, which is our control goal, i.e., reduce the amplitude of this frequency using a scheme of control based on the PPF combined with SMC. In particular, PPF control has a limitation derived from the controller's tuning gain, which is very important to achieve the necessary control objectives related with the minimization of displacement and vibrations of the civil structure.

4 Optimization Algorithm

This section presents the optimization method used to tune the gains of the controllers. The authors consider the integrated absolute tracking error as the objective e_{IAE} to optimize; defined as

$$e_{IAE} = \int_0^{T_{ss}} |r(\hat{t}) - x(\hat{t})| d\hat{t}, \tag{4}$$

where $r(t)$ is a reference input and T_{ss} is the time when the response is close to reaching the steady state, resulting in $T_{ss} = 60$ s for our experiments and the numerical results of the system in open-loop.

4.1 Differential Evolution Method

The method used to perform the optimization is Differential Evolution (DE) (Rainer and Kenneth 1997), which is a well-known metaheuristic for real-valued black-box optimization, together with the Interior Point Algorithm (Byrd et al. 2000; Fenik and Starek 2008) (this method comes within the *fmincon* function in *Matlab®*). In particular, we used the DE/rand/1/bin version implemented in *Matlab®* (the source code is from (http://www1.icsi.berkeley.edu/~storn/code.html, 2016).

This method consists in the initialization of a random population of n individuals. DE uses two operations to generate new solutions. The first operation consists in a mutation process, which estimates new individuals using three random individuals using

$$v_{i,G+1} = x_{r1,G} + F\left(x_{r2,G} - x_{r3,G}\right), \tag{5}$$

where r_1, r_2 and r_3 are random indexes, G is the present estimation and $F > 0$ and $x_{i,G}$ represent the population x of the algorithm, considering i as the index of an individual. The next operation consists in a crossover operator, used to increase diversity during the search. This search operator is expressed as

$$u_{ji,G+1} = \begin{cases} v_{ji,G+1} \ if \, (rand(j) \leq CR) \ or \, j = rnbr(i), \\ x_{jiG} \ if \, (rand(j) > CR) \ or \, j \neq rnbr(i) \end{cases} \tag{6}$$

with

$$j = 1, 2, \ldots, D$$

where $rand(j)$ is a uniform random number in $[0, 1]$, $CR \in [0, 1]$ is the crossover constant and $rnbr(i) \in 1, 2, \ldots, D$ is a randomly chosen index. Finally, to decide which individuals will be part of the next estimation $u_{i,G+1}$ is compared to $x_{i,G}$ and the one with the smallest fitness is retained for the next estimation.

The parameters used for the DE method are given in Table 2.

Table 2. Parameters of the DE method.

Parameters	Values
Population size	50
Number of estimation	25
CR	0.8
F	0.95
Simulation time	60 s

5   Controller

The application of controllers based on the PPF combined with SMC, is a recently proposed method used in civil structures, which provide robustness to the controller allowing it to work with linear and nonlinear systems. Generally, the parameters of PPF control are tuned using a methodology that consists on establish a fixed or constant value for the damping factor ζ_f, which will attenuate the tuned resonant frequency and subsequently change the value of the gain g of the controller (Silva-Navarro and Hernández-Flores 2015). The correct tuning of the parameter will guarantee the stability of the system.

The SMC uses a sliding surface to attract the system and keep the states within a close neighborhood of the sliding surface. The SMC is a two-part controller design. The first part involves the design of a sliding surface so that the sliding motion satisfies the design specifications. The second is concerned with the selection of a control law that will make the switching surface attractive to the system states (Utkin 1992). The chattering property is considered into the dynamic of the system and therefore the tuning of the parameter's control is influence with it presence. In this work we consider a continuous approximation of *signum* function, which must be tuned as part of the sliding mode control. The quick or slow response of the control on the sliding surface depend of gain of the controller, which must be smooth with intention to reduce the chattering property. The chattering is a not desired characteristic in discontinuous control, which consist in parasitic dynamics, present in the sliding surface (Boiko 2011). The values of the parameters of the SMC are tuned in such a way that the controller ensures the dynamic response of the system on steady stable in close-loop (Sira-Ramirez 1993; Sira-Ramirez 2002).

5.1   PPF Control

The scheme of control based on the PPF is well-known as a class of modal control used in Experimental Modal Analysis (EMA), which can be implemented without considering a mathematical model of the system. This scheme of control is used to compensate the response of the system in the presence of external perturbations such a seismic force. This controller is widely used to reduce the vibration in structures (Utkin 1992; Baz and Poh 1996; Baz and Hong 1997; Friswell and Inman 1999; Inman et al. 2006; Cabrera-Amado and Silva-Navarro 2012). In the literature other kinds of schemes for modal control include Positive Velocity Feedback (PVF), Positive Acceleration Feedback (PAF), and others (Friswell and Inman 1999; Inman et al. 2006).

This scheme of control adds a degree of freedom to the original mechanical system, considered as a *virtual passive absorber* or as a filter of second order. This filter consists of two poles, which can be analyzed through transfer function of the system in close-loop. Generally, the poles can be classified as: real different, real equal and conjugated complexes, which depend of the value of natural frequency ω_n and the value of the damping factor ζ, which can behave like a underdamped system, critical damped and overdamped (Ogata 2010).

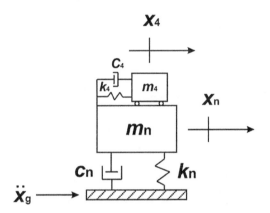

Fig. 2. Schematic diagram of a system of MDOF with AMD

The parameters of the controller can be obtained using experimental data, which make the technique of PPF control very useful in many applications of structures and control (Carlson et al. 1996). For this building model with an AMD mounted on the third floor (see Fig. 2), and the perturbation force at the base $\ddot{x}_g(t)$, and the active control force $u(t)$ supplied through the AMD, the dynamics of the system are represented by

$$M_4\ddot{x}(t) + C_4\dot{x}(t) + K_4x(t) = -M_4e_4\ddot{x}_g(t) + B_fu(t) \tag{7}$$

which is obtained considering the following characteristic equation of motion of a controlled system in control engineering

$$M\ddot{x}(t) + C\dot{x}(t) + Kx(t) = Ef(t) + Du(t)$$

where positions and numbers of the excitation or perturbation forces $f(t)$ and the control forces $u(t)$ are represented by the matrices E and D. The analysis process to obtain the Eq. (7) consider the relative displacement of the fourth masses or floors of the building, including the influence of the external forces or perturbation in terms of the acceleration $\ddot{x}_g$ in the dynamic of the overall system (Roy and Andrew 2006). In this case, the Eq. (7) is related with the Eq. (2), which represent the dynamic of the building of three floors in open-loop, and the degree of freedom of the system due to the presence of the AMD.

The control effort $u(t) \in R$, is implemented on the civil structure through of the mass, spring and damper m_4, k_4 and c_4, respectively (see Fig. 2). The parameters m_4, k_4 and c_4, are the elements of AMD. The vector $B_f = [0, 0, 0, 1]^T$ represent an input for the complete system and M, C and $K \in R^{4 \times 4}$, which denotes the matrices of mass, damping and stiffness respectively. The rest of the variables coincide with the previous definition in the open-loop scheme in a dimension of R^4.

The scheme of control for the PPF used in a building like structure with four degrees of freedom result in a closed-loop system described by Eq. (7) and

$$\ddot{\eta}(t) + 2\zeta_f \omega_f \dot{\eta}(t) + \omega_f^2 = g\omega_f^2 B_f^T y(t) \tag{8}$$

$$u(t) = g\omega_f^2 \eta(t) \tag{9}$$

where $y(t) \in R^4$ and $\ddot{x}_g(t) \in R$. The virtual passive absorber $\eta(t) \in R$, with a damping rate ζ_f and natural frequency ω_f, which is feedback with the primary system through of $g\omega_f^2 B_f^T y(t)$. The law of control PPF is represent via $u(t) \in R$, where g is a control gain.

The compact form of the system in closed loop is described as

$$\begin{bmatrix} M_4 & 0 \\ 0 & 1 \end{bmatrix} \begin{bmatrix} \ddot{y}(t) \\ \ddot{\eta}(t) \end{bmatrix} + \begin{bmatrix} C_4 & 0 \\ 0 & 2\zeta_f \omega_f \end{bmatrix} \begin{bmatrix} \dot{y}(t) \\ \dot{\eta}(t) \end{bmatrix}$$
$$+ \begin{bmatrix} K_4 & -B_f g\omega_f^2 \\ -g\omega_f^2 B_f^T & \omega_f^2 \end{bmatrix} \begin{bmatrix} y(t) \\ \eta(t) \end{bmatrix} = \begin{bmatrix} -M_4 e_4 \\ 0 \end{bmatrix} \ddot{x}_g(t) \tag{10}$$

5.2   Sliding Mode Control

The sliding mode control is a well-known control technique, which can be used to control linear and nonlinear systems. An important feature of SMC is their robustness against parametric uncertainties and unmodeled dynamics such as external disturbances (Enríquez-Zárate and Silva-Navarro 2004). To simultaneously compensate multiple vibration modes a PPF with SMC is synthesized as follows (Enríquez-Zárate and Silva-Navarro 2016)

$$\ddot{\eta}(t) + 2\zeta_f \omega_f \dot{\eta}(t) + \omega_f^2 \eta(t) = g\omega_f^2 B_f^T x(t) \tag{11}$$

$$\sigma\left(\eta - B_f^T q\right) = \alpha_1 \left(\eta - B_f^T x\right) + \alpha_2 \left(\dot{\eta} - B_f^T \dot{x}\right) \tag{12}$$

$$u(t) = g\omega_f^2 \eta(t) + W sign(\sigma) \tag{13}$$

where σ is the switching surface, established in terms of a virtual error $e = \left(\eta - B_f^T x\right)$, with positive parameters α_1 and α_2 and sign denotes the *signum* function, which for practical purposes is approximated by

$$\text{sign}(\sigma(e)) \approx \frac{\sigma(e)}{|\sigma(e)| + \varepsilon} \tag{14}$$

where ε is a sufficiently small positive parameter, which depends on the physical limitations of the actuator. The PPF and SMC parameters obtained using the criterion implemented from the optimization method were $\omega_f = \omega_2$, $\zeta_f = 0.10$, $g = 1.19$, $\varepsilon = 0.001$, $\alpha_1 = 0.02$, $\alpha_2 = 2.41$ and $W = 5.0$.

It's important to notice that effect of chattering property can be slightly compensating through of the parameter ε in terms of the switching more not eliminate it. In this case into the criterion of the optimization algorithm this smoothing parameter was considered and its effect is present in the dynamic of the numerical results.

6   Numerical Results

This section, presents numerical results of the controller used in the building-like structure. Figure 3 shows the displacement of the first floor considering the external force $\ddot{x}_g$ applied at the base or the civil structure. The Figs. 4 and 5 show the lateral displacement second and third floors of the structure, respectively.

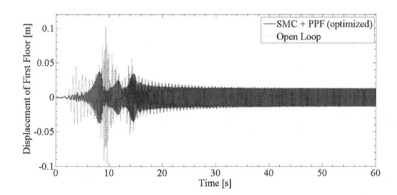

Fig. 3. Numerical result of the first floor with optimized parameters using an active control.

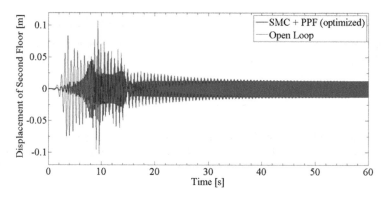

Fig. 4. Numerical result of the second floor with optimized parameters using an active control.

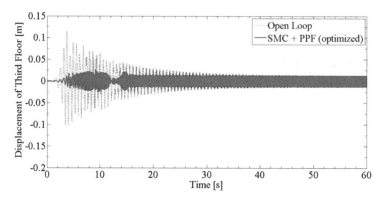

Fig. 5. Numerical result of the third floor with optimized parameters using an active control.

The FRF and effort control of the system are shown in the Figs. 6 and 7 respectively. The amplitude of the first resonant frequency of the system is reduced around of 90% with respect the open loop response. This reduction of the frequency is achieved using small control efforts, which is appropriate to the purpose controller. This scheme of control shows an acceptable performance with the goal of control. However, the other resonant frequencies of the system are affected by the action of the effort control. This effect known as *spill-over* can be improved considering the implementation an extension of the PPF controller, which consist on the design of a filter for each resonant frequency of the system known as Multiple Positive Position Feedback (MPPF) combined with SMC (Enríquez-Zárate and Silva-Navarro 2016).

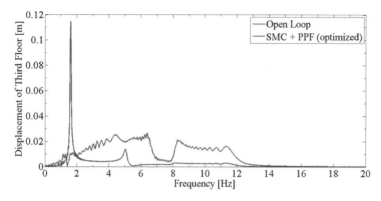

Fig. 6. Numerical FRF of the building-like structure using an optimized active control.

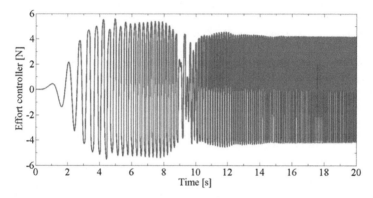

Fig. 7. The effort control of the SMC + PPF (optimized) used in the building-like structure

7 Discussion

The chattering property of SMC is a not desire effect into the dynamic of the controller of a system. However, it's possible to attenuate its presence, which occurs largely by the development of the discontinuous control in the sliding surface. The frequency of commutation of the controller can affect the response of the chattering, which can be adjusted with some schemes like one we present in this work. The consequence of chattering is present in the response of the system even the steady stable response, which in the numerical results not impact the propose of the controller due mainly to the robustness of the SMC and the optimal parameters obtained from the optimization algorithm. Finally, in the practice the chattering property must be considered in more detail, because its presence in electrical actuators (AMD scheme) can reduce its useful life and disturb the performance of the overall system.

8   Conclusions

This paper presents the performance of a Sliding Mode Control (SMC) combined with a modal control known as Positive Position Feedback (PPF). This controller is optimized tuning the parameters of the gains of both schemes of control. The PPF control is designed to reduce the amplitude of the first resonance frequency of the system, which was minimized around of 90% with small efforts control, reducing the lateral displacements of the third floor on the civil structure. The work present numerical results and the authors consider as future work the implementation of the controller in an experimental platform set-up.

References

Baz, A., Poh, S.: Optimal vibration control with modal positive position feedback. Optim. Control Appl. Method **17**(2), 141–149 (1996)

Baz, A., Hong, J.T.: Adaptive control of flexible structures using modal positive position feedback. Int. J. Adapt. Control Sign. Process. **11**(3), 231–253 (1997)

Cabrera-Amado, A., Silva-Navarro, G.: Semiactive vibration absorption in a rotor-bearing system using a PPF control scheme. In: Proceedings of International Conference on Noise and Vibration Engineering ISMA2012 + USD2012, pp. 209–221 (2012)

Carlson, J.D., Catanzarite, D.M., St Clair, K.A.: Commercial magneto-rheological fluid devices. Int. J. Mod. Phys. B **10**(23–24), 2857–2865 (1996)

Differential evolution code. http://www1.icsi.berkeley.edu/ ∼ storn/code.html. Accessed: 24 Feb 2016

Enríquez-Zárate, J., Silva-Navarro, G.: Active vibration control in a nonlinear mechanical system using a flatness based sliding-mode control: experimental results. In: Proceedings of International Conference on Electrical and Electronics Engineering (ICEEE) and X Conference on Electrical Engineering (CIE 2004), Acapulco, Guerrero, México, pp. 8–10 (2004)

Enríquez-Zárate, J., Silva-Navarro, G.: Ground-borne vibration control in a building-like structure using multi-positive position feedback combined with sliding mode control. Noise Control Eng. J. **64**(5), 668–676 (2016)

Enríquez-Zárate, J., Silva-Navarro, G.: Active vibration control in building-like structures submitted to earthquakes using multiple positive position feedback and sliding modes. DYNA **81**(1), 83–91 (2014)

Fenik, S., Starek, L.: Optimal PPF controller for multimodal vibration suppression. Eng. Mech. **15**(3), 153–173 (2008)

Friswell, M.I., Inman, D.J.: The relationship between positive position feedback and output feedback controllers. Smart Mater. Struct. **8**, 285–291 (1999)

Byrd, R.H., Charles Gilbert, J., Nocedal, J.: A trust region method based on interior point techniques for nonlinear programming. Math. Program. **89**(1), 145–185 (2000)

Inman, D.J., Tarazaga, P.A., Salehian, A.: Active and passive damping of structures. In: Proceedings of International Congress on Sound and Vibration, ICSV13, pp. 1–8 (2006)

Inman, D.J.: Vibration with Control. Wiley, New York (2006)

Boiko, I.M.: Analysis of chattering in sliding mode control systems with continuous boundary layer approximation of discontinuous control. In: 2011 American Control Conference on O'Farrell Street, San Francisco, CA, USA. 29 June–01 July 2011 (2011)

Nguyen, T.H., Kwok, N.M., Ha, Q.P., Li, J., Samali, B.: Adaptive sliding mode control for civil structures using magnetorheological dampers. In: ISARC 2006, pp. 636–641 (2006)

Ogata, K.: Ingeniería de control moderna. Prentice Hall – Pearson (2010). ISBN: 978-84-8322-660-5

Preumont, A., Seto, K.: Active Control of Structures. Wiley, New York (2008)

Silva-Navarro, G., Hernández-Flores, A.: On the passive and active lateral coupling of building-like structures under ground motion. In: The 22nd International Congress on Sound and Vibration, ICSV22, Florencia, Italy, 12–16 July 2015 (2015)

Rainer, S., Kenneth, P.: Differential evolution – a simple and efficient heuristic for global optimization over continuous spaces. J. Glob. Optim. **11**(4), 341–359 (1997)

Roy, R.C., Andrew, J.K.: Fundamentals of Structural Dynamics, 2nd edn. Wiley, New York (2006)

Sira-Ramirez, H.: On the dynamical sliding mode control of nonlinear systems. Int. J. Control **57**(5), 1039–1061 (1993)

Sira-Ramirez, H.: Dynamic second-order sliding mode control of the hovercraft vessel. IEEE Trans. Control Syst. Technol. **10**(6), 860–865 (2002)

Shtessel, Y., Edwards, C., Fridman, L., Levant, A.: Sliding Mode Control and Observation. Birkhäuser Basel (2014). https://doi.org/10.1007/978-0-8176-4893-0

Utkin, V.I.: Sliding Modes in Control and Optimization. Springer, Berlin (1992)

Real-World Applications

Biot's Parameters Estimation in Ultrasound Propagation Through Cancellous Bone

Miguel Angel Moreles[1,2(✉)], Joaquin Peña[2], and Jose Angel Neria[2]

[1] Mathematics Department, Universidad de Guadalajara,
CUCEI, Guadalajara, Mexico
[2] CIMAT, Jalisco S/N Valenciana, 36240 Guanajuato, GTO, Mexico
{moreles,joaquin,jose.neria}@cimat.mx

Abstract. Of interest is the characterization of a cancellous bone immersed in an acoustic fluid. The bone is placed between an ultrasonic point source and a receiver. Cancellous bone is regarded as a porous medium saturated with fluid according to Biot's theory. This model is coupled with the fluid in an open pore configuration and solved by means of the Finite Volume Method. Characterization is posed as a Bayesian parameter estimation problem in Biot's model given pressure data collected at the receiver. As a first step we present numerical results in 2D for signal recovery. It is shown that as point estimators, the Conditional Mean outperforms the classical PDE-constrained minimization solution.

Keywords: Cancellous bone · Ultrasound propagation
Biot's model · Inverse problem · Bayesian estimation

1 Introduction

Analysis of initial boundary value problems (IBVPs) usually consider well-posed problems, that is, problems where uniqueness and existence of the solution, as well as continuous dependence on the input data can be establish. Problems which lack any of these properties are ill-posed or inverse problems [1]. For example, problems arising in geophysics and medicine concern the determination of properties of some inaccessible region. The problem of interest in this work is of this sort. The properties to estimate are parameters of a saturated porous medium immersed in a fluid, a so called Biot's medium. The parameters are to be recovered from an ultrasound noisy signal.

The Biot's medium of concern is a medulla saturated cancellous bone. The bone is placed between an acoustic source and receiver. A potential application of the estimated parameters is as an aid in the diagnostic of osteoporosis.

Research on the problem is very active. In Buchanan et al. [2–4] the problem of inversion of parameters for a two-dimensional sample of trabecular bone is considered in a low frequency range ($f < 100$ KHz). In these works the recovered

© Springer International Publishing AG, part of Springer Nature 2019
L. Trujillo et al. (Eds.): NEO 2017, SCI 785, pp. 203–224, 2019.
https://doi.org/10.1007/978-3-319-96104-0_11

parameters are ϕ (porosity), α (solid tortuosity), K_b (bulk modulus of the porous skeletal frame) and N (solid shear modulus). In Sebaa et al. [17] trabecular bone samples are characterized by solving the inverse problem using experimentally acquired signals. In this case the direct problem is solved by a modified Biot's model, for the case a one-dimensional block of trabecular bone saturated with water. The recovered parameters are ϕ, α, ν_b (Poisson ratio of the porous skeletal frame) and E_b (Young's modulus of the porous skeletal frame).

In these cases, the inversion problem is posed as minimization problem in the least squares sense. So the maximum likelihood estimator is obtained. The reported results are somewhat unsatisfactory. In Buchanan et al. [4], an extensive exploration of minimization algorithms is carried out. In particular the Nelder-Mead simplex method. In their own words, mostly *subcorrect answers* are obtained. Namely, the target minimum for the objective function is attained, irrespective of the accuracy of the answers. In their exhaustive exploration, only porosity can be determined with an acceptable percentage error.

On the other hand, as pointed out in Sebaa et al. [17], the solution of the inverse problem for all model parameters using only data from the transmitted signal is difficult, if not impossible. This in part due to the high computational cost of the optimization of the objective function and partly because more experimental data is needed to obtain a unique solution.

In this setting, previous information such as physically acceptable ranges, results of other experiments, etc., may not be incorporated in a natural way. Also, noise in data and parameter uncertainty, require complementary techniques. In the works above, these issues are not fully addressed.

In the present work we follow an alternative approach. We pose the problem as one of Bayesian estimation. We show that in this approach, it is possible to estimate the parameters involved in the Biot's model, and previous information can be incorporated. Noisy data and parameter uncertainty, are an integral part of the formulation.

The outline of the chapter is as follows.

In Sect. 2 we introduce the Biot's model and the parameter estimation problem of interest. The underlying mathematical model is comprised by the coupling of two wave equations, acoustic and poroelastic. For simulation, the Finite Volume Method is used for solving this model of ultrasound propagation trough a heterogeneous, anisotropic, porous material. We argue that with this method, the physical boundary conditions are posed naturally at the interface. There is a vast literature on a variety of models and numerical methods for the Biot's problem. A short review is included.

Recovery of Biot's parameters by means of bayesian estimation, is the content of Sect. 3. To make our exposition self contained, we review the basics of the theory. Our contribution is a methodology to recover accurately the Biot's parameters, and fitting a given noisy signal. One may argue that the solution to a bayesian estimation problem, is the determination of the posterior density. From the latter we provide as point estimator the conditional mean. The posterior provides many tools to quantify parameter uncertainty, here we are content with showing confidence intervals with 0.9 probability for each estimated parameter. It will become apparent that estimation is robust.

Motivated by the work of Buchanan et al. [4], in Sect. 4 we carry out a minimization approach by using the Nelder-Mead method. For the reader convenience, a description of the method is presented. Our findings are similar to theirs. The target for the objective function is attained, but the approximated parameters have high percentage errors. In contrast to the conditional mean estimators. Also, we show, that as expected, the data signal is not recovered.

In Sect. 5 we draw conclusions and point out some venues of future work.

2   The Biot's Model and the Parameter Estimation Problem

2.1   A Clinical Motivation

Noninvasive techniques for assessing bone fracture risk, as well as bone fragility are of current interest. Bone can be characterized in two types, cancellous (spongy or trabecular) and cortical. There is an ongoing discussion on trabecular changes due to osteoporosis, in particular, thinning of the trabeculae. Consequently, early detection of these changes is a potential aid for diagnostics of osteoporosis. A promising noninvasive technique is by ultrasound propagation through cancellous bone, see Sebaa et al. [17], and references therein.

In accord with the *axial transmission* (AT) technique, Lowet and Van der [12], the configuration is a cancellous bone placed between an acoustic source and receiver. A schematic is presented in Fig. 1.

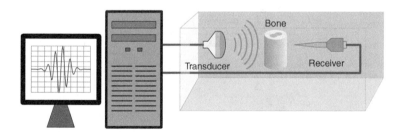

Fig. 1. Schematic showing the environment to study samples of cancellous bone using ultrasonic axial transmission.

An ultrasonic pulse is emitted from the transducer then propagated through the bone. The problem of interest, is to determine physical characteristics of the bone given a noisy signal collected at the receiver.

2.2   Biot's Model for a Fluid Saturated Porous Solid

The cancellous bone is modeled as a Biot medium, that is, a fluid saturated porous solid. Biot's theory yields the following governing equations. We follow the description in Buchanan et al. [2–4].

$$\rho_{11}\frac{\partial^2 \mathbf{u}}{\partial t^2} + \rho_{12}\frac{\partial^2 \mathbf{U}}{\partial t^2} = \nabla \cdot \sigma - b\frac{\partial}{\partial t}(\mathbf{u} - \mathbf{U}), \tag{1a}$$

$$\rho_{12}\frac{\partial^2 \mathbf{u}}{\partial t^2} + \rho_{22}\frac{\partial^2 \mathbf{U}}{\partial t^2} = \nabla s + b\frac{\partial}{\partial t}(\mathbf{u} - \mathbf{U}), \tag{1b}$$

$$\sigma = [(P - 2N)e + Q\varepsilon]I + 2N\bar{e}, \tag{1c}$$

$$s = Qe + R\varepsilon, \tag{1d}$$

where σ and s represent the forces acting on the solid and fluid portions of each side of a unit cube of the Biot medium, respectively, $\mathbf{u}$ and $\mathbf{U}$ are solid and fluid displacements, and

$$e = \nabla \cdot \mathbf{u},$$

$$\varepsilon = \nabla \cdot \mathbf{U},$$

$$\bar{e}_{i,j} = \frac{2 - \delta_{i,j}}{2}\left(\frac{\partial u_i}{\partial x_j} + \frac{\partial u_j}{\partial x_i}\right),$$

$$I_{ij} = \delta_{ij} = \begin{cases} 0, & \text{if } i \neq j, \\ 1, & \text{if } i = j. \end{cases}$$

Also P, Q, R are generalized elastic constants given by

$$P = \frac{(1 - \phi)\left(1 - \phi - \frac{K_b}{K_s}\right)K_s + \phi\frac{K_f}{K_f}K_b}{\Delta} + \frac{4N}{3}, \tag{2a}$$

$$Q = \frac{\left(1 - \phi - \frac{K_b}{K_s}\right)\phi K_s}{\Delta}, \tag{2b}$$

$$R = \frac{\phi^2 K_s}{\Delta}, \tag{2c}$$

$$\Delta = 1 - \phi - \frac{K_b}{K_s} + \phi\frac{K_s}{f}. \tag{2d}$$

The measurable quantities in these expressions are ϕ (porosity), K_f (bulk modulus of the pore fluid), K_s (bulk modulus of elastic solid) and K_b (bulk modulus of the porous skeletal frame). N is the solid shear modulus.

The remaining parameters are the mass coupling coefficients, namely

$$\rho_{11} + \rho_{12} = (1 - \phi)\rho_s, \tag{3a}$$

$$\rho_{22} + \rho_{12} = \phi\rho_f, \tag{3b}$$

$$\rho_{12} = -(\alpha - 1)\phi\rho_f. \tag{3c}$$

where ρ_s, ρ_f are the solid and fluid densities, α is the solid tortuosity and b is a parameter depending on the frequency of the incident wave and accounts for energy losses in the solid-fluid structure. In this first study, this parameter is regarded as a constant.

Below we give numerical values to all parameters when describing a synthetic example.

2.3 Acoustic Fluid

It is assumed that the Biot medium Ω^b (cancellous bone) is immersed in a fluid as shown in Figure 2.

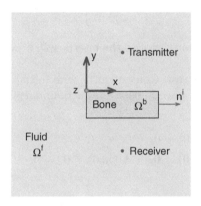

Fig. 2. Basic configuration of the elements in the domain.

The fluid within Ω^f is acoustic. Its density and speed are ρ, c, respectively. Consequently, in terms of the pressure $p(\mathbf{x},t)$ we have

$$\frac{1}{c^2}\frac{\partial^2 p}{\partial t^2} - \nabla^2 p = \frac{\partial Q}{\partial t}, \quad \forall \mathbf{x} \in \Omega^f, \tag{4}$$

where $Q(\mathbf{x},t)$ is the point source density located at $\mathbf{x}^s$ and given by

$$\frac{\partial Q}{\partial t} = \rho F(t)\delta(x - x^s)\delta(y - y^s), \tag{5}$$

where $F(t)$ is a scalar real function and $\delta(\cdot)$ is the Dirac's delta function.

In Ω^f, the velocity vector $\mathbf{v}(\mathbf{x},t)$, is related with the pressure gradient by means of the Euler's equation

$$\rho\frac{\partial \mathbf{v}}{\partial t} + \nabla p = \mathbf{0}, \quad \forall \mathbf{x} \in \Omega^f. \tag{6}$$

2.4   Initial and Boundary Conditions

According to Fig. 2, Ω^f is the domain occupied by the fluid whereas the fluid saturated porous medium is Ω^b. Null Dirichlet boundary conditions are prescribed in the outer boundary,

$$p = 0, \quad \forall \mathbf{x} \in \partial \Omega^f \setminus \partial \Omega^b. \tag{7}$$

The configuration is chosen so that at the receiver, the waves propagating in the fluid are not affected.

As derived in Lovera [11], in the Biot medium-fluid interface $\partial \Omega^b$, the boundary conditions are

$$\left.\begin{aligned} s &= -\phi p \\ \sigma \, \mathbf{n}^i &= -(1 - \phi) p \, \mathbf{n}^i \end{aligned}\right\} \quad \forall \, \mathbf{x} \in \partial \Omega^b, \tag{8}$$

where $\mathbf{n}^i$ is the normal unit vector to the interface $\partial \Omega^b$ pointing from Ω^b towards the fluid.

The system starts at rest. Consequently, zero initial conditions are added to the PDE system to have a well posed Initial Boundary Value Problem (IBVP). Namely

$$p(\mathbf{x}, 0) = 0, \quad \forall \mathbf{x} \in \Omega^f. \tag{9}$$
$$\mathbf{u}(\mathbf{x}, 0) = \mathbf{0}, \quad \mathbf{U}(\mathbf{x}, 0) = \mathbf{0}, \quad \forall \mathbf{x} \in \Omega^s. \tag{10}$$

2.5   Numerical Solution

In parameter estimation problems, the solution of the so called forward map, in this case involving the solution of the IBVP, is taken for granted. We have made an implementation of the finite volume method [13]. Our choice is based on the easy handling of the boundary conditions at the interface. Let us illustrate this fact.

Define

$$\mathbf{s} = s\mathbf{I}.$$

Integrating Biot's equations in a finite volume V (Fig. 3) and considering the terms involving σ and s, we have

$$\int_V \nabla \cdot \sigma = \int_{\partial V} \sigma \mathbf{n}$$

and

$$\int_V \nabla s = \int_V \nabla \cdot \mathbf{s} = \int_{\partial V} s\mathbf{n}.$$

Hence, in the part of the boundary of V contained in the interface, the boundary conditions (8) are straightforward.

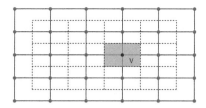

Fig. 3. Schematic representation of finite volume discretization of a rectangular domain by regular cells for each inner node.

Synthetic Example. We consider a water saturated porous medium, also immersed in water. The bone specimen is 4 mm thick and 10 mm long. The physical parameters of water are $\rho = 1000 \, \mathrm{Kg/m^3}$, $K = 2.2 \times 10^9 \, \mathrm{Pa}$. The source term is as in Nguyen and Naili [14], namely

$$F(t) = F_0 e^{-4(f_c t - 1)^2} \sin\left(2\pi f_c t\right), \tag{11}$$

where $f_c = 1 \, \mathrm{MHz}$ and $F_0 = 1 \, \mathrm{m/s^2}$. The transmitter is 2 mm above the specimen opposite to the receiver. For all experiments a time interval of length $T = 7 \times 10^{-5} \, \mathrm{s}$ is used. Following Buchanan et al. [4], the numerical value for the parameter b is 4.1.

Physical parameters for the porous medium are given in Table 1.

Table 1. Physical parameters of porous medium.

Parameter	Value
Porosity (ϕ)	0.5
Tortuosity (α)	1.4
Solid bulk modulus (K_s)	20×10^9 Pa
Squeletal frame bulk modulus (K_b)	3.3×10^9 Pa
Shear modulus (N)	2.6×10^9 Pa
Solid density (ρ_s)	1960 Kg/m^3

We contend that our numerical solution is robust. In the case reported here we used a time step of 10^{-7} s. In Fig. 4 there are some snapshots of the numerical solution obtained by the finite volume method.

Remark. Numerical modeling of ultrasound propagation trough a heterogeneous, anisotropic, porous material, such as a bone, is a research problem in itself. There is a vast literature on a variety of models and numerical methods. Finite Element in time and frequency domain have been used, as well as Finite Differences. For instance, see Nguyen and Naili [14] for a hybrid Spectral/FEM approach, Nguyen et al. [15] for FEM in time domain, and Chiavassa

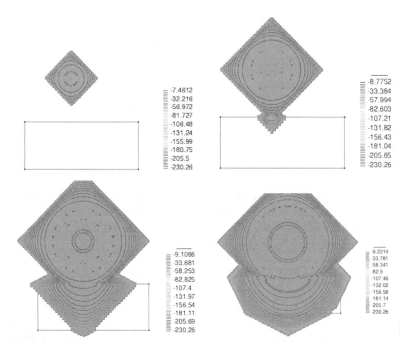

Fig. 4. Evolution of pressure field. Snapshots at $t = 10.8$ µs, $t = 18$ µs, $t = 25.2$ µs, $t = 32.8$ µs.

and Lombard [5] for a Finite Difference implementation. The model above is somewhat simple and does not include all the mechanical complexities. Nevertheless, it is realistic enough to test our methodology of parameter estimation. It will become apparent that it is straightforward to replace the underlying forward map (namely, parameters to solution of Biot's model to pressure data) with more complex models.

2.6   The Parameter Estimation Problem

For practical motivations, most studies are conducted on water-saturated specimens rather than medulla, the actual fluid saturating trabecular bone. Consequently, it is customary as a first approximation to consider the fluid saturating the porous medium in Biot's model as known.

We are led to the inverse problem: Given m pressure measurements at the receiver point (x^r, y^r),

$$p_i \sim p(x^r, y^r, t_i), \quad i = 1, 2, \ldots, m,$$

determine the Biot's parameters

$$\theta = (\phi, \alpha, K_s, K_b, N, \rho_s). \tag{12}$$

Remark. It is critical to have realistic reference values in Biot's model, regardless of the chosen methodology for estimation. Obtaining data from lab measurements is complex and costly. For instance, porosity can be obtain from 3D microtomography (μCT) Wear et al. [19], Pakula et al. [16]. Typical values for human trabecular bone vary between 0.55 and 0.95 depending on anatomic location and bone situation. Tortuosity values are also scarce. It can be measured using electric spectroscopy, wave reflectometry or estimated from porosity, Hosokawa and Otani [7,8]. Reported values are in the interval [1.01, 1.5] Laugier and Haeiat [10]. Elastic properties of bone tissue are required to estimate macroscopic elastic properties of the saturated skeletal frame. These have been measured using atom strength microscopy, nanoindentation or acoustic microscopy. Then, micro mechanic models can be used to determine volume and shear modulus of the solid.

3   Bayesian Parameter Estimation in Cancellous Bone

In the sense of Hadamard, a problem is well posed, if existence, uniqueness and continuity with respect to data (stability), can be established. For instance, in differential equations, continuity with respect to initial and/or boundary conditions. A problem is ill-posed if any of the conditions fails.

Classical well posed problems for differential equations are commonly referred as direct problems, in our case, the Initial-Boundary Value Problem for Biot's model (1), (4), (7), (8), (9), (10).

In practice, one is interested in a property of a system, to be determined from indirect information. The ill-posed Biot's parameters estimation problem is of this sort, and can be regarded as an inverse problem.

In general, a quantity $\mathbf{y} \in \mathbb{R}^m$ is measured to obtain information about another quantity $\theta \in \mathbb{R}^n$. A model is constructed that relates these quantities. The data $\mathbf{y}$ is usually corrupted by noise. Consequently, the inverse problem can be written as: Given the measurements $\mathbf{y}$ and the noise $\mathbf{e}$ find θ such that

$$\mathbf{y} = f(\theta, \mathbf{e}). \tag{13}$$

where $f : \mathbb{R}^n \times \mathbb{R}^k \rightarrow \mathbb{R}^m$ is a function of the model and $\mathbf{e} \in \mathbb{R}^k$ is the noise vector.

3.1   Bayesian Framework

Let us develop the Bayesian methodology for statistical inversion. This paragraph is deliberately terse, for details see Kaipio and Sommersalo [9] and Stuart [18].

The aim of statistical inversion is to extract information on θ, and quantify the uncertainty from the knowledge of **y** and the underlying model. It is based on the principles:

1. All variables are regarded as *random variables.*
2. Information is on realizations.
3. This information is coded in probability distributions.
4. The solution to the inverse problem is the posterior probability distribution.

As customary in statistical notation, random variables are capital letters, thus (13) reads

$$Y = f(\Theta, E). \tag{14}$$

In this context, the data y is a realization of Y.

In Bayesian estimation all we know about Θ it is encompassed in a probability density function, the *prior*, $\pi_0(\theta)$. The conditional probability density function $\pi(y|\theta)$ is the *likelihood function*, whereas the conditional probability density function $\pi^y(\theta) \equiv \pi(\theta|y)$ is the *posterior*. All densities are related by the Bayes' formula

$$\pi^y(\theta) = \frac{\pi_0(\theta)\pi(y|\theta)}{\pi(y)}. \tag{15}$$

Summarizing, solving an inverse problem in the Bayesian framework, consists on the following:

1. With all available information on Θ, propose a prior $\pi_0(\theta)$. This is essentially a modeling problem.
2. Find the likelihood $\pi(y|\theta)$.
3. Develop methods to explore the posterior.

For the last step we use a Markov Chain Monte Carlo (MCMC) method. More precisely, we use emcee, an affine invariant MCMC ensemble sampler. Foreman-Mackey et al. [6].

It is well known that this sampling methodology does not depend on the normalizing constant $\pi(y)$ and we write,

$$\pi^y(\theta) \propto \pi_0(\theta)\pi(y|\theta).$$

Point Estimators

Given the posterior, a suitable value for the unknown variable Θ is needed. That is, a point estimator.

A maximizer of the posterior distribution is called a *Maximum A Posteriori estimator*, or MAP estimator. This amounts to the solution of a global optimization problem.

$$\theta_{MAP} = \arg\max_{\theta \in \mathbb{R}^n} \pi(\theta|y), \tag{16}$$

Also of interest is the *conditional mean* (MC) estimator, namely

$$\theta_{MC} = E(\theta|y) = \int_{\mathbb{R}^n} \theta\pi(\theta|y)dx, \tag{17}$$

This is an integration problem usually in high dimensions, computationally costly. Classical quadrature rules are prohibited.

From MAP to Tikhonov

For Bayesian estimation we consider θ as a random vector distributed as π_0, a given prior density. Thus $\mathbf{y}$ is given by

$$\mathbf{y} = \mathscr{G}(\theta) + \eta$$

where η is random noise with density ρ. Here $\mathscr{G}$ is the observation operator.

From Bayes' formula, the posterior distribution $\pi^y(\theta)$ satisfies

$$\pi^y(\theta) \propto \rho(\mathbf{y} - \mathscr{G}(\theta))\pi_0(\theta).$$

Assuming Gaussian prior $\theta \sim \mathcal{N}(\theta_0, \sigma^2 \mathbf{I})$, and Gaussian noise $\eta \sim \mathcal{N}(\mathbf{0}, \gamma^2 \mathbf{I})$ we have

$$\pi_0(\theta) \propto \exp\left(-\frac{1}{\sigma^2}\|\theta - \theta_0\|^2\right)$$

and

$$\rho(\eta) \propto \exp\left(-\frac{1}{\gamma^2}\|\eta\|^2\right).$$

We are led to

$$\theta_{MAP} = \arg\max\left[\exp\left(-\frac{1}{\gamma^2}\|\mathbf{y} - \mathscr{G}(\theta)\|^2\right)\exp\left(-\frac{1}{\sigma^2}\|\theta - \theta_0\|^2\right)\right]$$

$$= \arg\min\left[\|\mathbf{y} - \mathscr{G}(\theta)\|^2 + \left(\frac{\gamma}{\sigma}\right)^2\|\theta - \theta_0\|^2\right].$$

Consequently, θ_{MAP} coincides with Tikhonov' solution with regularization parameter $\alpha = \left(\frac{\gamma}{\sigma}\right)^2$.

3.2   Application to Biot's Problem

We consider the problem described in Sect. 2.4 and assume as input data the collection of $m = 80$ pressure measurements. See Fig. 5. For the normalized signal we add Gaussian noise with a standard deviation of 0.2. A large value in this context.

Likelihood Function

In our synthetic example we have considered additive noise, a common assumption. Thus the model becomes,

$$Y = f(\Theta) + E. \tag{18}$$

Also, it is assumed that the random variables Θ and E are independent. Consequently, the likelihood function is

$$\pi(y|\theta) = \rho(y - f(\theta)), \tag{19}$$

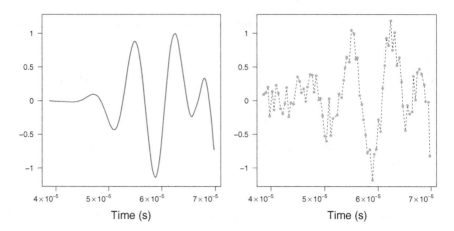

Fig. 5. Left: Solution of the Biot's model on the position of the receiver. Right: Input data are generated by sampling the receiver signal and adding Gaussian noise.

where ρ is a Gaussian probability density function of E.

We are led to explore the posterior,

$$\pi(\theta|y) \propto \pi_0(\theta)\rho(y - f(\theta)).$$

Prior Densities

We consider two cases, Gaussian and uniform (uninformative) priors. In both cases the posterior is sampled and the conditional mean estimator is computed. A comparison is made with the MAP estimator.

Gaussian prior

In Table 2 we list the parameters for the prior Gaussian densities. Notice that the mean of the Gaussian distribution for each parameter is *far away* form the true value.

Table 2. Parameters for Gaussian priors.

Property	True value	Mean (θ_i^0)	Standard deviation (γ_i)
Porosity (ϕ)	0.50	0.8	0.10
Tortuosity (α)	1.4	1.6	1.5
Solid bulk modulus (K_s)	20×10^9 Pa	25×10^9 Pa	9×10^9 Pa
Squeletal frame bulk modulus (K_b)	3.3×10^9 Pa	3.8×10^9 Pa	2.5×10^9 Pa
Shear modulus (N)	2.6×10^9 Pa	4.5×10^9 Pa	5.5×10^9 Pa
Solid density (ρ_s)	1960 Kg/m^3	1940 Kg/m^3	250 Kg/m^3

It is remarkable that the conditional mean estimator is capable of recovering the noisy signal. See Fig. 6.

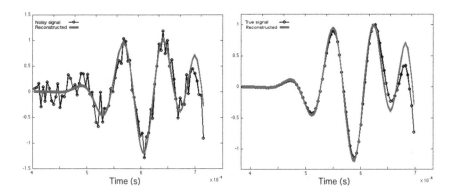

Fig. 6. Left: Comparison between the generated signal (solid line) using θ_{CM} and the noisy input (dashed line); Right: Comparison between the true signal (dashed line) and the recovered signal (solid line).

Uniform Prior

Let us consider uniform priors, namely

$$\pi_0(u) \ \propto \ \chi_{[a,b]}(u) = \begin{cases} 1 & \text{if } a \le u \le b, \\ 0 & \text{elsewhere,} \end{cases} \tag{20}$$

where the parameter of interest is believed to belong to the interval $[a,b]$.
The intervals are chosen to be physically meaningful, see Table 3.

Table 3. Parameters for uniform priors.

Property	Interval
Porosity (ϕ)	$[0.3,\ 0.95]$
Tortuosity (α)	$[1,\ \infty)$
Solid bulk modulus (K_s)	$[1.5 \times 10^{10},\ 3.0 \times 10^{10}]$ Pa
Squeletal frame bulk modulus (K_b)	$[2.0 \times 10^{9},\ 4.5 \times 10^{9}]$ Pa
Shear modulus (N)	$[2.0 \times 10^{9},\ 3 \times 10^{9}]$ Pa
Solid density (ρ_s)	$[1000,\ 3000]$ Kg/m^3

Again, the conditional mean estimator fits satisfactorily even the noisy signal. See Fig. 7.

Confidence Intervals

Having the posterior density, allows to quantify the uncertainty of the estimated parameters. Here we just provide confidence intervals with a 0.9 probability. Results are shown in Tables 4 and 5 for Gaussian and uniform priors respectively.

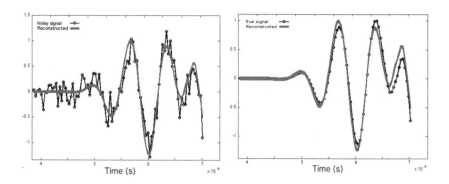

Fig. 7. Left: Comparison between the reconstructed signal using θ_{CM} (solid line) and the noisy input (dashed line); Right: Comparison between the reconstructed signal (solid line) and the true signal (dashed line).

Table 4. Comparison of estimated parameters.

Parameter	Gaussian prior density		
	θ_{TRUE}	θ_{CM}	Interval
ϕ	0.5	0.536	[0.478, 0.614]
α	1.4	1.421	[1.350, 1.505]
K_s ($\times 10^{10}$ Pa)	2.0	2.000	[1.9999922, 2.0000079]
K_b ($\times 10^9$ Pa)	3.3	3.299	[3.299, 3.300]
N ($\times 10^9$ Pa)	2.6	2.544	[2.226, 2.862]
ρ_s ($\times 10^3$ Kg/m^3)	1.96	1.955	[1.943, 1.969]

Table 5. Comparison of estimated parameters.

Parameter	Uniform prior density		
	θ_{TRUE}	θ_{CM}	Interval
ϕ	0.5	0.549	[0.505, 0.642]
α	1.4	1.432	[1.321, 1.540]
K_s ($\times 10^{10}$ Pa)	2.0	2.085	[1.477, 2.451]
K_b ($\times 10^9$ Pa)	3.3	3.270	[2.847, 3.808]
N ($\times 10^9$ Pa)	2.6	2.682	[1.645, 3.482]
ρ_s ($\times 10^3$ Kg/m^3)	1.96	1.949	[1.815, 2.077]

Remark. A drawback of Bayesian estimation is its computational cost. At each step of the random walk of the MCMC method, the forward map involves the solution of the Biot's model. Nevertheless, the results above show that the conditional mean is a reliable point estimator. We shall see below that in this case other approaches of estimation may not suffice.

4   A PDE-Constrained Optimization Approach

Let $\pi_0(\theta)$ denote the prior density for the Biot's parameters. As pointed out above, a classical approach is to consider the problem of estimation as one of optimization by means of the MAP estimator. Namely,

$$\theta_{MAP} = \arg\max_{\theta} \ \exp\left\{-\frac{1}{\sigma}||y - \mathscr{G}(\theta)||^2\right\} \pi_0(\theta). \tag{21}$$

Applying natural logarithm

$$\theta_{MAP} = \arg\min_{\theta} ||y - \mathscr{G}(\theta)||^2 - \sigma \log\left(\pi_0(\theta)\right). \tag{22}$$

This minimization problem is constrained by the IBVP for Biot's model (1), (4), (7), (8), (9), (10). Consequently any evaluation of the objective function requires the solution of this IBVP. Derivative based algorithms are computationally expensive. A reasonable alternative is the Nelder-Mead method, a description below.

4.1   The Nelder–Mead Method

The optimization problem to calculate the MAP estimator is solved by using the derivative free Nelder–Mead method [20, 23].

The method only requires evaluations of the objective function in (22). It is based on the iterative update of a *simplex*, which in this case is a set of $n + 1$ points in n-dimensional space and they do not lie in a space of lower dimension. Each point of the simplex is called a *vertex*. The shape and size of the simplex is modified according to the values of the objective function $f : \mathbb{R}^n \to \mathbb{R}$ at each vertex.

The algorithm starts with an initial guess of the vertices. In reference to Fig. 8, let L and S be the vertices where the objective function has its largest and smallest values, respectively. The algorithm tries to modify the vertex L to find a new point, such that the value of the objective function at this point will be smallest than $f(S)$. To illustrate the complexity of the method we delve a little further.

The Nelder-Mead algorithm has four parameters:

- the reflection coefficient $\tau_r > 0$ (usually τ_r is set to 1),
- the expansion factor $\tau_e > \max\{1, \tau_r\}$,
- the contraction parameter $\tau_c \in (0, 1)$, and
- the shrinkage factor $\tau_s \in (0, 1)$.

In each iteration all the vertices are indexed according to the values of the function $f(x)$. Thus the simplex is composed by the vertices

$$S = v_1, v_2, ..., v_{n+1} = L \quad \text{if} \quad f_i = f(v_i) \leq f_{i+1} = f(v_{i+1}), \quad i = 1, 2, ..., n. \tag{23}$$

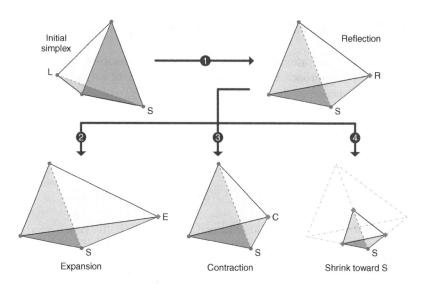

Fig. 8. Steps of the Nelder–Mead method.

To define the transformations of the simplex, we need to calculate the centroid $\bar{v}$ of the n first vertices,

$$\bar{v} = \frac{1}{n} \sum_{i=1}^{n} v_i, \tag{24}$$

and the point

$$v(\tau) = \bar{v} + \tau\,(\bar{v} - S),$$

for $\tau \in \{\tau_r, \tau_e, \tau_c, \tau_s\}$. Each parameter is associated to an operation (see Fig. 8):

- *Reflection* produces a movement of the simplex towards regions where f is getting smaller values.
- *Expansion* increases the size of the simplex to advance more quickly in search of the local minimum.
- *Contraction* is applied when reflection and expansion fails, and it allows to get an inner point of the simplex in which f takes a value lower than $f(L)$ at least.
- *Shrink toward S* moves all the vertices in the direction of the current best point to reduce the size of the simplex when it is in a valley of the objective function. This allows previous operations can continue to be applied in the following iterations.

The algorithm applies the following steps in each iteration to find a local minimum of the function $f(x)$:

1. Indexing the vertices of the simplex according to the objective function values (23).

2. Calculate the centroid $\bar{v}$ of the first n vertices (24).
3. Transform the simplex by the following operations:
   a. (Reflection) Calculate $R = v(\tau_r)$. If $f(S) < f(R) < f(L)$, L is replaced by R and we move to the step 1 to start the next iteration.
   b. (Expansion) If $f(R) < f(S)$, we calculate $E = v(\tau_e)$. If $f(E) < f(S)$, L is replaced by E. Otherwise L is replaced with R. The process is restarted. This
   c. (Contraction) If $f(v_i) < f(R)$ for $i = 1, ..., n$, the point $C = v(\tau_c)$ is calculated.

      If $f(C) < \min\{f(L), f(R)\}$, L is replaced with C and the process is restarted. Otherwise a shrink toward S is applied.
   d. (Shrink toward S) For $i = 2, 3, ..., n + 1$, the vertices v_i are modified by

$$v_i = S + \tau_s\,(v_i - S).$$

The process continues until a maximum number of iterations is reached, or when the simplex reaches some minimum size, or the best vertex S becoming less than a given value or failing to change its position between successive iterations.

Figure 9 shows a 2D example of the transformations applied to an initial simplex. First a reflection of the L vertex is applied to reach the point R. Next a contraction replaces the vertex R by the point C. Then a contraction followed by a shrink contraction reduces the size of simplex to allow it reaches a valley floor and finally an expansion is applied.

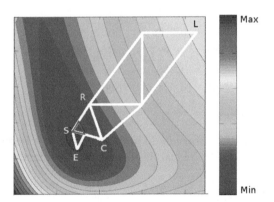

Fig. 9. Illustration of the evolution of a simplex to find the minimum value of a function $f : \mathbb{R}^2 \to \mathbb{R}$.

Remark.

(i) We contend that the Nelder-Mead method is appropriate for the problem at hand. It does not require the objective function to be smooth, hence it does not require computation of derivatives.

(ii) On the down side, it is well known that its performance decreases significantly in problems with more than 10 variables Han and Neumann [21]. Also, in the case of problems with few variables it may fail to converge to a critical point of the objective function Mckinnon [22].

4.2 Nelder–Mead Solutions to Biot's Problem and Percentage Errors

Chronologically, we posed the estimation problem as one of PDE-Constrained optimization. In the Biot's problem the number of parameters to be calculated using the MAP estimator (22) is at most six. Thus Nelder-Mead is appropriate. The prior information of the variables was used to build the initial simplex.

First assuming a Gaussian prior, or equivalently a regularized least square problem, the MAP estimators in Table 6 are obtained. The true and recovered signal are shown in Fig. 10. Starting with estimating the full set of six parameters, it was observed that the method failed to recover the noiseless signal. The problem was simplified one parameter at time. For instance, Fig. 10(c) show the estimated signal assuming ρ_s known, and so on.

Table 6. Comparison of estimated parameters for a Gaussian prior.

Parameter	θ_{TRUE}	θ_{MAP}
ϕ	0.5	0.541
α	1.4	4.171
K_s ($\times 10^{10}$ Pa)	2.0	2.369
K_b ($\times 10^9$ Pa)	3.3	3.241
N ($\times 10^9$ Pa)	2.6	9.651
ρ_s ($\times 10^3$ Kg/m^3)	1.96	2.149

Next uniform priors are considered and the same experiment is carried out. As shown in Table 7 and Fig. 11, no improvement is attained.

Table 7. Comparison of estimated parameters for a uniform prior.

Parameter	u_{TRUE}	u_{MAP}
ϕ	0.5	0.613
α	1.4	1.350
K_s ($\times 10^{10}$ Pa)	2.0	1.382
K_b ($\times 10^9$ Pa)	3.3	2.301
N ($\times 10^9$ Pa)	2.6	4.212
ρ_s ($\times 10^3$ Kg/m^3)	1.96	2.351

As remarked above, Buchanan et al. [4], call a solution subcorrect if the objective function reaches a desired tolerance. Therein, percentage errors are unsatisfactory. Our results for the MAP estimators are consistent. Noteworthy, the percentage errors of the Conditional Mean estimators show their accuracy as point estimators. See Table 8.

Table 8. Percentage errors of estimated parameters.

Parameter	Gaussian priors		Uniform priors	
	θ_{MAP}	θ_{CM}	θ_{MAP}	θ_{CM}
ϕ	8.2	7.2	22.6	9.8
α	197.92	1.42	3.57	2.28
K_s	18.45	1.1×10^{-6}	30.9	4.25
K_b	1.81	0.03	30.27	0.9
N	271.19	2.15	62	3.15
ρ_s	9.64	0.25	19.94	0.56

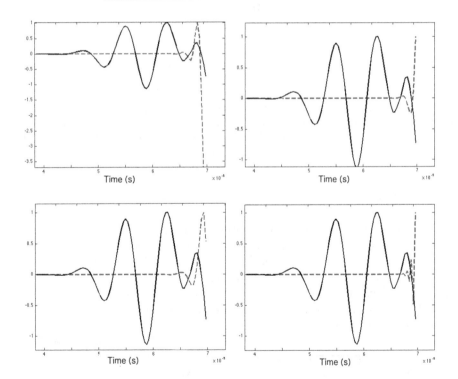

Fig. 10. Comparison between four reconstructed signals using θ_{MAP} with uniform priors (dashed line) and the true signal (solid line) for the estimated parameters: (a) $\theta = (\phi, \alpha, K_s)$, (b) $\theta = (\phi, \alpha, K_s, K_b)$, (c) $\theta = (\phi, \alpha, K_s, K_b, N)$, (d) $\theta = (\phi, \alpha, K_s, K_b, N, \rho_s)$.

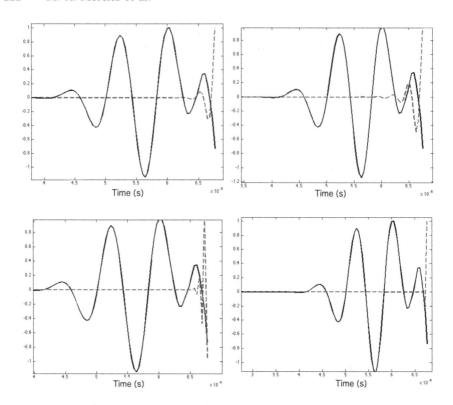

Fig. 11. Comparison between four reconstructed signals using θ_{MAP} with Gaussian priors (dashed line) and the true signal (solid line) for the estimated parameters: (a) $\theta = (\phi, \alpha, K_s)$, (b) $\theta = (\phi, \alpha, K_s, K_b)$, (c) $\theta = (\phi, \alpha, K_s, K_b, N)$, (d) $\theta = (\phi, \alpha, K_s, K_b, N, \rho_s)$.

5   Conclusions

We have introduced the problem of parameter estimation for a Biot's medium modeling a cancellous bone. The problem has been posed for both a minimization problem and a posterior density estimation in the Bayesian framework. The MAP estimator is shown as the solution of the minimization problem. We carried out extensive experiments with a variety of methods, classical descent methods as well as derivative free methods. All led to the same conclusion, the MAP estimator carries an unsatisfactory percentage error and is not capable of recovering the given signal. We show results only for the derivative free method Nelder Mead. In contrast, the conditional mean recovers the noisy signal satisfactorily. Moreover, the point estimator is accurate and robust.

Consequently, although the computation of the conditional mean is costly because of the sampling of the posterior, its use is advisable for diagnostics of the bone properties.

One may argue that the conditional mean is better suited to represent an intrinsically heterogeneous porous medium, a query worth of an in depth study. Also of interest, is to consider the initial geophysical phenomena where the Biot's model applies.

Acknowledgments. M. A. Moreles would like to acknowledge to ECOS-NORD project number 000000000263116/M15M01 for financial support during this research. Also, part of this research was carried out while M. A. Moreles was a visiting professor at the mathematics Department of the Universidad de Guadalajara. Their hospitality is greatly appreciated.

References

1. Tarantola, A.: Inverse Problem Theory and Methods for Model Parameter Estimation. SIAM, Philadelphia (2005)
2. Buchanan, J.L., Gilbert, R.P.: Determination of the parameters of cancellous bone using high frequency acoustic measurements. Math. Comput. Mode. **45**(3), 281–308 (2007)
3. Buchanan, J.L., Gilbert, R.P.: Measuring osteoporosis using ultrasound. In: Fotiadis, D.I., Massalas, C.V. (eds.) Advances in Scattering and Biomedical Engineering, pp. 484–494. World Scientific, Singapore (2004)
4. Buchanan, J.L., Gilbert, R.P.: Determination of the parameters of cancellous bone using high frequency acoustic measurements II: inverse problems. J. Comput. Acoust. **15**(02), 199–220 (2007)
5. Chiavassa, G., Lombard, B.: Wave propagation across acoustic/Biot's media: a finite-difference method. Commun. Comput. Phys. **13**(4), 985–1012 (2013). https://doi.org/10.4208/cicp.140911.050412a
6. Foreman-Mackey, D., Hogg, D.W., Lang, D., Goodman, J.: emcee: the MCMC hammer. Publ. Astron. Soc. Pac. **125**(925), 306 (2013)
7. Hosokawa, A., Otani, T.: Ultrasonic wave propagation in bovine cancellous bone. J. Acoust. Soc. Am. **101**(1), 558–562 (1997)
8. Hosokawa, A., Otani, T.: Acoustic anisotropy in bovine cancellous bone. J. Acoust. Soc. Am. **103**(5), 2718–2722 (1998)
9. Kaipio, J., Somersalo, E.: Statistical and Computational Inverse Problems, vol. 160. Springer, New York (2006)
10. Laugier, P., Haeiat, G. (eds.): Bone Quantitative Ultrasound. Springer, Amsterdam (2011)
11. Lovera, O.M.: Boundary conditions for a fluid-saturated porous solid. Geophysics **52**(2), 174–178 (1987)
12. Lowet, G., Van der Perre, G.: Ultrasound velocity measurements in long bones: measurement method and simulation of ultrasound wave propagation. J. Biomech. **29**, 1255–1262 (1996)
13. Mazumder, S.: Numerical Methods for Partial Differential Equations: Finite Difference and Finite Volume Methods. Academic Press, Cambridge (2016)
14. Nguyen, V.H., Naili, S.: Simulation of ultrasonic wave propagation in anisotropic poroelastic bone plate using hybrid spectral/finite element method. Int. J. Numer. Methods Biomed. Eng. **28**(8), 861–876 (2012)
15. Nguyen, V.H., Naili, S., Sansalone, V.: Simulation of ultrasonic wave propagation in anisotropic cancellous bones immersed in fluid. Wave Motion **47**(2), 117–129 (2010)

16. Pakula, M., Padilla, F., Laugier, P., Kaczmarek, M.: Application of Biot's theory to ultrasonic characterization of human cancellous bones: determination of structural, material, and mechanical properties. J. Acoust. Soc. Am. **123**(4), 2415–2423 (2008)
17. Sebaa, N., Fellah, Z.E.A., Fellah, M., Ogam, E., Wirgin, A., Mitri, F.G., Lauriks, W.: Ultrasonic characterization of human cancellous bone using the Biot theory: inverse problem. J. Acoust. Soc. Am. **120**(4), 1816–1824 (2006)
18. Stuart, A.M.: Inverse problems: a Bayesian perspective. Acta Numer. **19**, 451–559 (2010)
19. Wear, K.A., Laib, A., Stuber, A.P., Reynolds, J.C.: Comparison of measurements of phase velocity in human calcaneus to Biot theory. J. Acoust. Soc. Am. **117**(5), 3319–3324 (2005)
20. Olsson, D.M., Nelson, L.S.: The Nelder-Mead simplex procedure for function minimization. Technometrics **17**(1), 45–51 (1975)
21. Han, L., Neumann, M.: Effect of dimensionality on the Nelder-Mead simplex method. Optim. Methods Softw. **21**(1), 1–16 (2006)
22. Mckinnon, K.I.M.: Convergence of the Nelder-Mead simplex method to a nonstationary point. SIAM J. Optim. **9**, 148–158 (1998)
23. Nelder, J.A., Mead, R.: A simplex method for function minimization. Comput. J. **7**, 308–313 (1965)

Optimal Sizing of Low-DropOut Voltage Regulators by NSGA-II and PVT Analysis

Jesus Lopez-Arredondo[1], Esteban Tlelo-Cuautle[1(⊠)],
Luis Gerardo de la Fraga[2], Victor Hugo Carbajal-Gomez[3],
and Miguel Aurelio Duarte-Villaseñor[4]

[1] INAOE, Luis Enrique Erro No. 1, 72840 Tonantzintla, Puebla, Mexico
jesus.arredondo28@gmail.com, etlelo@inaoep.mx
[2] Computer Science Department,
Cinvestav, Av. IPN 2508, 07360 Mexico City, Mexico
fraga@cs.cinvestav.mx
[3] Universidad Autónoma de Tlaxcala, Tlaxcala, Mexico
victhug26@gmail.com
[4] Catedrático CONACYT en el Instituto Tecnológico de Tijuana, Tijuana, Mexico
miauduvi@gmail.com

Abstract. The optimization of analog integrated circuits has been a challenge due to the fact that there are not rules or systematic guidelines to bias and size the transistors and other elements in the circuit under design. This Chapter reviews the design of generic operational amplifiers by using complementary metal-oxide-semiconductor (CMOS) integrated circuit technology and shows the optimization of three different low-dropout (LDO) voltage regulators that consists of an operational amplifier and passive circuit elements. We highlight that if one performs a sensitivity analysis for each LDO, then a reduced set of design variables can be selected to create the chromosome for performing a multi-objective optimization by the well-known Non-Dominated Sorting Genetic Algorithm II (NSGA-II). In addition, the computed sensitivities are used to reduce the search spaces for the design variables of the MOS transistors to accelerate the optimization process. Finally, from the feasible solutions of the three LDOs, a process-voltage and temperature (PVT) analysis is performed to guarantee that the designed LDO is robust to variations.

Keywords: Low-dropout voltage regulator
Multi-objective optimization · Circuit sizing · NSGA-II
Operational transconductance amplifier · MOS transistor · SPICE

1 Introduction

Analog integrated circuit design is considered the bottleneck of the electronic industry because the manual design depends on the experience of the designer

© Springer International Publishing AG, part of Springer Nature 2019
L. Trujillo et al. (Eds.): NEO 2017, SCI 785, pp. 225–247, 2019.
https://doi.org/10.1007/978-3-319-96104-0_12

and very few design automation tools have been developed. For instance, the invention and synthesis of new amplifier topologies has been a research subject during the last years [1–7]. In this Chapter, three different operational amplifier topologies are used to design a low-dropout (LDO) voltage regulator, which is very useful to design applications in the area of power management. In fact, due to the exponential growth in the use of electronic portable devices, the design of high performance power management circuits like LDOs has become essential. However, due to the nonlinear behavior of the LDO circuits, the performance metrics are in conflict and then an analog designer has the challenge to discover the design trade-offs. In this manner, due to the large number of performances of the LDO that must be evaluated, the large number of design variables, and the design trade-offs, then multi-objective evolutionary algorithms (MOEAs) are a good option to optimize LDOs. It is worth mentioning that MOEAs have already been applied in the optimal sizing of integrated circuits [1, 8–12]. In this Chapter, the non-dominated sorting genetic algorithm (NSGA-II), is applied to perform multi-objective optimization of three LDO topologies that are designed with complementary metal-oxide-semiconductor (CMOS) technology.

The Non-Dominated Sorting Genetic Algorithm II (NSGA-II) has been applied in [13] to size CMOS operational amplifiers, and it is also applied herein linking the simulation program with integrated circuit emphasis (SPICE) to evaluate electrical characteristics of the amplifiers. The feasible solutions provided by NSGA-II are validated by performing a process-voltage and temperature (PVT) variation analysis [14], to guarantee robust design. As already shown in [13], one can estimate a reduced search space for the design variables of the MOS transistors, namely: widths (W) and lengths (L), which are multiples of lambda that is provided by the manufacturer. Estimated search spaces are obtained herein by applying sensitivity analysis to the CMOS LDO, which also helps to reduce the number of design variables [15]. This reduction accelerates the multi-objective optimization of the LDO, which target specifications or objectives are for example: power supply rejection (PSR), power consumption, layout area, and line regulation (LNR). The optimization includes as constraints: all MOS transistors must operate in a desirable design region, and W/L must be multiples of lambda that is provided by the manufacturer.

Section 2 describes the problem formulation for optimizing LDO voltage regulators. A short review on the design of operational amplifiers is given in Sect. 3. The sensitivity analysis is applied in Sect. 4 to establish a reduced set of design variables that will be included into the chromosome. Section 5 discusses the results provided by NSGA-II and performs PVT analysis to guarantee robust feasible solutions. Finally, the conclusions are listed in Sect. 6.

2   Problem Formulation

The multi-objective optimization problem for sizing LDO voltage regulators is formulated as follows [16, 17],

$$\text{Optimize} \quad F(\mathbf{x})$$

$$\text{Subject to: } \Omega = \{\mathbf{x} \in \Re^n | G(\mathbf{x}) \leq 0\} \tag{1}$$

where $\mathbf{x}$ is a multidimensional vector in $\Re^n$ that denotes the decision parameters, which are delimited by $x_{min}^i \leq x^i \leq x_{max}^i$. $F(\mathbf{x})$ represents a vector of m objectives $(f_1(\mathbf{x}), ..., f_m(\mathbf{x}))$ that are being minimized/maximized, and $G(\mathbf{x})$ is the vector of p constraints that must be satisfied to guarantee feasible solutions. When m equals to one, (1) is associated to a mono-objective problem, when $m = 2$ or 3, (1) is associated to a multi-objective problem. In both cases, one can have multiple constraints to deal with a global and multi-dimensional problem.

In this work, and to optimize LDOs, we propose the use of a reduced chromosome that consists of the most sensitive design variables, such as: W/L sizes of the MOS transistors, bias currents and the passive circuit element values. The following relationships are also used during the sizing of the LDOs,

$$y = \frac{x}{k_1} + k_2, \quad x \in \{0, 1, 2, ..., N - 1\}; \quad y \in [C_{min}, C_{max}]$$

where y represents the real values in the LDO voltage regulator, and x represents the values associated to the position in the chromosome. It can be inferred that each design variable can acquire N different values that can be restricted by the limits C_{min} and C_{max}. The constants k_1 and k_2 are computed to perform the conversion between x and y, and to scale the values of the design variables W/L that must be according to the parameters of integrated circuit technology.

3   CMOS Operational Amplifier Design

An LDO embeds an operational amplifier to measure the error that controls the output of the voltage regulator. However, the operational amplifier has undesirable nonlinear phenomena that must be mitigated, and it can be done by circuit sizing [13]. The main requirement in many applications is that the amplifier must have a linear transfer characteristic, so that linearization techniques are applied [18–20], and three major linearization approaches can be identified [21]: Ohmic transconductors, degenerated differential pairs, and square-law transconductors.

Ohmic transconductors exploit the operation of the MOS transistor in ohmic region and are well suited for high-frequency operation. Figure 1(a) illustrates the principle. Assuming that both transistors are perfectly matched and that their drain-to-source (DS) and saturation (DS_{sat}) voltages accomplishes $V_{DS1} = V_{DS2} \equiv V_{DS} < V_{DS_{sat}}$. Therefore, it follows that the difference of the output currents I_o depend on the technology parameter k, the sizes W/L, V_{DS} and input differential voltage V_{id},

$$I_{o1} - I_{o2} = \left(k \frac{W}{L} V_{DS} \right) V_{id} \tag{2}$$

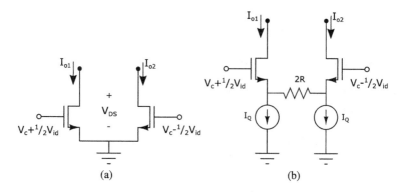

Fig. 1. (a) Concept of linearized ohmic transconductors, and (b) Concept of degenerated differential pair. V_c is the input common-mode voltage.

In (2) one can see that the voltage-to-current transfer characteristics and transconductance gain can electrically be tuned via V_{DS}. Besides, the major source of nonlinearity for all of them is due to the dependence of parameter k.

Figure 1(b) illustrates the principle of degenerated differential pair, which achieves large linearity at the cost of large silicon area and small transconductance gain. The circuit exploits feedback to maintain the gate-to-source V_{GS} voltage of the MOS transistors practically constant for large differential input voltage V_{id}. Proper operation requires that sufficiently large feedback is applied, which infers fulfilling the relation between the transconductance gain g'_m and resistance R as $g'_m R \gg 1$. Consequently, the output current difference can be given by (3), which shows that the external transconductance gain is determined by the feedback resistor.

$$I_{o1} - I_{o2} = \frac{g_m}{1 - g'_m R} V_{id} \approx R^{-1} V_{id} \tag{3}$$

The last group of linearized transconductors include structures that have linear operation by algebraic combination of square-law functions, as the following,

$$(a+b)^2 - (a-b)^2 = 4ab \tag{4}$$

These topologies exploit transfer characteristics of MOS transistors operating in strong inversion inside the saturation region. Figure 2 shows examples of some of them. As for ohmic region transconductors, a main limitation of these structures is the dependence of parameter k on terminal voltages, which produces deviation from the ideal square-law operation.

4   Sensitivity Analysis and Chromosome Construction

The three LDO topologies analyzed herein are based on different operational amplifiers that are designed with simple differential pairs and a second stage

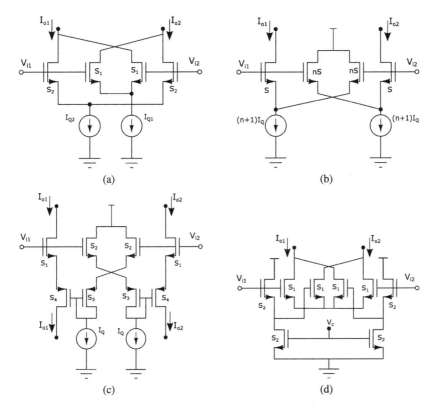

Fig. 2. Illustrative square-law MOS transconductors. (a) Cross-coupling, (b) Adaptive biasing, (c) Class AB, and (d) Voltage shifting.

to provide high voltage gain. From a manual design, sensitivity analysis is performed to identify the tolerances of the W/L sizes of the transistors and the passive elements R and C. That way, one can reduce the number of design variables by considering the most sensitive ones, and the search spaces can be estimated according to the target specifications. Reducing the number of design variables and their associated search spaces is a difficult task, but fortunately, from sensitivity analysis one can identify those W/L sizes or circuit elements that affect the response of the LDO with a minimum or maximum variation. The design variables that produce maximum variations in the performances of the LDO must be selected to create the chromosome for multi-objective optimization. Besides, the optimization of LDO voltage regulators is highly dependent on the topology and circuit elements. For instance, the compensation of an LDO can be done internally or externally, one can also add circuitry to enhance the transient response, among other design considerations.

Sensitivity analysis is performed to the three LDO voltage regulators shown in Fig. 3 by using SPICE and the results are listed in Table 1 for the LDO1 shown in Fig. 3(a), Table 2 for the LDO2 shown in Fig. 3(b), and Table 3 for the LDO3 shown in Fig. 3(c). As one sees, the dropout and output voltages (V_{DO} and V_{out}) are mainly affected by the variations of the width of the pass transistor W_{M_P}, the bias current I_{Bias}, and the compensation capacitor C in the three LDO topologies, so that they are the first candidates to create the chromosome for multi-objective optimization.

Table 1. Sensitivity analysis for the LDO1 from Fig. 3. The variations of the design variables (Var) are different and the maximum sensitivities produce different variations in the performances, such as: V_{DO}, V_{out}, power supply rejection PSR, open loop gain A_{OL}, phase margin (PM), line regulation LNR and load regulation LDR.

Var	Variation	V_{DO}	$\pm\Delta V_{out}$	PSR_{1kHz}	PSR_{10kHz}	A_{OL}	PM	LNR	LDR
W_{M_P}	±30%	±44%	±14%	±0.4%	±3.6%	±1.8%	±1.9%	±18%	±11%
I_{Bias}	±50%	±0%	±48%	±0.5%	±13.2%	±0.8%	±33.9%	±22.3%	±34.5%
$W_{M_{1-2}}$	±50%	±0%	±11.6%	±0.1%	±6.2%	±0.2%	±1.2%	±3.4%	±2.9%
$W_{M_{3-4}}$	±30%	±0%	±3.6%	±1.6%	±7.8%	±0.2%	±1.2%	±0.7%	±0.9%
W_{M_6}	±30%	±0%	±6.7%	±2.3%	±0.4%	±1.3%	±9.9%	±0.8%	±3.2%
W_{M_7}	±30%	±0%	±6.1%	±3.6%	±0.4%	±1.9%	±19.1%	±8.6%	±6.2%
R_Z	±50%	±0%	±4%	±0%	±0%	±0%	±0.6%	±0%	±0.1%
C_C	±80%	±0%	±62.2%	±0%	±12.4%	±0%	±85.3%	±0%	±0%
C_O	$1-5\mu F$	±0%	±61.7%	±0%	±1.2%	±13.3%	±58.9%	±28.2%	±36.8%
R_{ESR}	$50-500\,m\Omega$	±0%	±26.3%	±0%	±7.9%	±0%	±35.2%	±0%	±0%
R_{FB}	±50%	±0%	±1.1%	±0%	±0%	±0%	±0%	±0%	±0.1%

Tables 1, 2 and 3, show the tolerances of the circuit elements represented by the widths (W) of the MOS transistors W_M, the bias currents I_{Bias}, and the passive elements like capacitors C and resistors R. The variations of the values of the circuit elements is performed to see the variation in the dropout voltage V_{DO}, output voltage V_{out}, power supply rejection PSR, open-loop gain of the error amplifier A_{OL}, phase margin of the voltage regulator PM, line regulation LNR and load regulation LDR. Observing the variations on these parameters one can select the more sensitive design variables. This leads us to propose the general chromosome given in (5), where the genes are associated to the width of the pass transistor W_P, I_{Bias}, compensation capacitors C, external R_{Ext} and internal R_{Int} resistances, and the widths of the MOS transistors in the error amplifier to measure the open-loop gain A_{OL}. In all cases, A_{OL} has a big impact in the open loop response of the LDO voltage regulator, and the worst case is when the error amplifier gain is below 40 dB, a fact that makes the open loop response of the regulator being the more sensitive target objective.

$$\text{Chromosome:} \qquad W_P \quad I_{Bias} \quad C \quad R_{Ext} \quad R_{Int} \quad A_{OL} \tag{5}$$

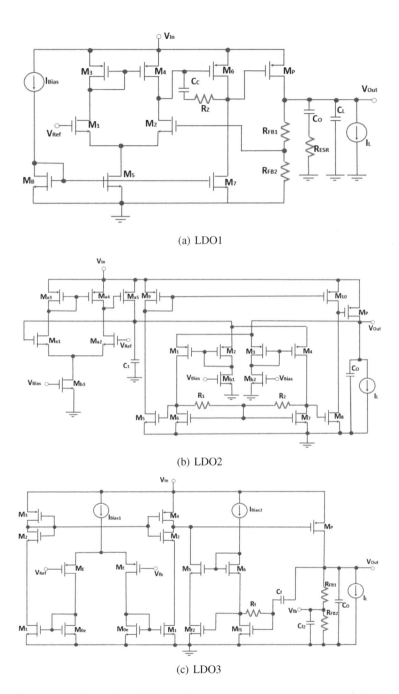

(a) LDO1

(b) LDO2

(c) LDO3

Fig. 3. Voltage regulator: (a) LDO1 with an error amplifier including Miller compensation, (b) LDO2 taken from [22], and (c) LDO3 taken from [23].

Table 2. Sensitivity analysis for the LDO2 shown in Fig. 3. The variations of the design variables (Var) are different and the maximum sensitivities produce different variations in the performances, such as: V_{DO}, V_{out}, power supply rejection PSR, open loop gain A_{OL}, phase margin (PM), line regulation LNR and load regulation LDR.

Var	Variation	V_{DO}	$\pm\Delta V_{out}$	PSR_{1kHz}	PSR_{10kHz}	A_{OL}	PM	LNR	LDR
W_{M_P}	±30%	±42.2%	±31.1%	±0.5%	±0.5%	±3.8%	±0.2%	±40.6%	±14.4%
I_{Bias1}	±50%	±0%	±16.9%	±11.8%	±2.8%	±7.4%	±9.3%	±64.8%	±47.2%
W_{M_E}	±30%	±0%	±21.5%	±2.3%	±0.5%	±1.6%	±2.2%	±46.9%	±11.4%
$W_{M_{0e}}$	±30%	±0%	±3%	±7.7%	±2.8%	±3.3%	±7.7%	±5.2%	±1.7%
W_{M_1}	±30%	±0%	±25.8%	±9.4%	±2.8%	±4.7%	±5.7%	±0%	±0%
W_{M_2}	±30%	±0%	±30.1%	±6.3%	±2.6%	±5.4%	±2.7%	±46%	±21%
W_{M_3}	±30%								
W_{M_4}	±30%	±0%	±29.9%	±7.2%	±2.3%	±4.1%	±5.4%	±28.1%	±30.5%
I_{Bias2}	±50%	±0%	±40%	±3.5%	±0.6%	±2.6%	±6%	±48.1%	±16.6%
W_{M_5},M_{f2}	±30%	±0%	±35.8%	±7.4%	±2.9%	±3.3%	±2.6%	±31.9%	±22.7%
C_f	±80%	±0%	±47.7%	±0%	±8.3%	±0.3%	±0.2%	±0%	±0%
R_f	±50%	±0%	±32.4%	±0%	±8.7%	±0.3%	±5.6%	±0%	±0%
C_{f2}	±80%	±0%	±13.2%	±0%	±0%	±0%	±0%	±0%	±0%

Table 3. Sensitivity analysis for the LDO3 shown in Fig. 3. The variations of the design variables (Var) are different and the maximum sensitivities produce different variations in the performances, such as: V_{DO}, V_{out}, power supply rejection PSR, open loop gain A_{OL}, phase margin (PM), line regulation LNR and load regulation LDR.

Var	Variation	V_{DO}	$\pm\Delta V_{out}$	PSR_{1kHz}	PSR_{10kHz}	A_{OL}	PM	LNR	LDR
W_{M_P}	±20%	±14.5%	±51.6%	±0%	±2.2%	±12.3%	±15.6%	±33.9%	±23.1%
I_{Bias}	±60%	±2.5%	±66.7%	±20.1%	±49.8%	±0%	±0%	±66%	±40.5%
$W_{M_{1-4}}$	±30%	±0%	±46.7%	±0.1%	±1.5%	±1.7%	±14.7%	±12.9%	±12.1%
$W_{M_{6-7}}$	±30%	±0%	±14.4%	±0.2%	±0.6%	±0.3%	±9.5%	±11.6%	±2.7%
W_{M_5},M_8	±30%	±0%	±11%	±0%	±5.3%	±0%	±5.6%	±2.6%	±0.3%
$W_{M_{9-10}}$	±30%	±0%	±23.3%	±0%	±0%	±0%	±10%	±0.9%	±0.3%
$W_{M_{a1-a2}}$	±30%	±0%	±46.6%	±8.2%	±5.2%	±0%	±0%	±6.7%	±0.4%
$W_{M_{a3-a4}}$	±30%	±0%	±13.3%	±23%	±17.8%	±0%	±0.6%	±26.7%	±0.2%
$W_{M_{a5}}$	±30%	±0.3%	±14.2%	±5.6%	±9%	±0%	±8.6%	±12.9%	±0.7%
C_1	±80%	±0%	±2.9%	±0%	±1%	±0%	±7.3%	±0%	±0%
R_{1-2}	±50%	±0.4%	±25.4%	±2.1%	±9.4%	±9.2%	±39.1%	±39.7%	±31.8%

5   LDO's Optimization by NSGA-II and PVT Analysis

The multi-objective optimization of the three LDOs shown in Fig. 3 was performed by NSGA-II with the following parameters: population size = 50 individuals, maximum number of generations = 100, crossover probability = 0.7, and mutation probability = 0.05. However, the values of the genes in (5) have different orders of magnitude, for example: the widths W and capacitors C are in micrometers μm (10^{-6}), and the resistors R near to kilo-ohms $k\Omega$ (10^3). In this case, the design variables are normalized to avoid the dispersion of the ranges of values to perform genetic operations within NSGA-II.

As already highlighted in [13], during the optimization process the evaluation of each individual (tentative solution) begins from the encoding of the design variable into an input text-file that is simulated within SPICE and contains the circuit description. The simulator requires analysis statements according to the target specifications being evaluated and at the end, SPICE generates an output text-file listing all simulation results, which are cleaned to extract the useful information and the conditions for the next generation (iteration) within NSGA-II.

Several applications of LDO voltage regulators require that they be connected as post-regulation stages, and in this case they must possess the highest power supply rejection (PSR) over a wide range of frequencies, as shown in [24]. At the same time, reducing the value of the compensation capacitor and the power consumption (efficiency) are a priori tasks in designing power management circuits [25]. From the issues mentioned above, the objectives being optimized in the three LDOs are the PSR, the external compensation capacitor value (C_O) and the quiescent current I_{Bias} also known as I_Q of the LDO voltage regulators.

NSGA-II provides a set of feasible solutions that are compared herein according to two figures of merit (FOM). The first one is given in (6) [26], and is widely used in the literature for comparing new topologies of LDO voltage regulators, which can be designed for different specifications of (I_Q), C_O, output voltage variations (ΔV_{out}), and maximum load current (I_L). The second FOM was proposed in [27], and it is generally used to evaluate the velocity of the voltage regulator in terms of the settling time (T_{settle}), current efficiency (I_Q/I_L), output capacitance value (C_O) and the chip area in terms of the compensation capacitor (C_C). In both FOM_1 and FOM_2, the smaller the value the better the LDO voltage regulator performance.

$$FOM_1 = \frac{\Delta V_{Out} C_O I_Q}{I_L^2} \tag{6}$$

$$FOM_2 = \frac{T_{settle} I_Q C_C}{I_L C_O} \tag{7}$$

5.1   Topology LDO1

The topology for the LDO1 is shown in Fig. 3(a), it is externally compensated with a Miller operational amplifier as error amplifier. This LDO was designed and optimized by using a 0.18 μm CMOS process in order to be capable of driving a load current up to 100 mA, while it is being connected to an input voltage between 2.2 V and 3.3 V. The objectives being optimized for this LDO1 topology are PSR and the output capacitor value (C_O).

The values for the different performance metrics (LDR, LNR, A_{OL}, ΔV_{Out} and t_{settle}) were normalized with respect to the values obtained by traditional design methods by evaluating (8), where the subscript I indicates the values of the solution found by NSGA-II, and the subscript N stands for the reference values obtained from the voltage regulator designed by traditional methods or manual design.

$$G = \left[\left(\frac{LDR_I}{LDR_N} \right) + \left(\frac{LNR_I}{LNR_N} \right) + \left(\frac{A_{OL,N}}{A_{OL,I}} \right) + \left(\frac{\Delta V_{Out,I}}{\Delta V_{Out,N}} \right) + \left(\frac{t_{set,I}}{t_{set,N}} \right) \right] \quad (8)$$

As for the case of the FOMs, the smaller G (representing the vector of constraints) the better the voltage regulator. The threshold value for G must be selected in order to increase the probabilities of finding solutions with at least one performance metric being better than those for the reference design. Among the values that are included into the FOMs, we choose PSR and the output capacitor value C_O to show the Pareto front in Fig. 4.

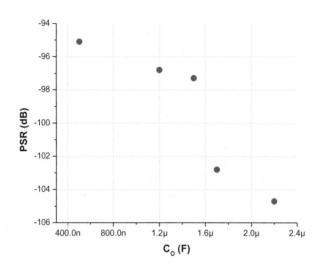

Fig. 4. Pareto front of the optimization of LDO1 from Fig. 3.

The characterization and comparison of the five solutions shown in the Pareto front with respect to the manual or human-made design is shown in Table 4, where it can be appreciated that although the PSR is better for almost all the feasible solutions provided by NSGA-II, also almost all the feasible solutions have a greater FOM. Nevertheless, the feasible solution 2 (Ind. 2) has the smaller FOM value, thus proving the better performance according to (6). Moreover, it has assigned the lowest C_O value, therefore it is considered the best size solution from Table 4.

Figure 5 shows the PSR response for the topology LDO1 under PVT variations (best, worse and typical cases). Also, Figs. 6 and 7 show the effects on LDR and LNR, respectively, for the regulator LDO1 under PVT variations.

Figure 8 shows the transient response for different process corners, temperature values and input voltages. According to the load transient analysis, the selected feasible solution 2 (Ind. 2) from Table 4, is stable even for the worst condition.

Table 4. Comparison between a manual or human-made design and the feasible solutions provided by NSGA-II for the LDO1 from Fig. 3.

Parameter	Trad. design	Ind. 1	Ind. 2	Ind. 3	Ind. 4	Ind. 5
Technology (μm)	0.18	0.18	0.18	0.18	0.18	0.18
V_{DO} (mV)	200	220	164	185	315	232
I_Q (μA)	65	62	109	185	88	128
LDR $(\mu V/mA)$	0.74	1	0.5	1.5	0.9	0.6
LNR $(\mu V/V)$						
$I_L = 0mA$	57.8	12	20	78	45	51
$I_L = 100mA$	44.5	435	13	19	38	72
PSR (dB)						
@$1kHz$	-95.3	-97.3	-95.1	-102.8	-96.8	-104.7
@$10kHz$	-90.4	-87.5	-74.9	-96.3	-82.5	-84.6
A_{OL} (dB)						
$I_L = 0mA$	87	95	105	83	89	92
$I_L = 100mA$	102	99	109	94	85	88
PM (Grades)						
$I_L = 0mA$	7	3	20	19	26	15
$I_L = 100mA$	65	11	24	18	39	29
ΔV_{Out} (mV)	15	40	186	19	37	23
T_{settle} (μS)	≈ 10	≈ 25	≈ 25	≈ 12	≈ 20	≈ 16
C_O (μF)	3	1.5	0.5	1.7	1.3	2.2
FOM_1 (ps)	293	372	142	597	324	647

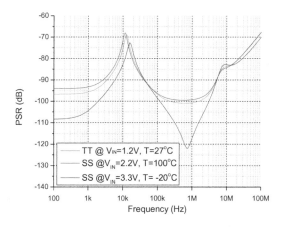

Fig. 5. PVT variations on the PSR of the feasible solution 2 (Ind. 2) from Table 4.

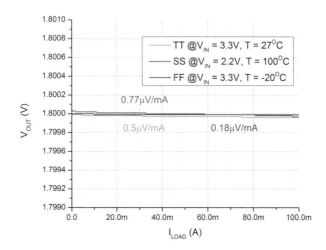

Fig. 6. PVT variations on the LDR of the feasible solution 2 (Ind. 2) from Table 4.

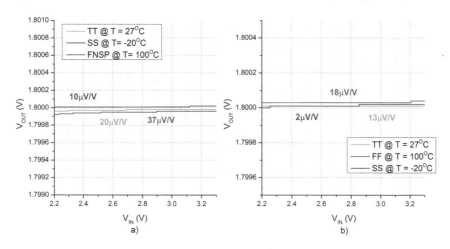

Fig. 7. PVT variations on the LNR under full load condition (a), and no load condition (b), of the feasible solution 2 (Ind. 2) from Table 4.

5.2   Topology LDO2

The LDO2 shown in Fig. 3(b), is a voltage regulator with a compensation scheme that provides a fast transient response and a full range AC stability [22]. This LDO provides a 2.8 V at its output without using an external compensation capacitor, and it is driven by a 3 V power supply. Its optimization is performed by NSGA-II by using a 0.35 μm CMOS technology. This LDO2 must be capable of driving a load current (I_L) as high as 50mA and an output capacitance (C_O) of 100pF.

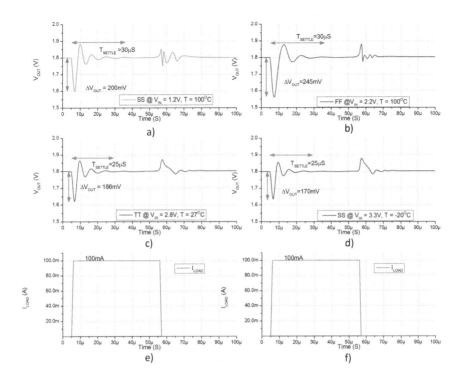

Fig. 8. PVT variations on the transient response of the feasible solution 2 (Ind. 2) from Table 4.

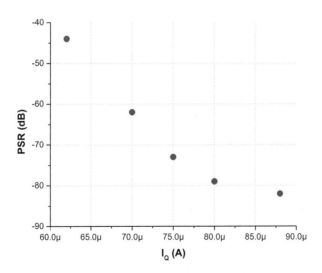

Fig. 9. Pareto front of the optimization of LDO2 from Fig. 3.

Table 5. Comparison between a manual or human-made design and the feasible solutions provided by NSGA-II for the LDO2 from Fig. 3.

Parameter	Trad. design [22]	Ind. 1	Ind. 2	Ind. 3	Ind. 4	Ind. 5
Technology (μm)	0.35	0.35	0.35	0.35	0.35	0.35
V_{DO} (mV)	200	365	177	146	274	132
I_Q (μA)	65	88	80	75	70	62
LDR (mV/mA)	0.62	1.2	0.21	0.83	0.51	0.43
LNR (mV/V)						
$I_L = 0mA$	23	0.9	0.16	3.77	0.91	6.9
$I_L = 50mA$	2.3	20.2	6.87	2.45	3.26	4.6
PSR (dB)						
@$1kHz$	−56	−82	−79	−73	−62	−44
@$10kHz$	−70	−72	−61	−60	−62	−43
GL (dB)						
$I_L = 0mA$	64	62	72	65	71	74
$I_L = 50mA$	56	30	54	8	37	45
PM (Grades)						
$I_L = 0mA$	>50	14	5	8	22	95
$I_L = 50mA$	80	98	108	103	86	98
ΔV_{Out} (mV)	90	200	460	348	270	33
T_{settle} (μS)	≈15	≈7	≈5	≈18	≈10	≈8
FOM_1 (ps)	0.23	0.7	1.47	1.04	0.75	0.08
FOM_2 (ns)	3.9	1.23	0.4	7.02	3.08	1.98

The Pareto front is shown in Fig. 9, where the trade-off between the two objective parameters can be clearly appreciated. Table 5 shows the comparison between a design taken from [22], and the five feasible solutions. From Table 5 one can see that NSGA-II provides different feasible solutions, from which the analog designer can choose the one more suitable for the desired application.

Based on the overall performance and the performance metric, the feasible solution 2 (Ind. 2) from Table 5, was selected in order to perform a PVT analysis and to prove the feasibility and robustness of the solutions provided by NSGA-II.

Table 6 lists the sizes associated to the circuit elements of the LDO2 topology. One column lists the sizes given in [22], and the last column lists the sizes provided by NSGA-II. One can appreciate that the MOS transistor's sizes are larger compared to [22]. However, the most important issue is that the pass transistor M_P and the compensation capacitor have smaller values after its optimization by NSGA-II, which leads to lower area consumption compared to manual design.

Figure 10 shows the simulated PSR frequency responses of the LDO2 at different process corners, temperature and input voltage values. The simulation

Table 6. Comparison between the values of the circuit elements provided by an analog designer and by NSGA-II for the LDO2.

Element	Traditional design from [22]	NSGA-II
M_E (λ)	$(100/10)$	$(120/4)$
M_{0e} (λ)	$(5/10)$	$(19/4)$
M_1 (λ)	$(10/10)$	$(80/4)$
M_2 (λ)	$(25/10)$	$(38/4)$
M_3 (λ)	$(14.5/10)$	$(202/4)$
M_4 (λ)	$(101.5/10)$	$(171/4)$
M_5, M_{f2} (λ)	$(15/2)$	$(15/2)$
M_6, M_{f1} (λ)	$(5/2)$	$(5/2)$
M_P (λ)	$(80000/2)$	$(52600/2)$
I_{Bias1} (μA)	5	11
I_{Bias2} (μA)	10	1
R_{FB1} $(k\Omega)$	156	156
R_{FB2} $(k\Omega)$	124	124
R_f $(k\Omega)$	200	237
C_f (pF)	20	5
C_{f2} (pF)	1	0.5

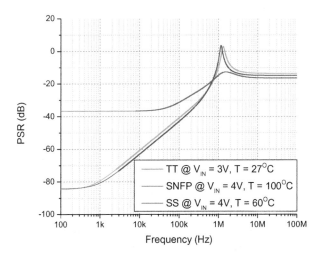

Fig. 10. PVT variations on the PSR of the feasible solution 2 (Ind. 2) from Table 6.

was performed under zero load condition. The green line is the typical case, while the red and blue line represent the worst case and the best case for the PVT variations, respectively. The effect of PVT variations on the load and line regulations are illustrated in Figs. 11 and 12, respectively, and Fig. 13 shows the simulated transient response. For a complete characterization, the cases higher and lower values of phase margin were added.

5.3 Topology LDO3

The LDO3 shown in Fig. 3(c) shows a low quiescent current (I_Q) internally compensated output capacitorless voltage regulator, with a current mode transconductance (CTA) error amplifier. The slew rate (SR) for the CTA is high in order to enhance the transient response at the gate of the pass device. This LDO3 topology was optimized by using a $0.18\,\mu m$ CMOS technology to supply a maximum load current of $100\,mA$, while driving an output capacitor of $100\,pF$, with

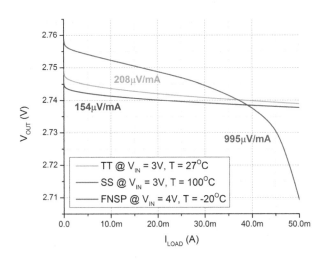

Fig. 11. PVT variations on the LDR of the feasible solution 2 (Ind. 2) from Table 6.

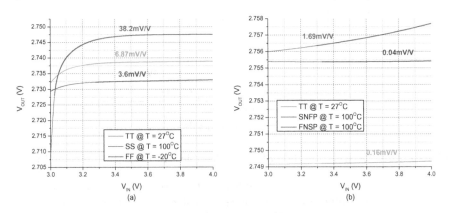

Fig. 12. PVT variations on the LNR under full load condition (a), and no load condition (b) of the feasible solution 2 (Ind. 2) from Table 6.

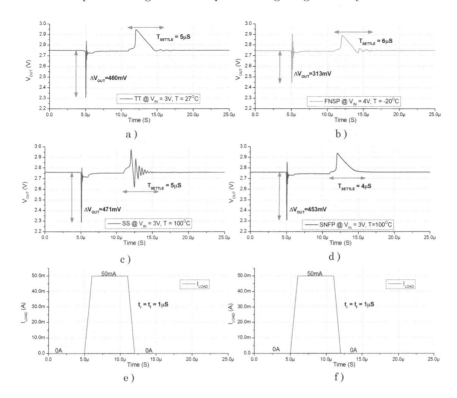

Fig. 13. PVT variations on the transient response of the feasible solution 2 (Ind. 2) from Table 6.

an input voltage between 1.2–2 V. The Pareto front is shown in Fig. 14, were one can see six feasible solutions, which are compared to [23], as shown in Table 7.

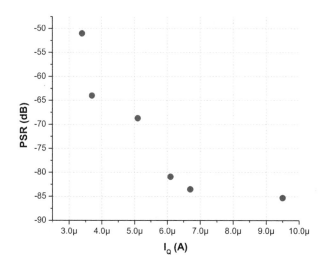

Fig. 14. Pareto front by optimizing the LDO3 voltage regulator shown in Fig. 3.

Table 7. Comparison between a human-made design and the feasible solutions found by NSGA-II for the LDO3 shown in Fig. 3.

Parameter	Trad. design [23]	Ind. 1	Ind. 2	Ind. 3	Ind. 4	Ind. 5	Ind. 6
Technology (μm)	0.18	0.18	0.18	0.18	0.18	0.18	0.18
V_{DO} (mV)	200	142	388	239	194	139	212
I_Q (μA)	3.7	9.5	6.7	6.1	5.1	3.7	3.4
LDR ($\mu V/mA$)	70	25.4	42	74	82	50	69
LNR (mV/V)							
$I_L = 0mA$	0.7	0.5	0.73	1	1.8	0.74	0.13
$I_L = 100mA$	1	0.87	1.27	2.3	4.2	1.48	0.8
PSR (dB)							
@1kHz	-65	-85.3	-83.5	-80.9	-68.7	-64	-51
@10kHz	-45	-72.1	-71.9	-60.9	-53.9	-49	-31
GL (dB)							
$I_L = 1\mu A$	68.6	75.3	78.8	65.3	63.2	67.2	64.9
$I_L = 100mA$	42.5	42.3	44.2	33.5	51.8	36.6	35.6
PM (Grados)							
$I_L = 1\mu A$	12	2	1	10	12	7	16
$I_L = 100mA$	88	80	32	77	88	90	91
ΔV_{Out} (mV)	277	88	226	138	199	126	374
T_{settle} (μS)	≈ 6	∞	∞	>20	>20	≈ 16	≈ 9
FOM (fs)	10.2	NA	NA	8.4	10.1	4.7	12.7

Table 8. Comparison between the elements values assigned by manual design and NSGA-II for the LDO3.

Element	Traditional Design [23]	GA
$M_1 - M_4$ (λ)	(55.55/5.55)	(45/6)
$M_5 - M_7$ (λ)	(44.44/3.33)	(85/4)
M_8 (λ)	(44.44/3.33)	(47/4)
M_9 (λ)	(222.22/5.55)	(270/6)
M_{10} (λ)	(222.22/5.55)	(148/6)
M_{a1}, M_{a2} (λ)	(11.11/3.33)	(43/4)
M_{a3}, M_{a4} (λ)	(5.55/3.33)	(53/4)
M_{a5} (λ)	(44.44/3.33)	(189/4)
M_P (λ)	(66666.66/2)	(37200/2)
$I_{Bias1} - I_{Bias3}$ (μA)	0.50	0.43
$R_1 - R_2$ $(k\Omega)$	66	140
C_1 (pF)	–	4
C_{Out} (pF)	100	100

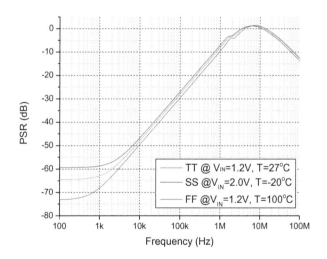

Fig. 15. PVT variations on the PSR of the feasible solution 3 (Ind. 3) from Table 8.

Table 8 shows the element values given in [23] and those values provided by NSGA-II. As for the previous LDO2, again most of the transistors' sizes found by NSGA-II are larger than the values provided by an analog designer. Nevertheless the size of the pass transistor M_P is much lower, thus leading to lower chip area consumption.

Figure 15 shows the PSR for LDO3 at $I_L = 1mA$ under PVT variations. It can be appreciated that the PSR exhibits larger variations below 10 kHz.

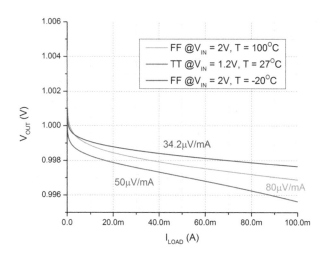

Fig. 16. PVT variations on the LDR of the feasible solution 5 from Table 8.

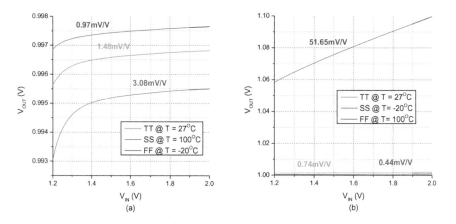

Fig. 17. PVT variations on the LNR under full load condition (a), and no load condition (b) of the feasible solution 5 from Table 8.

The effect of PVT variations on the load and line regulations are illustrated in Figs. 16 and 17, respectively.

Finally, Fig. 18 shows the transient response for different process corners, temperature values and input voltages. According to the load transient analysis, the feasible solution 5 is stable even for the worst condition.

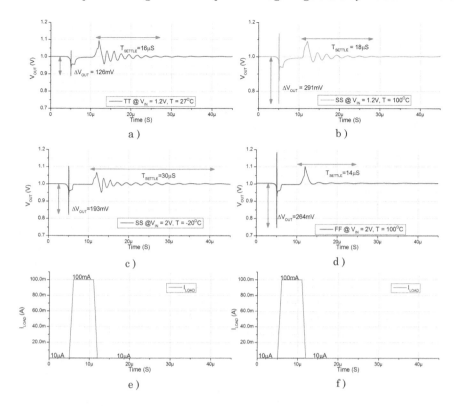

Fig. 18. PVT variations on the transient response of the feasible solution 5 from Table 8.

6 Conclusions

This Chapter summarized the design of operational amplifiers and their linearization to provide better performances in analog circuit design. It was highlighted that an LDO voltage regulator is implemented with one operational amplifier and external circuit elements. In this manner, three LDO topologies were the cases of study to show their optimization by applying the evolutionary algorithm known as NSGA-II. Before proceeding to the optimization process, a sensitivity analysis was performed for each LDO topology in order to identify the most sensitive circuit elements, which were used to create the chromosome. From the tolerances of the values for each LDO topology, a reduced number of design variables were included into the chromosome, and reduced search spaces were assigned to the design variables, after analyzing the sensitivity analysis results. For the three LDO topologies the optimization targets were PSR and the quiescent current (I_Q).

The results provided by NSGA-II were compared with the results obtained by traditional or manual design approaches. From the set of feasible solutions provided by NSGA-II, one solution was selected for each LDO topology to perform a PVT variation analysis, in order to guarantee that the feasible solution is robust to variations. According to the results that were compared with respect to two Figures of Merit, we conclude that NSGA-II is quite suitable for the optimization of LDO voltage regulators and that performing a sensitivity analysis helps to reduce the number of design variables, that are included into the chromosome. The statistical tests for more executions of the EA is the future work of this research.

Acknowledgements. This work is partially supported by CONACyT-Mexico under grant 237991.

References

1. Yang, Y., Zhu, H., Bi, Z., Yan, C., Zhou, D., Yangfeng, S., Zeng, X.: Smart-MSP: a self-adaptive multiple starting point optimization approach for analog circuit synthesis. IEEE Trans. Comput. Aided Des. Integr. Circuits Syst. **37**(3), 531–544 (2018)
2. Bhanja, M., Ray, B.: Synthesis of nonlinear analog functions. J. Circuits Syst. Comput. **27**(03), 1850040 (2018)
3. Dvorak, J., Langhammer, L., Jerabek, J., Koton, J., Sotner, R., Polak, J.: Synthesis and analysis of electronically adjustable fractional-order low-pass filter. J. Circuits Syst. Comput. **27**(02), 1850032 (2018)
4. Oliveira, V.A., Alzate, R., Bhattacharyya, S.P.: A measurement-based approach with accuracy evaluation and its applications to circuit analysis and synthesis. Int. J. Circuit Theory Appl. **45**(12), 1920–1941 (2017)
5. Bhanja, M., Ray, B.N.: Synthesis procedure of configurable building block-based linear and nonlinear analog circuits. IEEE Trans. Comput. Aided Des. Integr. Circuits Syst. **36**(12), 1940–1953 (2017)
6. Kourany, T., Ghoneima, M., Hegazi, E., Ismail, Y.: Passiot: a pareto-optimal multi-objective optimization approach for synthesis of analog circuits using sobol'indices-based engine. Integr. VLSI J. **58**, 9–21 (2017)
7. Duarte-Villaseñor, M.A., Tlelo-Cuautle, E., De la Fraga, L.G.: Binary genetic encoding for the synthesis of mixed-mode circuit topologies. Circuits, Syst. Signal Process. **31**(3), 849–863 (2012)
8. Singh, K., Jain, A., Mittal, A., Yadav, V., Singh, A.A., Jain, A.K., Gupta, M.: Optimum transistor sizing of CMOS logic circuits using logical effort theory and evolutionary algorithms. Integr. VLSI J. **60**, 25–38 (2018)
9. De, B.P., Maji, K.B., Kar, R., Mandal, D., Ghoshal, S.P.: Design of optimal cmos analog amplifier circuits using a hybrid evolutionary optimization technique. J. Circuits, Syst. Comput. **27**(02), 1850029 (2018)
10. Papadimitriou, A., Bucher, M.: Multi-objective low-noise amplifier optimization using analytical model and genetic computation. Circuits Syst. Signal Process. **36**(12), 4963–4993 (2017)
11. Maji, K.B., Kar, R., Mandal, D., Ghoshal, S.P.: An evolutionary approach based design automation of low power cmos two-stage comparator and folded cascode ota. AEU-Int. J. Electron. Commun. **70**(4), 398–408 (2016)
12. López-Arredondo, J., Tlelo-Cuautle, E., Fernández, F.V.: Optimization of IDO voltage regulators by nsga-ii. In: 2016 13th International Conference on Synthesis, Modeling, Analysis and Simulation Methods and Applications to Circuit Design (SMACD), pp. 1–4. IEEE (2016)

13. Sanabria-Borbón, A.C., Tlelo-Cuautle, E., de la Fraga, L.G.: Optimal sizing of amplifiers by evolutionary algorithms with integer encoding and g_m/i_d design method. In: NEO 2016, pp. 263–279. Springer (2018)
14. McConaghy, T., Breen, K., Dyck, J., Gupta, A.: Variation-aware design of custom integrated circuits: a hands-on field guide. Springer Science & Business Media (2012)
15. Lopez-Arredondo, J., Tlelo-Cuautle, E., Trejo-Guerra, R.: Optimizing an IDO voltage regulator by evolutionary algorithms considering tolerances of the circuit elements. In: 2015 16th Latin-American Test Symposium (LATS), pp. 1–5. IEEE (2015)
16. Nye, W., Riley, D.C., Sangiovanni-Vincentelli, A., Tits, A.L.: Delight. spice: an optimization-based system for the design of integrated circuits. IEEE Trans. Comput. Aided Des. Integr. Circuits Syst. **7**(4), 501–519 (1988)
17. Michalewicz, Z.: Evolutionary computation techniques for nonlinear programming problems. Int. Trans. Oper. Res. **1**(2), 223–240 (1994)
18. Allen, P.E., Holberg, D.R.: CMOS analog circuit design, New York. Holt, Rinehart, and Winston, (1987)
19. Silva-Martinez, J., Steyaert, M., Sansen, W.: High-performance CMOS continuous-time filters, vol. 223. Springer Science & Business Media (2013)
20. Pankiewicz, B., Szczepański, S., Wójcikowski, M.: Bulk linearized CMOS differential pair transconductor for continuous-time OTA-C filter design. Bull. Pol. Acad. Sci. Techn. Sci. **62**(1), 77–84 (2014)
21. Nauta, B.: Analog CMOS filters for very high frequencies, vol. 190. Springer Science & Business Media (2012)
22. Milliken, R.J., Silva-Martínez, J., Sánchez-Sinencio, E.: Full on-chip CMOS low-dropout voltage regulator. IEEE Trans. Circuits Syst. I: Regul. Pap. **54**(9), 1879–1890 (2007)
23. Fathipour, R., Saberkari, A., Martinez, H., Alarcón, E.: High slew rate current mode transconductance error amplifier for low quiescent current output-capacitorless cmos ldo regulator. Integr. VLSI J. **47**(2), 204–212 (2014)
24. Gupta, V., Rincon-Mora, G.A., Raha, P.: Analysis and design of monolithic, high PSR, linear regulators for SoC applications. In: Proceedings of the IEEE International SOC Conference, pp. 311–315, September 2004
25. Rincon-Mora, G.: Analog IC Design with Low-Dropout Regulators. McGraw-Hill, Inc. (2009)
26. Hazucha, P., Karnik, T., Bloechel, B.A., Parsons, C., Finan, D., Borkar, S.: Area-efficient linear regulator with ultra-fast load regulation. IEEE J. Solid-State Circuits **40**(4), 933–940 (2005)
27. Giustolisi, G., Palumbo, G., Spitale, E.: Robust miller compensation with current amplifiers applied to ldo voltage regulators. IEEE Trans. Circuits Syst. I: Regu. Pap. **59**(9), 1880–1893 (2012)

Genetic Optimization of Fuzzy Systems for the Classification of Treated Water Quality

Itzel G. Gaytan-Reyes, Nohe R. Cazarez-Castro[✉],
Selene L. Cardenas-Maciel, David A. Lara-Ochoa,
and Armando Martinez-Graciliano

Tecnológico Nacional de México-Instituto Tecnológico de Tijuana, Calz. del
Tecnológico S/N, Fracc. Tomás Aquino, 22414 Tijuana, Baja California, México
{itzel.gaytan,aamg_a}@tectijuana.edu.mx, {nohe,lilettecardenas}@ieee.org,
dr.davidlara@gmail.com

Abstract. The water quality problem is a world-wide challenge for human development. Traditionally, since the 1960's water quality evaluation has been promoted through the application of various quality or pollution indexes. However, these indexes present excessive divergence and little complementarity that have not allowed to advance toward a unified system of globalized application due to the quantity, the range and the complexity of parameters. We propose a Takagi-Sugeno type fuzzy system to classify and make decisions about treated water reuse for direct or indirect contact activities. The fuzzy system design is based on expert knowledge, and considers guidelines established in mexican and international standards about pollution and water quality indexes. The coefficients in some rule consequents of the fuzzy system were determined by solving an optimization problem through a genetic algorithm. The classification performance of the fuzzy system was verified using real data of a water treatment plant through a leave-one-out cross validation and with an analysis of variance for the pollution indexes assessments.

Keywords: Wastewater · Quality and pollution indexes
Fuzzy Logic · Genetic algorithm

1 Introduction

The water quality is a relevant issue in the world. Scientist, non-profits and government organizations of many countries have studied and alerted about water quality deterioration and its potential negative impacts in the ecosystems. Factors affecting the water quality are due to natural processes and human actions. The increasing water demand to support agriculture, industry, and human living activities along with inappropriate procedures for sewage management (since waste water is poured into water bodies such as lakes, rivers or oceans) are the predominant factors that increase abrupt changes in water composition and

© Springer International Publishing AG, part of Springer Nature 2019
L. Trujillo et al. (Eds.): NEO 2017, SCI 785, pp. 248–260, 2019.
https://doi.org/10.1007/978-3-319-96104-0_13

eventually induce contamination. As a result, many initiatives and projects have been launched with the objective to properly manage fresh water resources, these including sensitization and training about sustainable use. In addition, topics of sanitation and monitoring of water are taking relevance, several efforts have been done to implement sanitation-purification procedures of waste water through treatment plants and methodologies for water quality assessment.

The problem of water quality and its assessment have achieved attention since the 1960's. The water assessment is determined by comparing the physical-chemical features of a water sample with guidelines based on toxicity levels scientifically acceptable for humans and another organisms. Several indexes of quality and pollution have been developed around the world with the purpose of improving water valuations. The methods of water quality assessment based on indexes offer multiple advantages such as methodological simplicity, rapidity of results, easy interpretation and a retrospective of pollution events. Nevertheless, the excessive divergence and little complementarity between indexes have not allowed to advance toward a unified system of globalized application that integrates the appropriate variables, ranges, criteria and standards [7].

In this paper a Takagi-Sugeno type unified intelligent system is presented as proposal for the still open problem in hand, that allows an evaluation of water quality in accordance with mexican and international regulations for treated water. The fuzzy system called Local Treated Water Classifier (CATL, from Spanish - Clasificador de Agua Tratada Local -) was designed to realize water quality assessments and make decisions about treated water reuse for direct or indirect contact activities. The consequents rules in the CATL are modeled as linear functions in terms of variables which represents pollution indexes and a genetic algorithm was used to adjust the coefficients of these functions to minimize a mean square error criterion. Our design was verified with real data measurements corresponding to the effluent of treatment plant which belongs to State Public Services Commission of Tijuana (CESPT, from Spanish - Comisión Estatal de Servicios Públicos de Tijuana), local government agency responsible for water treatment. A statistical study was carried out to validate the effectiveness of the assessments.

The rest of the chapter is organized as follows. The previous results that motivated this work are mentioned in Sect. 2. Section 3 explains in detail the architecture of the proposed model to classify the effluent water of a treatment plant and how it was designed. Section 4 shows the response of our classifier tested with real data, the statistical study using leav-one-out cross validation and hypothesis test performed with the pollution indexes valuations of the fuzzy system in comparison with the mathematical formulas. Finally, the conclusion is stated in Sect. 5.

2   Theoretical Framework

Often the pollution of water bodies comes from waste discharges of domestic or industrial origin. The pollutant load in domestic waste is composed by high percentages of organic matter and microorganisms. The determination of basic bacteriological, organic and chemical parameters to identify water pollution require a series of analyzes, and these parameters are used to evaluate the quality of water bodies and sewage discharge [11] through a collection of formulas called indexes.

Today the formulas to calculate water-pollution indexes use one parameter and even more than thirty, which can be grouped into different categories such as: contamination by organic and inorganic matter, eutrophication, health aspects, suspended and dissolved substances, oxygen level, physicochemical characteristics and dissolved substances. The implementation of new methodologies for assessment of water quality involving more than two parameters became increasingly important because they allow the reduction of information to a simple expression. These indexes are known as water quality indexes (WQI) and water pollution indexes (WPI), which encompass several parameters mostly physicochemical and in some cases microbiological [10].

According to [12], monitoring results must allow for the resolution of different types of problems, for example water usage and the integrity of aquatic ecosystems, which imply socioeconomic aspects. WQI and WPI are important tools because their correct use allows for the evaluation of water-resources management programs.

Pioneers in generating a unified methodology for calculating the WQI were [10,13]. However, these were only used and accepted by water quality monitoring companies in the 1970's when the WQI became more important in the evaluation of water. The general index of water quality was developed by [14] and improved by Deininger for the National Academy of Sciences of the United States in 1975 [15]. With these studies, the Scottish Department for Development (SSD), in collaboration with regional institutions for water preservation, carried out extensive research to evaluate the quality of water in scottish rivers [10]. The water quality assessments are made through indexes of different type because of their lack of unification in [15,18].

The AMOEBA methodology [12] (A General Method of Ecological and Biological Assessment), developed by the Netherlands, uses physico-chemical and biological parameters that allow ecological and biological assessment of aquatic systems. Its development was carried out by the Dutch Ministry of Transport, Public Works and Water Resources Management considering the agricultural production and yield, sustainable species diversity and sustainable regulations.

Another important work was done by the Regional Corporation of Valle del Cauca (CVC) and Universidad del Valle, which developed the quality index for the Cauca river called ICAUCA in the Cauca River (Cauca) mathematical modeling and characterization project, the index uses Biochemical Oxygen Demand from 5 days (BOD5), total nitrogen, total phosphorus, total solids, total suspended solids and fecal coliforms [17]. In addition, some of these parameters also are used to build a unified index to assess water of the Pamplonita River [7].

On the other hand, Fuzzy Logic (FL) has been booming since the 1990's in this topic due to the need to find better techniques for the management of natural resources, such as water quantity and quality aspects [7]. Regarding water assessments, FL has been proven to contribute with its synthetic evaluation [5], to realize the trophic evaluation of reservoir waters through a fuzzy index model [2], as well as comparative assessments with other classical indicators [3], Artificial Neural Networks (ANN) have been used to solve a water assessment problem, which results in a huge computational expense and low effectiveness according to [7].

3   Methodological Aspects

Considering that water quality assessment is based on experts knowledge and information interpretation, and since FL [20] is an effective technique to represent imprecise values and make human-like inferences, a system based on FL is proposed to assess effluent water from a treatment plant.

The system is devised to perform a classification task via pollution indexes and a quality index, where the last index makes the water quality assessment. For the fuzzy inference system (FIS) design for classification of treated water it was necessary to consult experts in the area of toxicology, chemical reactors, statisticians and consequently programming experts for graphic representations of the results. The methodology used for the design is described in more detail below.

3.1   Selection of Water Quality Indexes

In accordance with the AMOEBA strategy the characterization of domestic water is done mainly by concentrated organic matter, where the amount of oxygen available in the water is a relevant variable, therefore, the organic pollution index (OPI) was selected. Also another important characteristics cited in [18,19] for water evaluations are turbidity and conductivity, for these reasons, the Suspended Solids Index (ICOSUS) and the Conductivity Indexes (IC) were chosen.

Table 1. Permissible pollution limits

Parameter	Permissible limit
BOD (mg/L)	20
DOC (mg/L)	40
Conductivity ($\mu S/cm$)	3500
Total suspended solids (mg/L)	20

The waste water treatment plant must supply water guaranteeing that it will not pose any risk to the community during indirect or direct usage. Direct water usage means water contact through activities such as filling ponds and recreational artificial channels with boat rides, rowing, canoeing, among others; otherwise, indirect usage stands for water contact through activities such as irrigation of green areas (roundabouts, ridges, gardens, recreational centers, parks, sports fields, ornamental fountains) and for industrial uses (services for washing yards and industrial premises, washing vehicle fleet, toilets, heat exchangers, boilers, water curtains, etc.) [4,8]. Additionally, from the pollutants of domestic water exposed in the AMOEBA strategy and the indexes mentioned above, 4 parameters were selected: Biochemical Oxygen Demand (BOD), Chemical Oxygen Demand (COD), Conductivity and Total Suspended Solids (TSS). The measurement limits for each of them are shown in Table 1.

3.2   Architectural Design

Based on the information contained in mexican norm NOM 003 SEMARNAT [21][1], formulas and interpretations of the indexes provided in [6,12,16,19], and expert recommendations, the overall fuzzy system to classify treated water is shown in Fig. 1. The fuzzy model takes measurements of BOD, COD, conductivity and TSS parameters as input and processes them by means of three FIS to determine values for WPI's (OPI, CI and ICOSUS) which are passed by a fourth FIS providing as final output the water assessment.

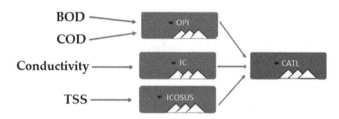

Fig. 1. Architecture of the water classifier model.

3.3   Fuzzification and Knowledgement Base

Fuzzfication consists of transforming input values of each parameter (BOD, COD, conductivity, TSS) into linguistic tags characterized by membership functions. The knowledge base is structured with fuzzy rules defined by expert knowledge. According to the class of rules, the classifier whose architecture are shown in Fig. 1 has two MISO-type FIS (Multiple Input Single Output) and two SISO-type FIS (Single Input Single Output) named as follows: locally treated water classifier (CATL acronym of Clasificador de Agua Tratada Local, in spanish), OPI for the Organic Pollution Index, for Conductivity Index CI and for the Suspended Solids Index ICOSUS, respectively.

---

[1] Norm that indicates pollution thresholds of treated-waste water for reuse in public services, it was published on *Diario Oficial de la Federacion* in 1998.

The literature offers formulas to calculate pollution indexes. The knowledge provided in [1,6,12,16] was incorporated for characterization of membership functions and to derive the rule base of the OPI, CI and ICOSUS fuzzy systems. The formulas (1)–(3) represent the usual way which the OPI, CI and ICOSUS indexes are calculated. The OPI is computed with

$$OPI = e^{\sum_{i=1}^{n} ln(PDQ_i) \cdot w_i}, \tag{1}$$

where PQD is the quality index for the i-th parameter, a dimensionless number between 0 and 100, which are extracted from functions (curves showed in Fig. 2, for example), W_i is the weighting factor for each parameter, $W_i = \frac{1}{n}$, here the factor is $W_i = 0.5$ because only BOD and COD parameters are available. In case of unregistered values, the weight must be calculated according to the number of missing parameters.

$$IC = 10^{2+0.45 \cdot log_{10} C}, \tag{2}$$

where C is the conductivity measured in $\mu S/cm$.

$$ICOSUS = \frac{1}{30TSS}, \tag{3}$$

where TSS is the total suspended solids measured in mg/l.

For the OPI FIS, the linguistic tags for the input parameters are named of the form "dXLow" and "dXHigh", the X character is replaced by numbers 1 or 2 to make a difference between first and second parameter tags, the membership functions and universe of discourse are presented in Fig. 3. The linguistic tags for the CI FIS and ICOSUS FIS are shown in Figs. 4 and 5, respectively.

The rule base design of each FIS takes the format of Takagi-Sugeno-type Fuzzy Inference System (T-S FIS) [9], this means that rule antecedent use linguistic tags (defined through membership functions) and rule consequent is a linear function of the inputs. With the knowledge in hand about quasi-linear or piecewise linear behavior of the formulas (1)–(3), the rule consequents of OPI, IC

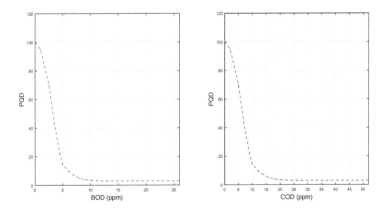

Fig. 2. Typical behavior of BOD and COD parameters.

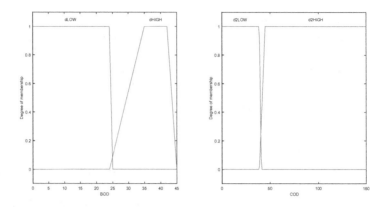

Fig. 3. Definition of linguistic tags for inputs of OPI FIS: BOD parameter (left) COD parameter (right).

and ICOSUS fuzzy systems were designed such that they replicate each piecewise linear function. Examples of rules for the OPI, ICOSUS and CI fuzzy systems are respectively

- If BOD is $d1LOW$ and COD is $d2HIGH$ then OPI is $a_1 \times d1LOW + b_1 \times d2HIGH + c_1$.
- If SST $s1$ then ICOSUS is $a_2 \times s1 + b_2$.
- If Conductivity is $c1$ then IC is $a_3 \times c1 + b_3$.

On the other hand, the CATL FIS takes as inputs organic pollution, conductivity and suspended solids indexes, all membership functions are defined in $[0, 1]$ as universe of discourse. The rule format is:

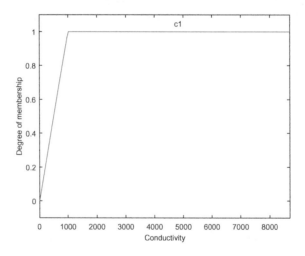

Fig. 4. Definition of linguistic tag for input of CI FIS.

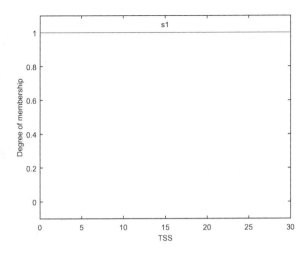

Fig. 5. Definition of linguistic tag for input of ICOSUS FIS.

- If OPI is x_1 and CI is x_2 and ICOSUS is x_3 then Quality is $A \times x_1 + B \times x_2 + C \times x_3$.

The rule consequent coefficients are obtained by a genetic algorithm, this is shown in the next section.

The output of the overall fuzzy system is calculated with [23]

$$y_0 = \frac{\sum_{i=1}^{m} MV_i \cdot h_i}{\sum_{i=1}^{m} h_i}, \qquad (4)$$

where h_i is the maximum membership in the domain of the input variable for each value MV_i, that is the input.

3.4 Fuzzy System Design by Genetic Optimization

The usual form to assess water quality is through a formula based on a weighted linear sum (WLS) involving different indexes or parameters, for example [12] proposed a water quality index (WQI) defined as

$$WQI = 0.22DOI + 0.19BOD + 0.16COD + 0.15AI + 0.16SSI + 0.12pHI \quad (5)$$

where DOI is Disolved Oxigen Index, BOD is Biochemical Oxigen Demand, COD is Chemical Oxigen Demand, AI is Ammoniac Index and pHI is de pH Index and SSI is the Suspended Solids Index.

Considering our first approach to classify treated water through a Mamdani-type FIS (Mamdani FIS) which was reported in [22], the next step of this work is motivated for the creation of a WQI through of a T-S FIS since the rule consequents and its output calculation is similar to a WLS. A genetic algorithm (GA) was used to find the coefficients in the rule consequent of the CATL FIS such that the assessments will be almost equal to Mamdani FIS assessments. In order to reach this requirement a minimization problem is formulated, whose objective function is defined by

$$f(e) = \frac{1}{m} \sum_{i=1}^{m} e_i^2,$$ (6)

where the error of each water evaluation realized by Mamdani FIS and T-S FIS, (using the same inputs) is represented as $e_i = WQI_i^{Mamdani} - WQI_i^{T-S}$.

The WQI proposed through CATL FIS has one rule with consequent of the form $A \times OPI + B \times CI + C \times ICOSUS$, this means that only three coefficients must be determined. Therefore, each individual in the population consists of three real numbers, the search domain for each coefficient is bounded.

The executions of the GA was performed using its implementation in MATLAB® 8.5 with the configuration and parameters as shown in Table 2.

Table 2. GA configuration and its parameters.

Population	Fitness scaling	Number of variables	Selection	Lower limits and upper limits
Population size:100 Type: double vector Initial scores, range and population: default	Rank	3	Stochastic uniform	$[0.40, -0.2, 0.1]$ and $[0.60, -0.14, 0.27]$
Reproduction	Mutation and crossover	Migration	Constraint parameters	Stop criteria
Elite count: 0.05× population size crossover fraction: 0.8	Constraint dependents	Foward, fraction: 0.2, interval: 20	Nonlinear constraint algorithm: Augmented Lagrangian, Initial penalty: 10, Penalty factor: 100	Generations: 150 time limit: inf., fitness limit: inf, stall time limit: inf., stall test: average change, function tolerance: 1×10^{-6} and constraint tolerance: 1×10^{-3}

4   Results

The overall fuzzy system T-S FIS described in Sect. 3 was tested with data of an operating treatment plant managed by CESPT. The data set shown in Fig. 6 corresponds to tuples of DBO, DQO, Conductivity and TSS measurements collected from treated water throughout thirty one days between the months of December of 2015 and February of 2016.

The T-S FIS use the rule-consequent coefficients of the best individual obtained after thirty executions of the GA. The water assessments are showed by a graphical representation. The plots use a colorimetric scale based on the Purple Project [8], whose color meaning is described in Table 3.

The results from Mamdani FIS [22] and T-S FIS are shown in the Fig. 7(a) and (b), respectively. The Mamdani FIS evaluations shows water with good conditions for reuse only in two of the thirty one days, meanwhile the T-S FIS only in nine days indicates water with marginally acceptable conditions for reuse. The water evaluations correspond to water with characteristics mostly not suitable for reuse, because in several days COD values exceeds the limit of 40 mg/L which indicates that water is contaminated according with Prati index [18]. It is important to mention that COD is not specified in the mexican norm NOM 003 SEMARNAT [21], but in our design it is considered.

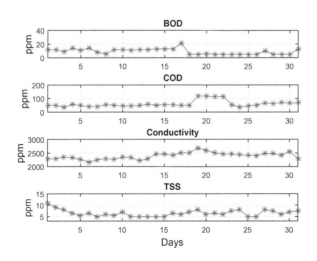

Fig. 6. Raw data from thirty one days.

The statistical hypothesis test was performed with the purpose to evaluate through variance analysis at 95% confidence level, if exist the same performance between OPI, CI and ICOSUS FIS (which determine pollution indexes) and the indexes calculated by formulas (1)–(3). The statistical report presented in Table 4 allows to support the hypothesis that there is no significant difference between the values of the pollution indexes obtained through fuzzy inference

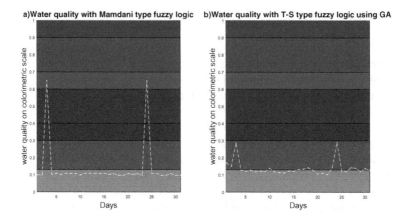

Fig. 7. Graphic representation of treated water assessment with: (a) Mamdani FIS, (b) T-S FIS.

Table 3. Description of water-quality colorimetric scale.

	Gray	Grayish purple	Purple	Turquiose	Blue
Description	Poor Conditions to reuse	Marginally acceptable conditions to reuse	Acceptable conditions to Mexican standars	Good conditions to reuse	Excellent conditions to reuse
Range	0-0.13	0.13-0.3	0.3-0.6	0.6-0.9	0.9-1.0

Table 4. Results of the statistical F for pollution indexes.

Index	Mean square	F	Probability	Critical value for F
OPI	0.0848	2.2214	0.14135	4.0012
CI	1.210E-07	0.00112	0.97335	4.0011
ICOSUS	0.00031	0.1389	0.710682415	4.0011

systems and those calculated using the traditional scheme based on formulas. This provides evidence that fuzzy systems valuations concerning to pollution indexes resembling the results calculated with usual form, i.e., T-S FIS makes an appropriate valuation for water pollution indexes.

To evaluate the classifier a leave-one-out cross validation (LOOCV) was carried out and a confusion matrix is presented in Table 5, the results of LOOCV show that the classifier reach 93.54% of accuracy and 6.45% of misclassification rate.

Table 5. Confusion matrix using LOOCV.

						Total
Poor conditions	27	0	0	0	0	27
Marginally acceptable conditions	1	2	1	0	0	4
Marginally acceptable conditions to Mexican norms	0	0	0	0	0	0
Good conditions	0	0	0	0	0	0
Excellent conditions	0	0	0	0	0	0
Total	28	2	1	0	0	31

5 Conclusion

A Takagi- Sugeno type fuzzy system was designed to evaluate effluent water from a treatment plant. The design of the rule base knowledge was carried out by modeling pollution indexes using mexican - international standards, accepted formulas and expert knowledge. Also, fuzzy rules representing water quality index were modeled through a genetic algorithm in order to determine the consequent coefficients such that its assessments approximated those of a by Mamdani-type fuzzy system.

The architecture of the fuzzy model showed good performance to support decision making in the assessment of water quality, since the statistical hypothesis validates that pollution indexes calculations with fuzzy systems are close to results done with traditional calculations and the results of cross validation show a good response of the fuzzy system related with correctly classification in the reuse scale of treated water.

The monitoring of effluent water with the proposed fuzzy system showed low reuse conditions because the COD values went above the upper limit of $40\,\mathrm{mg/L}$. This means that our system is stricter due the mexican standard NOM 003-SEMARNAT which does not considers COD, which explains the reasons of institutional declaration that treated water has good conditions for reuse. This should motivate more research and investments to get a deeper insigh of the problem and make proposals to adapt the actual purple project implementation.

Aknowledgments. The authors thank to CESPT for providing data for this research. This research was partially funded by project number 6351.17-P and "Estabilización orbital de sistemas mecánicos - Parte II: Estudio de aspectos de estabilidad"? from "Tecnológico Nacional de México".

References

1. Mitchell, M.K., Stapp, W.B., Beebe, A.: Field manual for water quality monitoring: an environmental education program for schools (1994)
2. Liou, Y.-T., Lo, S.-L.: A fuzzy index model for trophic status evaluation of reservoir waters. Water Res. **39**(7), 1415–1423 (2005)

3. Lioua, C.Y.-T., Lob, S.-L.: A fuzzy index model for trophic tatus evaluation of reservoir waters. Water Res. **39**, 1415–1423 (2005)
4. Silvert, W.: Ecological impact classification with fuzzy sets. Ecol. Model. **96**(1), 1–10 (1997)
5. Melcher, D., Matthies, M.: Application of fuzzy clustering to data dealing with phytotoxicity. Ecol. Model. **85**(1), 41–49 (1996)
6. Silvert, W.: Fuzzy indices of environmental conditions. Ecol. Model. **130**(1), 111–119 (2000)
7. Fernández, N., Carvajal, L., Colina, E.: Sistema difuso tipo Mamdani para la determinación genérica de la calidad del agua. Bistua: Revista de la Facultad de Ciencias Básicas, vol. 8(1) (2010)
8. Navarro-Chaparro, K., Rivera, P., Sánchez, R.: Análisis del manejo de agua en la ciudad de Tijuana, Baja California: Factores críticos y retos. Estudios Fronterizos **17**(33), 53–82 (2016)
9. Jang, J., Sun, C., Mizutani, E.: Neuro-Fuzzy and Soft Computing. Prentice Hall, Upper Saddle River (1997)
10. Liebman, H.: Atlas of Water Quality: Methods and Practical Conditions. R. Oldenbourgh, Munich (1969)
11. Ramalho, R.S., Beltran, D.J., de Lora, F.: Tratamiento de aguas residuales, Reverté (1990)
12. Samboni Ruiz, N.E., Carvajal Escobar, Y., Escobar, J.C.: A review of physical-chemical parameters as water quality and contamination indicators. Ingeniería e Investigación **27**(3), 172–181 (2007)
13. Horton, R.K.: An index number system for rating water quality. J. Water Pollut. Control Fed. **37**(3), 300–306 (1965)
14. Brown, R.M., McClelland, N.I., Deininger, R.A., Tozer, R.G.: A water quality index- do we dare (1970)
15. Cude, C. G.: Oregon water quality index a tool for evaluating water quality management effectiveness. J. Am. Water Resour. Assoc. **37**, 125–137 (2001)
16. Fernández, N., Solano, F.: Índices de calidad y de contaminacion del agua. Universidad de Pamplona, pp. 43–53 (2005)
17. Callejas, C.R.: Caracterización y modelación matemática del río Cauca. Revista Ingeniería y Competitividad **4**(1), 7–18 (2011)
18. Liou, S.-M., Lo, S.-L., Wang, S.-H.: A generalized water quality index for Taiwan. Environ. Monit. Assess. **96**(1–3), 35–52 (2004)
19. Ramírez, A., Restrepo, R., Viña, G.: Cuatro índices de contaminación para caracterización de aguas continentales. Formulaciones y aplicación. Ciencia, Tecnología y Futuro **1**(3), 135-153 (1997)
20. Zadeh, L.A.: Fuzzy Sets, Fuzzy Logic, and Fuzzy Systems: Selected Papers by Lofti A. Zadeh. World Scientific (1996)
21. Mexicana, D.N.O.: Normas Oficiales Mexicanas: NOM-003-SEMARNAT-1997. http://www.conagua.gob.mx/CONAGUA07/Publicaciones/Publicaciones/SGAA-15-13.pdf
22. Gaytan, I.-G., Cazarez-Castro, N.-R., Lara, D., Cardenas-Maciel, S.-L.: Sistema difuso tipo Mamdani para la determinación de calidad del agua tratada de origen doméstico. Avances recientes en ciencias computacionales CiComp 2016, pp. 206–213 (2016)
23. Cordon, O., Herrera, F., Hoffmann, F., Magdalena, L.: Genetic fuzzy systems: Evolutionary tuning and learning of fuzzy knowledge bases. World Scientific (2001)

Stabilization Based on Fuzzy System for Structures Affected by External Disturbances

Marco A. Alcaraz-Rodriguez, Nohe R. Cazarez-Castro$^{(\boxtimes)}$,
Selene L. Cardenas-Maciel, Luis N. Coria, and Sergio Contreras-Hernandez

Tecnológico Nacional de Mexico - Instituto Tecnológico de Tijuana,
22414 Tijuana, BC, Mexico
{nohe,lilettecardenas}@ieee.org,
{marco.alcaraz,luis.coria,sergio.contreras}@tectijuana.edu.mx

Abstract. Vertical structures such as buildings, bridges and trusses are subjected to strong changes with respect to their original state of design due to different external disturbances. Seismic waves are leading sources of disturbances giving rise to lateral displacements, as well as vertical and angular deformations, which increase the risk of structural failure compromising the structure integrity. In order to prevent catastrophic failures in structures and its subsequent side effects, in this work we propose a scheme for the attenuation of the vibration effects in vertical structures by means of a control system based on a Mamdani-type fuzzy inference system. The fuzzy rules of the controller were designed such that the close loop system is guaranteed to satisfy the Lyapunov stability criterion. Numerical simulations were performed to evaluate the best performance and effectiveness of fuzzy control, considering that the controller is placed at different levels of the building and inducing as perturbations signals that approximate a seismic event. The results show that the proposed controller attenuates the vibration in the structure accomplishing the control objective.

Keywords: Structures · Fuzzy system · Stabilization

1 Introduction

The protection of civil structures and their occupants is undoubtedly a priority. Minimizing damage in buildings caused by external perturbations such as earthquakes is very important to maintain urban functions. The area where most seismic energy is released is known as the circumpolar belt. Baja California is considered as one of the places that stands out for its seismic activity [1]. Accordingly, it is necessary to implement strategies for safety improvements in buildings to avoid risk along seismic events. There are passive insulation systems in civil structures, but these are ineffective if a large magnitude earthquake

© Springer International Publishing AG, part of Springer Nature 2019
L. Trujillo et al. (Eds.): NEO 2017, SCI 785, pp. 261–274, 2019.
https://doi.org/10.1007/978-3-319-96104-0_14

occurs. Recently, significant progress has been made to make the active structural control technology for active isolation of vibration as a useful strategy to improve structural functionality and security against natural hazards [2].

There is an active community which studies the modeling of vertical structures and develops methodologies for isolation of vibrations through control techniques and material technologies. In [3], the authors proposed a parametric identification for a building of three floors to scale is presented. Besides authors of [4] proposed a vector parameterizations that allows to obtain measurements about the floors and the ground acceleration. In [5], the authors analyze an adaptive observer which estimates parameters and states of a building excited by an earthquake. In [6], the authors propose an estimation of parameters for a building model, this is relevant for damage detection, life cycle, design, maintenance of the building and for the design of a vibration control. They authors employ Gershgorin circles and properties of triangular matrices establishing conditions for a stable model of a building during the process of identification.

On the other hand, in [7], the authors provide an overview of modeling and control of buildings structures, focused on different types of devices and control strategies. In [8], the authors obtain a model for a structure of n levels structure using the D'Alembert principle and a PID-fuzzy driver to counteract vibrations is proposed. In [9], the authors deal with the problem of robust reliable energy-to-peak controller design for seismic-excited buildings with actuator faults and parameter uncertainties, it is assumed that uncertainties mainly exist in damping and stiffness of the buildings because they are difficult to be precisely measured. To control the vibrations in a building under the excitation induced by an earthquake, authors of [10] proposes a variable structure controller and a fuzzy sliding controller. Regarding experimental studies in active vibration control, in [11] the authors report an active vibration suppression through positive acceleration feedback on a building-like structure.

This chapter considers a mathematical model that represents a structure of n levels. The proposed control technique is based on fuzzy logic. This control technique generally depends on the empirical knowledge of the process or the plant, from which a base of linguistic rules of the type IF-THEN is formulated. One of its advantages is that it can successfully model nonlinear functions and facilitates the constructions of an inference system [12]. The system stability is analyzed following the Lyapunov stability criterion from which are determined conditions to construct the fuzzy linguistic rule base for the fuzzy control system. The synthesis of the controller is determined by analyzing conditions that demonstrate that a Lyapunov candidate function can guarantee stability in that sense.

The rest of the chapter is organized as follows: Sect. 2 describes the problem and the control objective. Section 3 presents the methodology to design a fuzzy controller based on the Lyapunov stability criterion. Section 4 presents simulation results and the conclusions are presented in Sect. 5.

2   Problem Statement

Structural control mainly concerns with the protection of buildings against vertical and horizontal displacements produced by external disturbances. The vibrations produced by an earthquake cause horizontal, vertical and rotational deflec-
tions in the structure and could severely damage the building. It is necessary to
determine an internal force that allows to attenuate or reduce the vibrations in
the structure due to the excitation induced by a disturbance.

2.1   Mathematical Model of a Structure

Consider a n level building, where its structure is shaped by a set of intercon-
nected bars and joined by articulated knots, forming frames in the building. The
model of a structure of n levels is based on the representation of the building as
a multiple mass-spring-damper system as shown in Fig. 1 [8]. The mathematical
model for each level is given by

$$m_i \ddot{q}_i(t) + b_{m_i} \dot{q}_i(t) + k_i q_i(t) = m_i \ddot{q}_g(t), \tag{1}$$

where q_i is the angular displacement; $\dot{q}_i$ is the angular velocity; $\ddot{q}_i$ is the corresponding acceleration; m_i is the mass of each of the levels of the structure;
b_{m_i} represents damping; k_i represents rigidity and $\ddot{q}_g$ is the acceleration of the
medium (soil) caused by an external disturbance (an earthquake for example), it
can be considered with the same magnitude for each level or through a function
of energy dissipation. The full model for a structure of five levels is presented
in (2), where the dynamics of the structure depends on interactions between
adjacent floors.

$$
\begin{aligned}
&m_1 \ddot{q}_1(t) + (b_{m1} + b_{m2})\dot{q}_1(t) + (k_1 + k_2)q_1(t) - b_{m2}\dot{q}_2(t) - k_2 q_2(t) = m_1 \ddot{q}_g(t) \\
&m_2 \ddot{q}_2(t) + (b_{m2} + b_{m3})\dot{q}_2(t) + (k_2 + k_3)q_2(t) - b_{m2}\dot{q}_1(t) - k_2 q_1(t) - b_{m3}\dot{q}_3(t) - k_3 q_3(t) = m_2 \ddot{q}_g(t) \\
&m_3 \ddot{q}_3(t) + (b_{m3} + b_{m4})\dot{q}_3(t) + (k_3 + k_4)q_3(t) - b_{m3}\dot{q}_2(t) - k_3 q_2(t) - b_{m4}\dot{q}_4(t) - k_4 q_4(t) = m_3 \ddot{q}_g(t) \\
&m_4 \ddot{q}_4(t) + (b_{m4} + b_{m5})\dot{q}_4(t) + (k_4 + k_5)q_4(t) - b_{m4}\dot{q}_3(t) - k_4 q_3(t) - b_{m5}\dot{q}_5(t) - k_5 q_5(t) = m_4 \ddot{q}_g(t) \\
&m_5 \ddot{q}_5(t) + (b_{m5})\dot{q}_5(t) + (k_5)q_5(t) - b_{m5}\dot{q}_4(t) - k_5 q_4(t) = m_5 \ddot{q}_g(t)
\end{aligned}
\tag{2}
$$

2.2   Control Objective

The model (1) is considered without control input. Introducing the control signal
τ in model (1), the model can be written as:

$$m_i \ddot{q}_i(t) + b_{m_i} \dot{q}_i(t) + k_i q_i(t) = m_i \ddot{q}_{i_g}(t) + \tau. \tag{3}$$

Defining q_d as the desired position, then the error can be defined as

$$x_1 = q - q_d, \tag{4}$$

and

$$x_2 = \dot{x}_1 = \dot{q}. \tag{5}$$

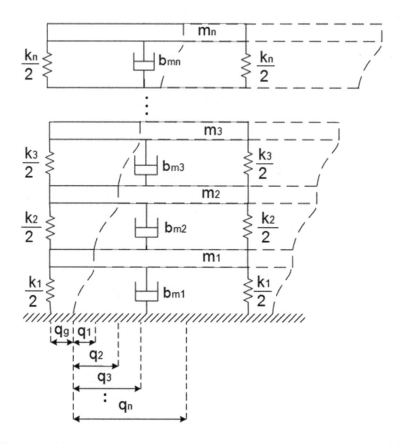

Fig. 1. Building representation as a multiple mass-spring-damper systems [8].

The problem consists in designing a fuzzy control law τ for a n level building with the mathematical model is given in (2) such that it attenuates the horizontal deflection induced by external disturbances, i.e., reduces the displacements in the building produced by seismic vibrations. This is defined through the control objective

$$\lim_{t \to \infty} \left\| (x_1, x_2)^T \right\| = 0. \tag{6}$$

where $x_1 \in \mathbb{R}$ is the angular error and $x_2 \in \mathbb{R}$ is the angular error derivative.

3 Fuzzy Controller Synthesis

Lyapunov theory is very important in the stability analysis of linear and non-linear dynamic systems, regardless of the order of the system it allows obtaining information about the stability of the equilibrium point of the system without solving the differential equation that describes the dynamical system. This theory offers an accurate characterization through functions (V) that are qualifying as energy functions in the vicinity of the equilibrium point and the main notion about the stability properties of the system is based on the energy function (named Lyapunov candidate function) decreasing along the trajectories of the dynamic system, i. e., the temporal derivative of the energy (power of the system) must be negative definite ($\dot{V} < 0$) or negative semi-definite ($\dot{V} \leq 0$) until reaching a state of equilibrium [13].

To define the control objective, a novel fuzzy control law is proposed that leans elements from [14], where the authors design is based on Lyapunov's theory to guarantee the stability of the closed loop system. The proposed Lyapunov candidate function is

$$V(x_1, x_2) = \frac{1}{2}x_1^2 + \frac{1}{2}x_2^2, \tag{7}$$

this is a positive definite and radially unbounded function. The time derivative of the Lyapunov candidate function results in

$$\dot{V}(x_1, x_2) = x_1\dot{x}_1 + x_2\dot{x}_2 = x_1x_2 + x_2\dot{x}_2. \tag{8}$$

Considering that for a second-order system the second derivative is directly related to the form of the input signal, for the system (3) should be $\dot{x}_2$ and it is proportional to $-\tau$, so it is possible to change the $\dot{x}_2$ by $-\tau$ at (8) resulting on

$$\dot{V}(x_1, x_2) = x_1x_2 + x_2(-\tau). \tag{9}$$

Now, to guarantee stability of the equilibrium point it is necessary to satisfy $\dot{V} \leq 0$. On the other hand it is possible to guarantee asymptotic stability looking to satisfy $\dot{V} < 0$, however, the algorithm is more complex computationally. Therefore, more conditions are required on x_1 and x_2 in (9) to guarantee $\dot{V} < 0$, implying an increase of the number of fuzzy rules and greater number of calculations for inference results, which decrease performance or even precludes an implementations to carry out experiments in real time. For such reasons, it is decided a fuzzy controller design such that

$$\dot{V}(x_1, x_2) = x_1x_2 + x_2(-\tau) \leq 0. \tag{10}$$

By inspection, testing the signs for each state variable schematically can be deduced with the qualitative state of the system for each situation, and hence to find conditions to satisfy (10). Table 1 represents the position error x_1 and the derivative of the error x_2, in column three we can see τ represents the control signal in the closed loop system. This signal is generated through the fuzzy inference system.

From Table 1 the rule-base for the fuzzy inference system (fuzzy controller) can be derived as follows:

1. IF x_1 is Negative and x_2 is Negative, Then τ is Large Negative.
2. IF x_1 is Negative and x_2 is Positive, Then τ is Zero.
3. IF x_1 is Positive and x_2 is Negative, Then τ is Zero.
4. IF x_1 is Positive and x_2 is Positive, Then τ is Large Positive.

Table 1. Conditions to satisfy (10)

x_1	x_2	τ
Negative	Negative	Negative such that $\dot{V} = (x_1, x_2) \leq 0$
Negative	Positive	Zero
Negative	Positive	Zero
Positive	Positive	Positive such that $\dot{V} = (x_1, x_2) \leq 0$

The type and form of the membership functions have been obtained heuristically, and nonlinear functions are selected over lineal ones to obtain a soft response. Both inputs for the fuzzy controller x_1 and x_2 are considered as two linguistic labels called *Negative* and *Positive*. For this particular case the sigmoidal membership functions x_{neg}, x_{pos} are selected, respectively defined

$$x_{Negative} = \frac{1}{1 + (\exp +1.45x)} \tag{11}$$

and

$$x_{Positive} = \frac{1}{1 + (\exp -1.45x)}, \tag{12}$$

representing the sign that variables x_1 and x_2 can take. The representation of the membership functions are shown in the Fig. 2.

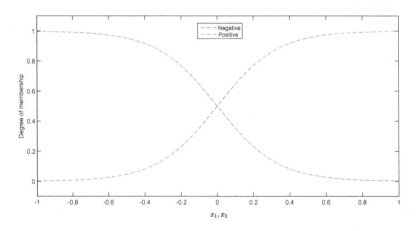

Fig. 2. Membership functions for the inputs x_1 and x_2.

The domain of inputs and outputs are normalized to the range $[-1,1]$, this is done with the purpose of easy adapt the fuzzy system to each particular structure. This means that the fuzzy system will not undergo any change in the ranges of input and output, and the necessary changes can be made by

modifying factors of the input or output. Gaussian membership functions are chosen to describe the output of the fuzzy controller which is granulated in three linguistic labels defined respectively as:
Negative large,

$$\tau_{Large-negative} = \exp[-(x-5)^2], \tag{13}$$

Zero, defined as

$$\tau_{Zero} = \exp[-(x)^2], \tag{14}$$

and *Large positive,* defined as

$$\tau_{Large-positive} = \exp[-(x+5)^2]. \tag{15}$$

In the Fig. 3 shows a visual representation of the output membership functions.

4   Results

To verify that the fuzzy control law designed through the methodology described in Sect. 3 can carry out the control objective, a numerical study of the closed-loop system has been performed. The 5-level building modeled by Eq. 2 is used for testing by applying three perturbation signals which represent different seismic events. At the beginning of each simulation it is considered that the building is in a position of equilibrium, there is no horizontal deflection, this means, the angular error and error derivative in each level are $x_1(0) = 0$ and $x_2(0) = 0$, respectively. The model parameters used are shown in Table 2 and are taken from [8].

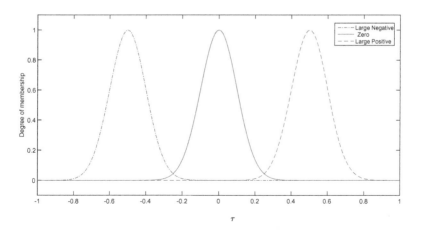

Fig. 3. Membership functions of the output τ

Table 2. The parameters of the building model

level [i]	$m_i[kg]$	$b_{m_i}[\frac{n}{m}]$	$k_i[\frac{N_s}{m}]$
1	6800	33,732,000	67,000
2	5897	33,732,000	67,000
3	5897	29,093,000	58,000
4	5897	28,621,000	57,000
5	5897	24,954,000	50,000

4.1 Study Case 1

A signal defined as a pulse is injected in each level of the building at the beginning of simulation. In the first level the signal amplitude corresponds to a 3 Richter scale degrees earthquake. The amplitude values of the signals applied to higher levels in the building decrease in linear form, for the second level the amplitude is 2.4, for the third level it is 1.8, at the fourth level it is 1.2 and for the fifth level it is 0.6.

Figure 4 shows the behavior of the angular error in each level of the building when no controller is operating. The trajectories represent that building has a horizontal deflection, if it does not exceed a predetermined threshold the civil structure is going to eventually attenuate the vibrations, but this implies possible structural failure caused by persistent and sustained deformations. On the other hand, Fig. 5 presents the angular error in each level considering the same perturbations and a control input applied at each level. The error trajectories decreasing in amplitude demonstrates that the building returns to the state of equilibrium at around 7 s.

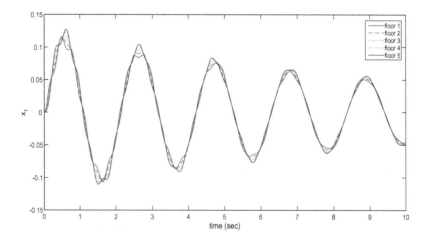

Fig. 4. Behavior of angular errors x_1 for each floor considering Case 1: system without control system.

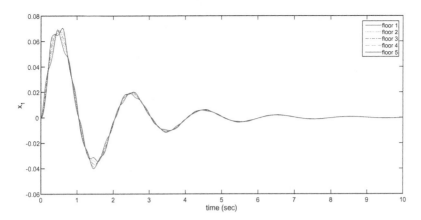

Fig. 5. Behavior of angular errors x_1 for each floor considering Case 1: system with fuzzy control system.

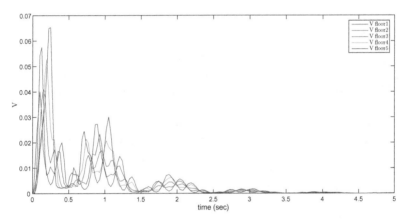

Fig. 6. Lyapunov candidate function V trajectories for each floor considering Case 1.

Figures 6 and 7 show the evolution of the Lyapunov candidate function V and its derivative $\dot{V}$ along the closed-loop system simulation. The function V shows to be positive definite and $\dot{V}$ satisfies the necessary design condition (10) to guarantee that the system is stable in the Lyapunov sense.

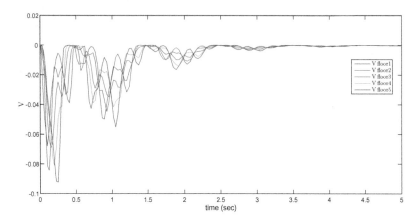

Fig. 7. Lyapunov candidate function derivative $\dot{V}$ trajectories for each floor considering Case 1.

4.2   Study Case 2

In this case six pulses of 3 Richter scale degrees are applied to the building. The first impulse happens in the first second, subsequent impulses are applied after one second. The angular errors without using the fuzzy controllers and using a fuzzy controller at each level of the building are shown in Figs. 8 and 9, respectively. The trajectories of the figures allow us to conclude that the fuzzy controllers attenuate the destructive effects of vibration because the maximum amplitude of the error signals does no exceeds the interval $[-0.02, 0.04]$ radians. Instead, without the controller the amplitude of the error signals imply that the building could be severely damaged or destroyed.

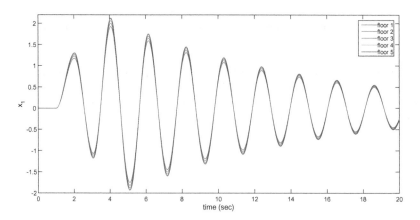

Fig. 8. Behavior of angular errors x_1 for each floor considering Case 2: system without a control system.

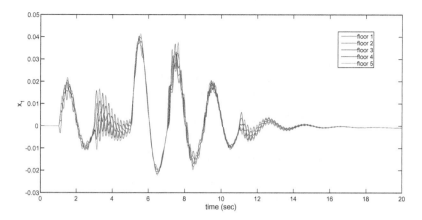

Fig. 9. Behavior of angular errors x_1 for each floor considering Case 2: system with fuzzy control system.

4.3   Study Case 3

A random signal is considered as perturbation and it is injected to the building. In this case the signal is continuous in all time and a control input is applied at all levels. Figure 10 shows that the amplitude of angular errors trajectories have a damping behavior and the maximum deflection does no exceeds the 0.3 radians, which will be acceptable under certain considerations. Figure 11 verifies that the stability in the Lyapunov sense is guaranteed since $\dot{V}$ is semi-negative definite and satisfies the conditions given in (10) and V showed in Fig. 12 is positive definite along the simulation.

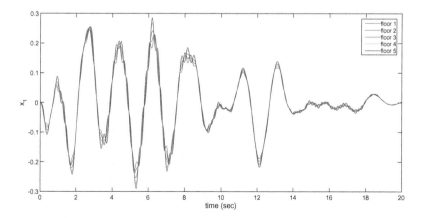

Fig. 10. Behavior of x_1 for each floor considering Case 3: the system with a random function as disturbance with control fuzzy.

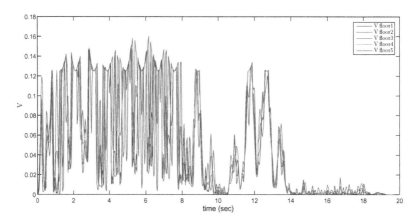

Fig. 11. Lyapunov candidate function V trajectories for each floor considering Case 3.

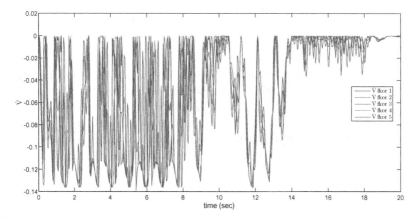

Fig. 12. Lyapunov candidate function $\dot{V}$ trajectories for each floor considering Case 3.

5   Conclusions

1. It can be verified that fuzzy systems are a good option to solve this type of complex problem, the solution presented here is is simple and effective.
2. The results obtained verify that the methodology used for the fuzzy control synthesis based on the stability criterion of Lyapunov for system type (3) is favorable. They show that the disturbance is efficiently damped when they have a control system at each level of the building in the closed loop system in comparison when it is in its natural state. Another important factor to consider is the time in which it manages to dissipate the energy within the system since for a case of this type if the disturbance persists for a long time in the system it is critical and the deformations can be irreparable and catastrophic for the structure.

3. Establishing necessary and sufficient conditions on the convergence of solutions via the Lyapunov stability theory guarantees that the results are satisfactory and generalizable.

5.1   Future Work

1. A neural network will be used to reproduce a real world seismic vibration behaviour of greater magnitude to test the robustness of the control system under high impact perturbation. From the data obtained of these simulations, the goal is to obtain structural deformation diagrams which include final moments and shears of design, this will verify whether the proposed control system has a physical influence on the building.
2. A prototype of a building with specific characteristics will be designed and instrumented in order to perform experiments with the fuzzy control system.

Acknowledgements. This research was partially funded by project number 6351.17-P and "Estabilización orbital de sistemas mecánicos - Parte II: Estudio de aspectos de estabilidad" from "Tecnológico Nacional de México".

References

1. Bazán, E., Meli, R.: Diseño Sismico de Edificio. Limusa (2004)
2. Forrai, A., Hashimoto, S., Funato, H., Kamiyama, K.: Structural control technology: system identification and control of flexible structures. Comput. Control Eng. J. **12**(6), 257–262 (2001)
3. Angeles, J.M., Alvarez-Icaza, L.: Identificacion paramétrica de un edificio con falla estructural durante la excitación sísmica. In: Congreso Nacional de Control Automático (2015)
4. Concha, A., Alvarez-Icaza, L.: Identificación de edificios acoplados torsionalmente usando una parametrización vectorial y filtros integrales lineales. In: Congreso Nacional de Control Automático (2015)
5. Concha, A., Alvarez-Icaza, L., Garrido, R.: Observador adaptable para edificios basado en proyeccion paramétrica.Memorias del XVI Congreso Latinoamericano de Control Automático (2017)
6. García-Illescas, M., Alvarez-Icaza, L.: Identificación en línea de modelos estables de edificios en 3d. In: Congreso Nacional de Control Automático (2015)
7. Thenozhi, S., Yu, W.: Advances in modeling and vibration control of building structures. Ann. Rev. Control **37**(2), 346–364 (2013)
8. González-Padilla, M.: Modelado y control difuso de una estructura de edificio sometida a las vibraciones de un temblor. Master's thesis, Centro de Investigación y de Estudios Avanzados del Instituto Politécnico Nacional, Unidad Zacatenco, 1 (2012)
9. Zhang, W., Chen, Y., Gao, H.: Energy-to-peak control for seismic-excited buildings with actuator faults and parameter uncertainties. J. Sound Vibr. **330**(4), 581–602 (2011)
10. Wang, A.-P., Lin, Y.-H.: Vibration control of a tall building subjected to earthquake excitation. J. Sound Vibr. **299**(4), 757–773 (2007)

11. Enríquez-Zàrate, J., Silva-Navarro, G., Abundis-Fong, H.: Active vibration suppression through positive acceleration feedback on a building-like structure: an experimental study. Mech. Syst. Signal Process. **72–73**, 451–461 (2016)
12. Salas, F.G., Juárez, R.: Controlador de seguimiento p-pi difuso auto-organizable aplicado a un robot paralelo. In: Congreso Nacional de Control Automático (2015)
13. Reyes, F.: Robótica-Control de robots manipuladores. Alfaomega Grupo Editor (2011)
14. Cazarez-Castro, N.R., Aguilar, L.T., Cardenas-Maciel, S.L., Goribar-Jimenez, C.A., Odreman-Vera, M.: Diseño de un controlador difuso mediante la síntesis difusa de lyapunov para la estabilización de un péndulo de rueda inercial. Rev. Iberoam. Autom. Inform. Ind. RIAI **14**(2), 133–140 (2017)

Comparison of Two Methods for I/Q Imbalance Compensation Applied in RF Power Amplifiers

S. A. Juárez-Cázares[1], E. Allende-Chávez[2], Y. Sandoval-Ibarra[2],
J. R. Cárdenas-Valdez[2], E. Tlelo-Cuautle[3], and J. C. Nuñez-Pérez[1(✉)]

[1] CITEDI, Instituto Politécnico Nacional, Tijuana, B.C., Mexico
sjuarez@citedi.mx, jnunez@ipn.mx
[2] Instituto Tecnológico de Tijuana,
Tecnológico Nacional de México, Tijuana, B.C., Mexico
edgar.allende@tectijuana.edu.mx, jumasaniba@gmail.com
[3] Instituto Nacional de Astrofísica, Óptica y Electrónica, Tonatzintla, Puebla, Mexico
etlelo@inaoep.mx

Abstract. In this paper, the design and implementation of two methods for I/Q imbalance compensation is presented, based on a low cost phase measurement approach. The design methodology for an I/Q imbalance correction system is presented based on a DSP-FPGA board. The first method employs some trigonometric properties. The second employs Volterra series to model the non-linear behavior of the I/Q imbalance. The system performance is verified using a complex signal with phase and amplitude imbalance. The implemented systems have the advantage of having low implementation cost and a high design flexibility, which allows for future revisions or enhancement. The Stratix III FPGA board from Altera is employed for the practical implementation and results verification of the system. A comparison between methods is introduced for correcting (I or Q) branches respectively to guarantee amplitude and phase balancing condition in the modulator output. Experimental results are implemented employing an FPGA by using DSP-Builder to bit true VHDL hardware description of proposed model. This work can be considered as a low cost alternative for I/Q imbalance correction given that it does not require additional measurement equipment nor uses complex algorithms.

Keywords: Compensation · DSP/FPGA · I/Q imbalance · LSE
Power Amplifier · Volterra series

1 Introduction

The unstoppable growth in the telecommunications field demands for complex digital modulation schemes capable of a high accuracy. The progress in the manufacture process allowed high density integrated circuits (IC), thus making

© Springer International Publishing AG, part of Springer Nature 2019
L. Trujillo et al. (Eds.): NEO 2017, SCI 785, pp. 275–294, 2019.
https://doi.org/10.1007/978-3-319-96104-0_15

possible the development of high performance digital systems in the form of Application Specific Integrated Circuits (ASIC), System on Chip (SoC), Field Programmable Gate Arrays (FPGA), among others [1]. In this context, modern FPGA devices have advantages in the field of hardware implementation and deployment, making it an attractive alternative which allows for an extensive design verification thus avoiding undependable design implementations.

The digital modulation systems available in the market, must have a low spurious production, otherwise errors in the signal spectrum will appear both in transmitter and receiver causing a high bit error rate (BER). These flaws increase in high bandwidth communications systems [2] and in complex modulation scenarios I/Q imbalance occurs. The I/Q imbalance is produced when there is a difference in amplitude and phase between channels I and Q. Several works are presented in the fields of measurement and compensation of the I/Q imbalance [2].

The method described in [3] presents a methodology to compensate the quadrature employing an envelope detector in the down-converter. However, it requires a previous estimation of the imbalance and employs an iterative algorithm which has a high computing cost. In [4] a simple algorithm is presented which cancels spurious produced by the local oscillator (LO). Nevertheless, it requires a spectrum analyzer which is expensive. In [5] a methodology for QAM modulator is introduced. Where an additional block in the feedback chain is introduced to compensate the I/Q imbalance.

Summarizing, the current tendency is to compensate the I/Q imbalance employing digital systems [6–9]. Following, FPGA devices become an attractive solution for digital system development. They also allow for a low cost debug, when compared to application specific tools and equipment [10].

The Power Amplifier (PA) is a crucial component for the signal transmission to be received without any difficulty or distortion. Nonlinearity is an intrinsic property for PAs by inheritance this generates three main consequences: spectral regrowth, memory effects and intermodulation products generation, resulting in adjacent channel interference and poor efficiency. Different algorithms are reported to represent behavioral modeling with or without memory. For example polynomial model [11], neural network [12], among other models based on Volterra series [13,14]. For this reason, the I/Q imbalance behavior can also be modeled in the same way. Thus, in this paper the comparison of two I/Q imbalance correction schemes is presented the first is based on a trigonometrical approach and the second is based in Volterra series, both are emulated in an FPGA device.

This work is organized as follows: In Sect. 2 the I/Q imbalance fundamentals are explained. The Sect. 3 describes both correction methodologies and implementation. Section 4 presents the obtained results. Finally, in Sect. 5 conclusions are given.

2   I/Q Imbalance Fundamentals

In a modern communications systems [15], both analog and digital domains are employed in both transmitter and receiver stages. This system is known as transceiver and is depicted in Fig. 1. However, drawbacks over the interest signal are caused by the several nonlinearity factors causing mismatching at the received signals. A proposed solution to correct these unwanted effects consists in identify the multiple transfer stages between RF transmit and receive chains.

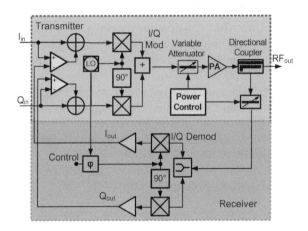

Fig. 1. Transceiver block diagram

In Fig. 2 the sources of errors present in I/Q transmitter are displayed, these stages play a key role in the system performance. Otherwise, they produce I/Q mismatch were phase and amplitude imbalances degrade the transmitted signal.

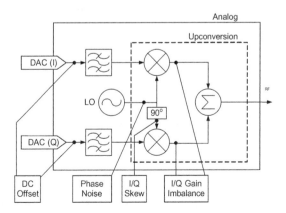

Fig. 2. Common sources of error in I/Q transmitter

These drawbacks are measured by means of DC offset, phase noise, quadrature skew and gain imbalance. These sources cause amplitude and phase imbalances in the baseband signal (in-phase and quadrature), propagating to the transmitted RF signal.

2.1   Phase Measurement

One of the most important parameters to be measured in any communications system is phase. Unfortunately, a Vector Network Analyzer is required, and such equipment is expensive. However, a low cost alternative for phase measurement is possible. In I/Q transmitters the quadrature conversion causes an unbalanced phase reflected in the signal amplitude. This effect should be taken into account when measuring the gain mismatch in the I/Q transmission using simple amplitude measurements by a power detector.

This technique reduces hardware requirements to a digital phase detector, developing in FPGA capabilities for measurement and compensation I/Q imbalance.

In this manner, the gain imbalance can be easily measured in I/Q transmission using a power detector, which is located generally in any communications system in the output of the power amplifier (PA). Thus, let $f(\theta, \phi)$ which is the input signal to the amplifier results in Eq. (1).

$$f(\theta, \phi) = I_t cos(\omega t) - Q_t sin(\omega t), \tag{1}$$

by using

$$I_t = cos(\omega_b t), \quad Q_t = sin(\omega_b t + \phi), \tag{2}$$

where

ϕ - represents the phase imbalance in rad
ω_b - is the message frequency to be transmitted in rad/s
ω - is the carrier frequency
θ - is the phase imbalance between baseband signals

Replacing Equation (2) in (1) results in:

$$f(\theta, \phi) = cos(\omega_b t)cos(\omega t) - sin(\omega_b t)sin(\omega t + \phi), \tag{3}$$

obtaining

$$f(\theta, \phi) = cos(\omega t)[cos(\omega_b t) - sin(\omega_b t + \phi)sin(\omega t)]$$
$$-sin(\omega t)[sin(\omega_b t + \phi)]cos(\omega t), \tag{4}$$

It must be noted in Eq. (4) the envelope has the I/Q imbalance information.

The phase transformation to the amplitude is an inherently property of the I/Q architecture. The direct approach in this paper is to use a power detector, so θ must be yield by the sensitivity at output power which is maximized for any small change in ϕ. Assuming that I_t and Q_t have the necessary amplitude and the PA drives into the linear region, the envelope detector equation can be written as follows in Eq. (5).

$$E(\theta, \phi) = envelope[G_{PA} \times f(\theta, \phi)], \tag{5}$$

$$E(\theta, \phi) = G_{PA}\sqrt{[cos(\omega_b t)cos(\omega t) - sin(\omega_b t)sin(\omega t + \phi)]^2 + [cos(\phi)sin(\omega_b t + \phi)]^2}, \tag{6}$$

Applying the quadratics terms into the root results in Eq. (7).

$$E(\theta, \phi) = G_{PA}\sqrt{[cos(\omega_b t)]^2 - 2sin(\omega_b t + \theta)sin(\phi)cos(\omega_b t) + [sin(\omega_b t + \phi)]^2}, \tag{7}$$

To compute the sensibility $S(\theta, \phi)$ is necessary to differentiate $E^2(\theta, \phi)$, respect to ϕ.

$$S(\theta, \phi) = \frac{d}{d\phi}E^2(\theta, \phi), \tag{8}$$

$$S(\theta, \phi) = -2G_{PA}^2 sin(\omega_b t + \phi)cos(\omega_b t)cos(\phi), \tag{9}$$

From Eq. (9) it can be demonstrated that power sensitivity reaches a maximum when $sin(\omega_b t + \phi)$ and $cos(\omega_b t)$ are equals or opposite, that happens when $\theta = 90^o$ or $\theta = 270^o$. Therefore, the sensitivity it is maximized to any output power and ϕ changes. The calculation can be evaluated when $I_t = cos(\omega_b t) = -Q_t$, for positives ϕ.

$$Tx_{out}(\phi) = G_{PA} \times cos(\omega_b t)[cos(\omega t) - cos(90^o + \omega t + \phi)], \tag{10}$$

Applying the trigonometric identity for the cosines remainder is obtained:

$$Tx_{out}(\phi) = G_{PA} \times cos(\omega_b t)\left[2sin\left(\omega t + 45^o + \frac{\phi}{2}\right)sin\left(45^o + \frac{\phi}{2}\right)\right], \tag{11}$$

To obtain the signal peaks the module is computed as shown in Eq. (5).

$$|Tx_{out}(\phi)| = 2G_{PA} \times K \times sin\left(45^o + \frac{\phi}{2}\right), \tag{12}$$

In Eq. (8) is shown that the signal peaks are directly related to phase imbalance of ϕ. In practical cases the proportionality constant depends not only of the amplifier gain also influences the mismatch impedance and the gain or attenuation of other circuits located in the transmitter.

Denoting the constant K in the Eq. (12) can be rewritten as:

$$|Tx_{out}(\phi)| = M \times sin\left(45^o + \frac{\phi}{2}\right), \tag{13}$$

where $M = 2GPA \times K$. Two steps are proposed to correct I/Q imbalance, in the first step gain imbalance between I and Q channels is detected and added to the Q channel. After adjusting the gain imbalance the phase imbalance is calculated as indicated in step 2.

Step 1:

1. $I_t = Acos(\omega_b t)$ and $Q_t = 0$ with $P = P_I$.
2. $Q_t = 0$ and $Q_t = Acos(\omega_b t)$ with $P = P_Q$.

Where P_I, P_Q represent the output power detector when it is active respectively I and Q. The mismatch gain is computed as:

$$G_{decouple} = \sqrt{\frac{P_I}{P_Q}}, \tag{14}$$

and K is expressed as:

$$K = \sqrt{\frac{2P_I}{G_{PA}^2}}, \tag{15}$$

Step 2:

By applying $I_t = Acos(\omega_b t)$, $Q_t = -G_{decouple} Acos(\omega_b t)$ to ensure that the signals are added with the same amplitude and therefore the phase to amplitude conversion shown in Eq. (12) is maintained. From the Eq. (12) can be determined ϕ:

$$\phi = 2 \left[arcsin \left(\frac{|Tx_{out}(\phi)|}{|M|} \right) - 45° \right], \tag{16}$$

3   I/Q Imbalance Correction

3.1   Method 1: The Trigonometrical Approach

This section introduces the model employed to compensate the I/Q phase imbalance. In Fig. 3 the proposed model is shown.

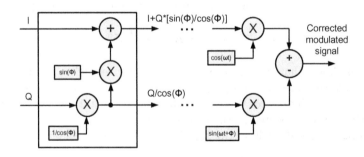

Fig. 3. I/Q imbalance compensation model

The I/Q phase imbalance corrector model ensures that when the signal mix occurs the phase difference between channels I and Q will be eliminated.

Figure 3 satisfies this premise. The mathematical development of this method is presented as follows:

Channel Q:

$$\frac{Q}{cos(\phi)} * [sin(\omega_c t + \phi)], \tag{17}$$

Applying the sine trigonometric identity for angle addition:

$$\frac{Q}{cos(\phi)} * [sin(\omega_c t * cos(\phi)] + \frac{Q}{cos(\phi)} * [sin(\phi) * cos(\omega_c t)], \tag{18}$$

Channel I:

$$cos(\omega_c t) * \left[I + Q * \frac{sin(\phi)}{cos(\phi)} \right], \tag{19}$$

Subtracting both I and Q channels to obtain the modulated output signal.

$$\left[I * cos(\omega_c t) + Q * cos(\omega_c t) * \frac{sin(\phi)}{cos(\phi)} \right] - \left[Q * sin(\omega_c t) + Q * cos(\omega_c t) * \frac{sin(\phi)}{cos(\phi)} \right], \tag{20}$$

Reducing terms:

$$I * cos(\omega_c t) - \frac{Q}{cos(\phi)} * sin(\omega_c t)cos(\phi) \tag{21}$$

Results in:

$$Sx_{modulated} = I * cos(\omega_c t) - Q * sin(\omega_c t) \tag{22}$$

Equation (20) demonstrates that the proposed model to correct the I/Q imbalance is correct given the fact that such imbalance is compensated in terms of phase ϕ. When substituting $I = cos(\omega_b t)$ and $Q = sin(\omega_b t)$ in Eq. (22) is obtained:

$$Sx_{modulated} = cos(\omega_b t) * cos(\omega_c t) - sin(\omega_b t) * sin(\omega t), \tag{23}$$

Following, sine and cosine multiplication trigonometric identity is applied as:

$$Sx_{modulated} = cos(\omega_b t + \omega_c t), \tag{24}$$

Finally, a I/Q modulated signal is obtained such that the signal spectrum should be a pure tone without any spurious tones or noise.

3.2   Method 2: The Numerical Approach

As described in the previous section, the I/Q imbalance can be compensated employing trigonometrical identities for the phase correction. In this section a solution based on numerical method is presented, where a function describes the relation between an input and output of a non-linear system. The block diagram of the proposed model is depicted in Fig. 4.

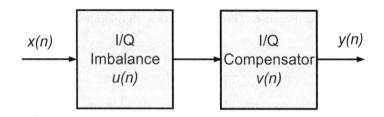

Fig. 4. I/Q imbalance compensation block diagram

As noted in Fig. 4, this methodology will create an inverse model of the I/Q imbalance in order to compensate such behavior. This technique is also known as predistortion, given the fact that the I/Q imbalance can be described as a non-linear function without memory effects truncated Volterra Series will be employed to model such behavior.

The Volterra series equation is described as follows:

$$x(n) = \sum_{k=0}^{L-1} u(k)s(n-k) + \sum_{k=0}^{L-1} v(k)s^*(n-k), \tag{25}$$

where

$x(n)$ - is the transmitted signal.
$y(n)$ - is the distorted signal.
$u(n)$ - is the non-linear behavior (I/Q imbalance) function.
$v(n)$ - are the coefficients of the system.

Rewriting the equation in its matrix form:

$$\mathbf{Y_{IQ}} = [\mathbf{S_{IQ}C_{IQ}}], \tag{26}$$

where

$\mathbf{Y_{IQ}}$ - is the output of the model.
$\mathbf{S_{IQ}}$ - is the input of the model.
$\mathbf{C_{IQ}}$ - are the model coefficients.

The correct estimation of the system coefficients is a crucial step, since the presicion of them will make the difference in terms of accuracy between the real I/Q imbalance behavior and the generated model.

A well trusted method for constant coefficients estimation is linear regression through Least Square Error (LSE) [16], this method has been successfully proven in previous research such as [16,17]. From, Eq. (26) the output data can be represented as shown.

$$\mathbf{Y_{IQ}} = [y(0), y(1), ..., y(N-1)], \tag{27}$$

where

Y_{IQ} - represent the vector of outputs
$y(n)$ - is the discrete output at time n
N - represents the number of observations

The observation matrix $\mathbf{H}$ is defined as:

$$\mathbf{H} = [\mathbf{H}_0, \mathbf{H}_1, ... \mathbf{H}_Q], \tag{28}$$

where:

$$\mathbf{H_q} = \begin{bmatrix} h_{1,q}(0), & h_{3,q}(0), & \cdots & h_{2k-1,q}(0) \\ h_{1,q}(1), & h_{3,q}(1), & \cdots & h_{2k-1,q}(1) \\ \vdots, & \vdots, & \ddots, & \vdots \\ h_{1,q}(N-1), & h_{3,q}(N-1), & \cdots & h_{2k-1,q}(N-1) \end{bmatrix}, \tag{29}$$

and

$$h_{2k-1} = |x(n-q)|^{2*(k-1)} * x(n-q). \tag{30}$$

Thus, the coefficients are expressed as

$$\mathbf{a} = [\mathbf{a_0}, \cdots, \mathbf{a_q}, \cdots, \mathbf{a_Q}]^T, \tag{31}$$

where a_q means:

$$\mathbf{a_q} = [\mathbf{a_{1,q}}, \mathbf{a_{3,q}}, \cdots, \mathbf{a_{2k-1,q}}], \tag{32}$$

So Eq. (1) is represented in matrix notation as:

$$\mathbf{Y} = \mathbf{H} * \mathbf{a}. \tag{33}$$

The coefficients that minimize the error between the output of the model and $\mathbf{Y}$ can be estimated using LSE through Eq. (33)

$$\hat{\mathbf{a}} = [\hat{\mathbf{a}}_0, \cdots, \hat{\mathbf{a}}_q, \cdots, \hat{\mathbf{a}}_Q] = \mathbf{H}^+ * \mathbf{Y}, \tag{34}$$

where $\mathbf{H}^+$ is the pseudo-inverse of the matrix $\mathbf{H}$ and can be calculated as

$$\mathbf{H}^+ = (\mathbf{H}^T * \mathbf{H}^{-1}) * \mathbf{H}^T. \tag{35}$$

The least square error is given by:

$$J_{min} = (\mathbf{Y} - \mathbf{H} * \hat{\mathbf{a}})^T * (\mathbf{Y} - \mathbf{H} * \hat{\mathbf{a}}). \tag{36}$$

Finally, the coefficients are calculated as:

$$\mathbf{C_{IQ}} = (\mathbf{S_{IQ}^H})^{-1} * (\mathbf{S_{IQ}^H} * \mathbf{y_{IQ}}), \tag{37}$$

4   Implementation, Test and Results

Both of techniques for I/Q imbalance compensation were simulated in Simulink
with DSP Builder, a Simulink toolkit from Altera which allows for a quick FPGA
Implementation since it generates synthetizable VHDL code. The methodology
described in Sect. 2 will be employed. The I/Q signal is shown in Fig. 5.

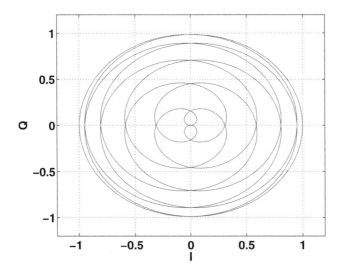

Fig. 5. I/Q Constellation

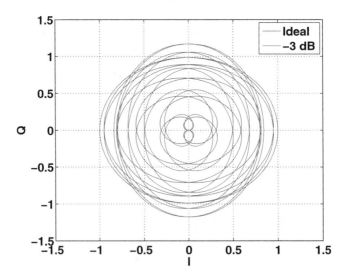

Fig. 6. I/Q ideal constellation (blue) and I/Q constellation compressed −3 dB (red)

To prove the compensation capabilities of each method 3 I/Q imbalance scenarios will be tested:

1. I/Q gain imbalance of −3 dB.
2. I/Q phase imbalance of 60°.
3. I/Q gain imbalance of +1 dB.

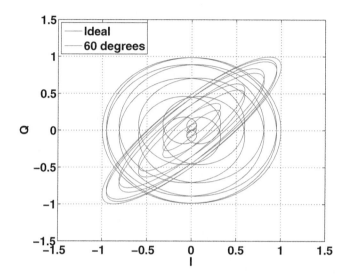

Fig. 7. I/Q ideal constellation (blue) and I/Q constellation dephased 60° (red)

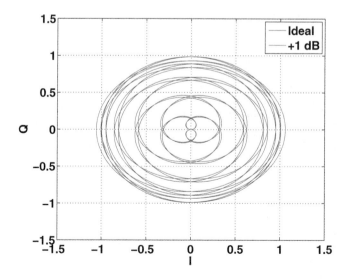

Fig. 8. I/Q ideal constellation (blue) and I/Q constellation saturated +1 dB (red)

As noted in Figs. 6, 7 and 8 a comparison between the imbalanced and ideal constellation is shown, where Fig. 6 shows the $-3\,\mathrm{dBm}$ compression, Fig. 7 presents the $60°$ dephase and finally Fig. 8 illustrates the $+1\,\mathrm{dB}$ saturation scenario. The DSP Builder implementation of model from Sect. 3.1 is shown in Fig. 9.

Figure 10 shows the model implementation from Sect. 3.2.

The Stratix III DSP Development Kit was employed with a clock frequency of 125 MHz, The frequency of the baseband signal is 1 MHz with a carrier frequency of 10 MHz (Fig. 11).

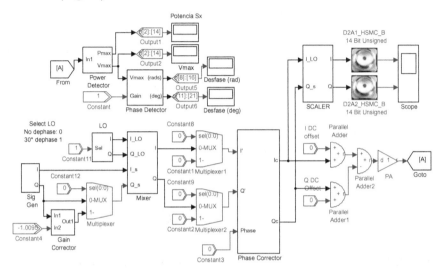

Fig. 9. Method 1 implemented in DSP Builder

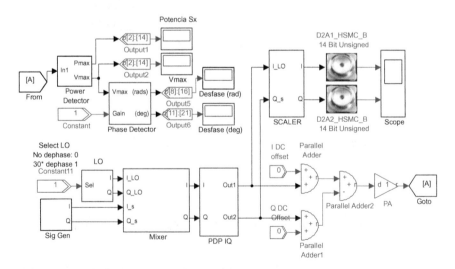

Fig. 10. Method 2 implemented in DSP Builder

Fig. 11. Experimental Testbed

1. I/Q gain imbalance of −3 dB.

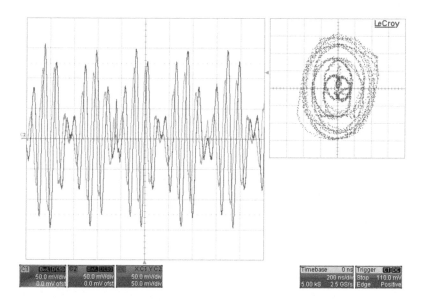

Fig. 12. Oscilloscope view and constellation with an I/Q gain imbalance of −3 dB

Method 1

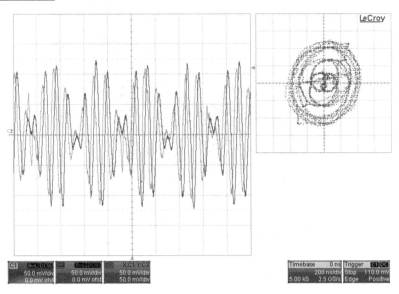

Fig. 13. Oscilloscope view and constellation with a corrected I/Q gain imbalance of −3 dB employing Method 1

Method 2

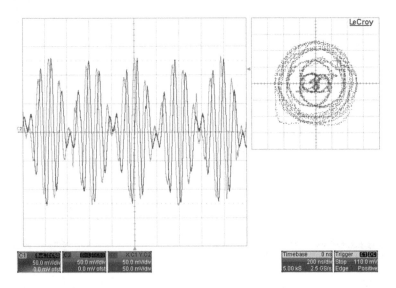

Fig. 14. Oscilloscope view and constellation with a corrected I/Q gain imbalance of −3 dB employing Method 2

2. I/Q phase imbalance of 60°.

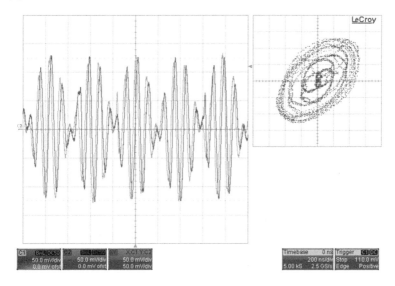

Fig. 15. Oscilloscope view and constellation with an I/Q phase imbalance of 60°

Method 1

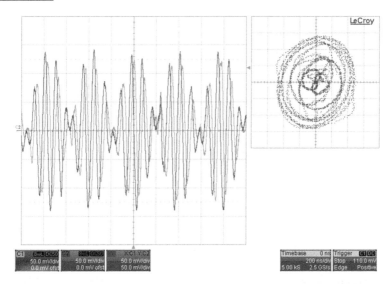

Fig. 16. Oscilloscope view and constellation with a corrected I/Q phase imbalance of 60° employing Method 1

Method 2

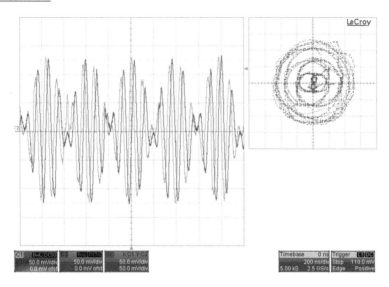

Fig. 17. Oscilloscope view and constellation with a corrected I/Q phase imbalance of 60° employing Method 2

3. I/Q gain imbalance of +1 dB.

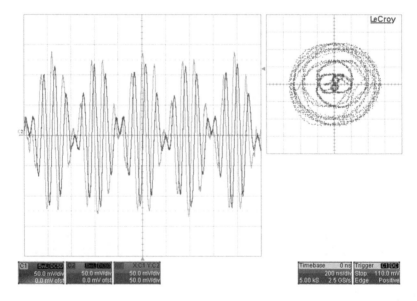

Fig. 18. Oscilloscope view and constellation with an I/Q gain imbalance of +1 dB

Method 1

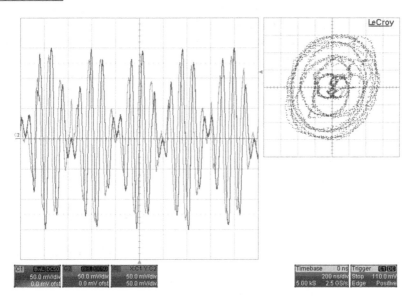

Fig. 19. Oscilloscope view and constellation with a corrected I/Q gain imbalance of +1 dB employing Method 1

Method 2

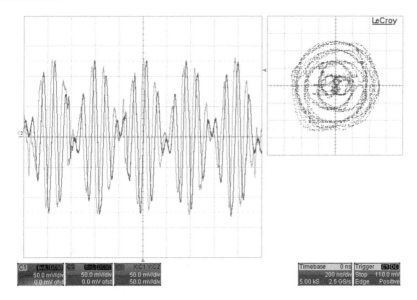

Fig. 20. Oscilloscope view and constellation with a corrected I/Q gain imbalance of +1 dB employing Method 2

As it can be confirmed in Figs. 12, 13, 14, 15, 16, 17, 18, 19 and 20 both methods of compensation have comparable results since they compensated to some degree the I/Q imbalance. Thus, in order to compare the performance of both methodologies in terms of accuracy and precision, the Normalized Mean Square Error (NMSE) of each is calculated. Table 1 indicates the respective NMSE for each model and scenario.

Table 1. Comparison of NMSE measurements in dBm

I/Q imbalance scenario	NMSE Method 1	NMSE Method 2
-3 dB compression	-25.76 dBm	-30.77 dBm
$60°$ dephase	-18.15 dBm	-27.77 dBm
$+1$ dB saturation	-23.53 dBm	-28.77 dBm

Considering the NMSE comparison from Table 1, it can be demonstrated that Method 2 is better than Method 1, given the fact that in this case a lower NMSE value, means a better compensation scheme. Thus, according to the NMSE measurements Method 2 is better for I/Q compensation.

Nevertheless, not only accuracy is a decisive factor, hardware requirements must be taken into account. Table 2 shows the hardware resources required by each compensation method.

Table 2. Logic resources used for each method

Component	Method 1	Method 2
Combinational ALUTs	$196/113,600$ ($< 1\%$)	$883/113,600$ ($< 1\%$)
Memory ALUTs	$32/56,800$ ($< 1\ \%$)	$32/56,800$ ($< 1\ \%$)
Logic registers	$199/113,600$ ($< 1\%$)	$1,206/113,600$ (1%)
Pins	$130/744$ (17%)	$130/744$ (17%)
Block memory bits	$1,581,056/5,630,976$ (28%)	$1,187,840/5,630,976$ (21%)
DSP block 18-bit	$38/384$ (10%)	$86/384$ (22%)
PLLs	$2/8$ (25%)	$2/8$ (25%)

The logic resources utilization described in Table 2 shows that Method 1 requires few DSP elements and much more memory bits than Method 2. This means that hardware implementation of Method 1 will be cheaper compared to Method 2, taking into account that memory is significantly cheaper to implement than High Speed DSP Blocks.

The results of both implementations show that there is a trade-off to be made, in terms of accuracy Method 2 is far better than Method 1. But, the fact that it requires an additional digital stage for coefficient calculation or to know in advance the I/Q imbalance behavior, makes the digital realization harder to implement. Thus, Method 1 has the upper hand since it requires no previous measurements, and is capable of compensating I/Q imbalance applying simple trigonometical properties.

5   Conclusions

In this work the comparison of two methods for I/Q imbalance compensation was showed and implemented, first a method for phase measurement is described, following two methods for I/Q imbalance compensation. The first method describes a simple way to compensate I/Q imbalance employing some of the theory described in the phase measurement scheme. Followed by a numerical analysis method based on Volterra series to obtain an inverse model capable of compensating I/Q imbalance, using linear regression and least squares for the system coefficient obtainment.

The viability of the two proposed methods has been proved through the simulation and implementation in FPGA. The obtained results showed that both methodologies were capable of compensating I/Q imbalance with differences in terms of accuracy. In order to prove which, was better the NMSE was measured, giving as result that the method based on Volterra series had a better accuracy, but with the disadvantage of requiring an additional stage for coefficient calculation or a previous I/Q imbalance modeling. In contrast Method 1 was better for scenarios were accuracy was not as important as effectiveness.

Finally, it can be said that the method based on Volterra series is better, given the fact that since is based on a behavioral model can compensate more complex signals such as 64, 256-QAM and its possible to do a prior measurement to detect any system I/Q imbalance. And the compensated signal is almost identical to the desired signal. Future work will analyze the possibility of implementing a coefficient calculation stage, in order to be capable of determining the I/Q imbalance behavioral model on the fly.

Acknowledgements. The authors wish to thank PhD. Patrick Roblin, Professor at Ohio State University, for its support provided through the measuring data. In addition, the authors would like to express their gratitude to the IPN for its financial support by the project SIP-20170588.

References

1. Dick, C., Harris, F., Rice, M.: FPGA implementation of carrier synchronization for QAM receivers. J. VLSI Signal Process. Syst. Signal Image Video Technol. **36**(1), 57–71 (2004)
2. Erdogan, E.S., Ozev, S.: Single-measurement diagnostic test method for parametric faults of I/Q modulating RF transceivers. In: Proceedings of the 26th IEEE VLSI Test Symposium, VTS 2008, pp. 209–214. IEEE Computer Society, Washington, DC (2008)

3. Asami, K.: An algorithm to evaluate wide-band quadrature mixers. In: 2007 IEEE International Test Conference, pp. 1–7, October 2007
4. Cavers, J.K.: New methods for adaptation of quadrature modulators and demodulators in amplifier linearization circuits. IEEE Trans. Veh. Technol. **46**(3), 707–716 (1997)
5. Nash, E.: Correcting imperfections in IQ modulators to improve RF signal fidelity. AN-1039, Application Note, Analog Device (2009)
6. Yu, L., Snelgrove, W.M.: A novel adaptive mismatch cancellation system for quadrature if radio receivers. IEEE Trans. Circ. Syst. II Analog Digit. Signal Process. **46**(6), 789–801 (1999)
7. Anttila, L., Valkama, M., Renfors, M.: Blind compensation of frequency-selective I/Q imbalances in quadrature radio receivers: circularity-based approach. In: 2007 IEEE International Conference on Acoustics, Speech and Signal Processing - ICASSP 2007, vol. 3, pp. III-245–III-248, April 2007
8. Anttila, L., Valkama, M., Renfors, M.: Circularity-based I/Q imbalance compensation in wideband direct-conversion receivers. IEEE Trans. Veh. Technol. **57**(4), 2099–2113 (2008)
9. Mattera, D., Paura, L., Sterle, F.: MMSE WL equalizer in presence of receiver IQ imbalance. Trans. Signal Process. **56**(4), 1735–1740 (2008)
10. Cong, J., Liu, B., Neuendorffer, S., Noguera, J., Vissers, K., Zhang, Z.: High-level synthesis for FPGAs: from prototyping to deployment. IEEE Trans. Comput. Aided Des. Integr. Circ. Syst. **30**(4), 473–491 (2011)
11. O'Droma, M., Yiming, L.: A new Bessel-Fourier memoryless nonlinear power amplifier behavioral model. IEEE Microw. Wirel. Compon. Lett. **23**(1), 25–27 (2013)
12. Bahoura, M., Park, C.W.: FPGA-implementation of an adaptive neural network for RF power amplifier modeling. In: 2011 IEEE 9th International New Circuits and Systems Conference, pp. 29–32, June 2011
13. Zhu, A., Brazil, T.J.: Behavioral modeling of RF power amplifiers based on pruned Volterra series. IEEE Microw. Wirel. Compon. Lett. **14**(12), 563–565 (2004)
14. Fehri, B., Boumaiza, S.: Baseband equivalent volterra series for behavioral modeling and digital predistortion of power amplifiers driven with wideband carrier aggregated signals. IEEE Trans. Microw. Theor. Tech. **62**(11), 2594–2603 (2014)
15. Sen, S., Devarakond, S., Chatterjee, A.: Phase distortion to amplitude conversion-based low-cost measurement of AM-AM and AM-PM effects in RF power amplifiers. IEEE Trans. Very Large Scale Integr. (VLSI) Syst. **20**(9), 1602–1614 (2012)
16. Ku, H., Kenney, J.S.: Behavioral modeling of nonlinear RF power amplifiers considering memory effects. IEEE Trans. Microw. Theor. Tech. **51**(12), 2495–2504 (2003)
17. Dvorak, J., Marsalek, R., Blumenstein, J.: Adaptive-order polynomial methods for power amplifier model estimation. In: 2013 23rd International Conference Radioelektronika (RADIOELEKTRONIKA), pp. 389–392, April 2013

An Application of Data Envelopment Analysis to the Performance Assessment of Online Social Networks Usage in Mazatlán Hotel Organizations

Manuel Cázares[1]([✉]) and Oliver Schütze[2]

[1] Research Institute for Economic and Social Growth, Mazatlán, Mexico
manuel.cazares@iices.mx
[2] Computer Science Department, Cinvestav-IPN, México City, Mexico
schuetze@cs.cinvestav.mx

Abstract. Best practices in social media have been recently a major concern for organizations, since they are an important key in online social networks to develop a comprehensive strategy both on e-commerce and on traditional business. This paper makes use of Data Envelopment Analysis (DEA) (Charnes et al. 1978) for measuring the economic performance of organizations that lead to best practices using online social networks in their business and strategic processes. In particular, we apply a DEA CRS input-oriented model on a dataset of thirteen hotel organizations in Mazatlán, México, to measure efficiency from the period of 2012 to 2013. The results show that management styles and technology adoption have a great impact on efficiency as well on the creation of competitive strategies. Also, hotel organizations that are willing to use the information obtained from online social networks to create new products or services and aggressively invest on these networks to reach new markets are increasing their market share and show the best performance.

Keywords: Competitiveness · Performance measurement
Data envelopment analysis · Social media · Strategy · Online social networks

1 Introduction

In recent years online, social networks have been recognized as a medium that companies use to sell their products and services to the market, issues regarding on how to find the best practices of these networks have become more and more important for managers. However, so far, little has been reported on these issues, and even now many organizations do not know how to evaluate the performance of their social media activities in order to find the best practices. Particular problems on this topic include the difficulties on choosing the right metrics and methods to evaluate the performance of these social media in organizations.

The lack of knowledge on how to evaluate social media can lead to misinformation on what are the best practices using this field. In this paper we make a first attempt in

© Springer International Publishing AG, part of Springer Nature 2019
L. Trujillo et al. (Eds.): NEO 2017, SCI 785, pp. 295–310, 2019.
https://doi.org/10.1007/978-3-319-96104-0_16

this direction via using a dataset from the Mazatlán hotels in México, but stress that the current methodology can be applied to any other organization or sector.

Social networks are defined as web-based sites that allow people to create a public profile within a system and provide a list of users with a nexus or connection (Boyd and Ellison 2007). In the previous definition, it is understood that sites like Facebook, Twitter, YouTube, etc., are clear examples of online social networks where users share information about their interests, friends, opinions, etc.

Since the inclusion of information technologies in the hospitality sector, hotel organizations have made efforts to understand the impact of these new approaches in efficiency. The latest efforts and discussions by hotel organizations worldwide are related to online social networks which have been presented at conferences where success stories are set. One of the most recent international conferences in this field is the "Cornell Hospitality Summit (CHRS)" which is held bi-annually since 2010 by the Cornell University in New York, USA.

Huge international chains like the Hilton Group have benefited from investing in information technology to increase market share, reducing the number of calls to support centers and thanks to the new online systems, information is always available to customers and to managers making the administrative process more efficient (Schwartz 2012).

It is also important to note that the new technologies have served as a basis to increase the satisfaction and loyalty of customers through personalized attention derived from the retrieval of the users' interests via social media. That said it is not surprising that the top 50 U.S. hotel brands have profiles on the major social networks like Facebook, Twitter, Flickr, and YouTube, along with big corporations such as the Hyatt Group, Best Western and Marriott which are visibly investing more and interacting more with their consumers to improve their sales (Withiam 2011).

Over the last five years[1], Mazatlán hotel organizations have been investing on online social networks due to the exponential growth of this technology, but the same issues remain, i.e., to know what the best practices are using online social networks. Without knowing these best practices hotel organizations cannot create effective strategies to attract online customers. The main goal of this research is to give a coherent methodology to evaluate the performance of the online social networks usage in Mazatlán hotel organizations in order to know the best practices that leads to competitive strategies. We adopted the data envelopment analysis (DEA) technique in this paper to achieve our goal.

The remainder of this paper is organized as follows: Sect. 2 describes the problem at hand and its background as well as a brief review of the related work in this field. Section 3 introduces the methodology used throughout this paper. Section 4 presents the results and the related analysis. Finally, conclusions are drawn in Sect. 5.

---

[1] The data for this study have been collected in the period of 2012 to 2013.

2 Problem

2.1 Background

Mazatlán is by far the most important tourist center in the region of Sinaloa, México, with larger numbers of tourists than any city in that state (INEGI 2011), and one of the six most visited beach centers in México (see Table 1). Currently, the city of Mazatlán is growing rapidly due to the creation of the Mazatlán-Durango super highway and the Baluarte Bridge by the government of México (Bojórquez 2013), with the intention of connecting the Pacific with the Gulf of México to create more trade routes in the country. Because of this, Mazatlán has experienced considerable growth since the first weekend after the opening of the bridge, tourism grew by 21% and this figure has been rising since the summer holiday period in which records show a total of 944,000 tourists (Sectur Sinaloa 2013). That is why it is considered that Mazatlán will become one of the most important beach centers in México.

Table 1. Most visited beach centers in México (January to July 2013). Source: (Sectur 2013)

Average number of occupied rooms				Variation		Available rooms
Rank	Beach centers	2011	2012	2013	2011–2013	2012–2013	2013
1	Riviera Maya	25,706	27,841	29,913	16.4%	7.4%	35,530
2	Cancún	18,122	19,984	22,488	24.1%	13%	30,036
3	Acapulco	7,505	7,830	7,934	5.7%	1.3%	18,494
4	Los Cabos	7,402	8,006	8,320	12.4%	3.9%	12,360
5	Puerto Vallarta	6,073	7,209	7,414	22.1%	2.8%	11,990
6	Mazatlán	4,554	4,593	4,834	6.1%	5.2%	9,218
7	Veracruz	4,053	3,975	4,277	5.5%	7.6%	8,712
8	Nuevo Vallarta	3,895	4,075	5,410	38.9%	32.8%	7,598
9	Ixtapa Zihuatanejo	2,644	2,785	2,563	−3.1%	−8%	5,092
10	Cozumel	2,303	2,494	2,387	3.6%	−4.3%	4,490
11	Manzanillo	1,740	1,741	1,643	−5.6%	−5.6%	3,549
12	Bahías de Huatulco	1,546	1,769	1,990	28.7%	12.5%	2,459
13	Playas Rosarito	380	451	530	39.4%	17.7%	2,272
14	Puerto Escondido	513	552	682	33.1%	23.6%	2,122
15	La Paz	699	784	781	11.6%	−0.5%	1,619

Mazatlán's hotel organizations are divided into two tourism associations, which delimit the area where they are located. This division also serves as a reference for tourists about the type and quality of the hotels; the Mazatlán Hotel Association, which usually includes hotels near the beach and the main tourist area of Mazatlán (also called the *Golden Zone*), only includes four and five stars hotels which are a mix of international hotel franchises and domestic investment (see Table 2).

Table 2. Basic information of hotels within the main tourist area from Mazatlán (Golden Zone). Source: own elaboration.

Hotel	Management	Rooms	Stars
Pueblo Bonito Emerald Bay	Local	162	6
Crowne Plaza Resort Mazatlán	International	90	5
The Inn at Mazatlán	Local	215	5
Emporio Mazatlán	National	133	5
Ramada Resort Mazatlán	Mixed	200	5
Hotel Playa Mazatlán	Local	400	5
El Cid Castilla Beach	Local	500	5
Pueblo Bonito Mazatlán	Local	247	5
Ocean Breeze	Local	280	5
Riu Emerald Bay	International	716	5
Royal Villas	Local	125	5
Torres Mazatlán	Local	126	4
Coral Island Hotel & Spa	Local	82	4
Hotel Quijote Inn	Local	101	4
City Express Mazatlán	International	110	4
The Palms Resort Mazatlán	Local	189	4
Olas Altas Inn	Local	80	4
Las Flores Beach Resort	Local	119	4
Costa de Oro	Local	230	4
Don Pelayo	Local	162	4
Quality Inn	International	89	4
Posada Freeman Zona Dorada	Local	50	4
Océano Palace	Local	200	4
Luna Palace	Local	71	4
Playa Bonita	Local	131	4
Villas El Rancho	Local	28	4
Torrenza Boutique	Local	30	4
Hacienda Blue Bay	Local	40	4
Isla Mazatlán Golden Resort	Local	25	4
Cerritos Resort	Local	33	4
Hotel Azteca Inn	Local	74	3
Hotel Misión Mazatlán	National	127	3
Hotel Amigo Plaza	Local	53	3
Hotel Las Jacarandas	Local	86	3
Del Real Hotel & Suites	Local	42	3
Margaritas Hotel & Tennis Club	Local	60	3
Hotel Marian Beach	Local	16	3
Hotel Mazatlán	Local	63	3
Suites Lindamar	Local	12	3

(*continued*)

Table 2. (*continued*)

Hotel	Management	Rooms	Stars
El Cid El Moro Beach	Local	294	3
El Cid Marina	Local	210	3
El Cid Granada	Local	120	3
Zona Dorada Inn	Local	n/a	3
Hotel Los Arcos	Local	24	3
Costa Bonita	Local	90	3
Solamar Inn	Local	52	3
Vistamar	Local	40	3
Hotel San Diego	Local	100	3
Blue Pacific Hotel	Local	17	3
Hotel Sombrero Suites	Local	18	3
La Costa Suite	Local	15	3
Motel Marley	Local	16	3
Girasoles	Local	18	3
Suites Nereidas	Local	16	3
La Casa Contenta	Local	8	3
Hotel Génesis	Local	30	2
Hotel Posada La Misión	Local	90	2
Hotel Villa Bonita Tecali	Local	42	n/a

Note: Mixed management are hotels owned by local investors
with an international franchise.

On the other hand, Mazatlán also has the 3 Islas Association, which usually contains two and three stars hotels. These hotels are located outside the boundaries of the main tourist area as well as near to the downtown. Commonly, these hotels are conformed from local investment only. By local investment we mean organizations that were founded by local investors without national or foreign capital and with no link to hotel chains or franchises. At present time there are about 87 hotels of various categories in Mazatlán. Out of those, 58 are in the area near the beach and the main tourist area, of which 51 are of local origin, two are of national origin and five are linked to international chains; similarly, the number of hotel rooms available in this area range from a minimum of 8 rooms to a maximum of 716 rooms.

From the hotels linked to international chains, we can count the Hotel Ramada Resort Mazatlán as a special case, because although this hotel has local capital, it pays annual royalties from implementing processes for quality standards in service to the Wyndham Group. Other hotel organizations are handled by groups of chains with global presence and foreign capital. From the hotels located outside of the main tourist area and near downtown (Table 3), 26 are of local origin, and 3 linked to international franchises. They are mostly small hotels compared to large hotels in the main tourist area.

Table 3. Basic information of hotels located outside the main tourist area in Mazatlán. Source: own elaboration.

Hotel	Management	Rooms	Stars
The Melville	Local	20	5
Casa Lucila Hotel Boutique	Local	8	4
Jonathon	International	18	4
Hotel Posada Freeman	International	72	4
Hacienda	Local	95	4
Hotel Aguamarina	Local	111	4
Wyndham Las Villas	International	71	4
Villa Serena	Local	20	4
Las 7 Maravillas	Local	7	3
Belmar	Local	110	3
Suitel 522 Echo Hostel	Local	9	3
Hotel de Cima	Local	140	3
Plaza Marina	Local	100	3
Sands Arenas	Local	97	3
Hotel del Sol	Local	20	3
Hotel Fiesta	Local	62	3
María Coral	Local	20	3
Hotel Celta	Local	34	3
Hotel Los Tabachines	Local	50	3
Hotel Emperador	Local	42	3
Santa María	Local	37	3
Hotel Machado Mazatlán	Local	6	3
La Siesta	Local	57	3
Hotel Acuario	Local	42	2
Hotel California	Local	37	2
Posada Playa Sur	Local	19	2
Hotel Bucanero	Local	30	2
Hotel Perlamar	Local	37	1
Hotel del Río	Local	n/a	n/a

2.2   Issues

Currently, online social networks are positioned as a new model of communication within the Internet; their impact has been reflected not only in communication systems, but also in organizations and how these interact with their customers and suppliers. Therefore, companies had to make changes within their organizations to adapt business processes to include this new communication tool.

However, it is recognized that there is still a debate about whether it is possible to evaluate the performance and profitability offered by online social networks to companies (Hoffman and Fodor 2010). Unfortunately, at present time organizations are still unaware of the best practices that can lead to a better economic performance by making use of online social networks in their business processes.

That is when it becomes important to know how to measure efficiency, because through efficiency performance evaluation we can know what the best practices are when using online social networks. Ignoring the best practices using online social networks may lead to a loss of efficiency and therefore a loss of competitiveness, not only in the hotel industry in Mazatlán but in any industry assiduously involved in the use of this networks in their processes.

If we want to know what are the best practices when using social networks, the main question arises, which hotel organizations in Mazatlán are more efficient and competitive by using online social networks? Hence, it is necessary to compare organizations among each other in order to see which ones have the best performance in order to clarify the strengths and weaknesses in formulating strategies. Therefore, this paper discusses the following issues:

1. What is the relative efficiency of hotels located in the main tourist area in Mazatlán when using online social networks in their business processes?
2. What are the strategies that help hotel organizations in Mazatlán to be more efficient and competitive by using social media?

To give answers to these questions we will have to define the input and output components to measure the (relative) efficiency.

2.3   Related Literature

Evaluating the performance becomes crucial because it allows organizations to know their strengths and weaknesses, which is necessary to have more control and to generate better strategies that serve as the basis for the growth of any organization. But using measures such as return on investment (ROI) or return on sales (ROS) may be misleading since they do not consider the interaction between different variables that directly affect performance.

Within the industrial context it is possible to develop standards to measure performance since most of the processes are developed automatically by machines, which can easily account for such production capacity in time; in the service sector standards simply does not exist, it is necessary to compare organizations between each other in order to see which ones have the better performance (Cook and Zhu 2008).

Currently, organizations from around all over the world are trying to use several variables to measure the performance of online social networks such as number of likes given to any post or publication inside a social network, number of fans attached to a firm profile, number of reposts, etc. Hoffman and Fodor (2010) suggest that many of the above indicators can be used to evaluate the performance of social media campaigns. On other hand, Porter (2001) says that these metrics are being used to see if someday they can be correlated with revenue, which is nothing more than the gap between price and cost; also, Porter mentions that this new array of performance metrics has a loose relationship with revenue.

However, in order to create good strategies, it is necessary to evaluate the performance of organizations to see which ones are the best to create and develop such strategies. Hwang and Chang (2002) suggest the use of data envelopment analysis (DEA) to evaluate the performance of managerial efficiency in hotel organizations in Taiwan and through this detect the failures in formulating competitive strategies.

A similar approach to evaluate the Meliá hotel franchise using DEA was made in Spain, trying to analyze which franchise unit has the best strategies (Alonso de Magdaleno et al. 2009). Cayón (2007) conducted also a DEA study in Spain including not only Meliá hotels, but the entire hospitality sector in that paper, Cayón shows that an organization structure has a significant impact in hotel efficiency, he also claims that innovation is an important factor to sustained efficiency.

Having said the above, it is fair to say that DEA is a popular technique to evaluate the performance between similar organizations because it can handle multiple inputs and outputs to measure efficiency (Charnes et al. 1978) and it has been widely used to evaluate the hospitality sector performance not only to get the best strategies, but to see the evolution of firms through time, and how this firms can be affected by the changes in the management style; One example of this kind of approach is proposed by Pestana and Almeida (2006), where they used DEA to analyze the changes in the management styles in the Portuguese hospitality sector.

Another DEA example in hotel industry is given by Alberca and Parte (2012), where the authors have used this technique to find out which communities have the hotels with the best performance in Spain. They suggest that hotels should diversify their offerings according to the market, because consumers demand new products and services, and it is through this strategy that hotels can reach higher levels of efficiency since their incomes will not be linked to seasonal revenue. Also, they suggest that hotel industry should direct their modernization process through competitiveness and not through generating new facilities.

3 Methodology

3.1 Defining Input-Output Components

According to Porter (2001), the economic value of an organization is represented by the gap between price and cost, which provides us a reliable metric to measure the profitability. When hotel organizations and companies in general use social media in their business processes, they usually pay people to create, maintain, and evaluate the performance of their online profiles. Hence, we can define as one input all the money that these companies are spending in hiring and training people to control their social media profiles in favor of their objectives.

Another expense should be the equipment for these social media managers use to achieve the companies objectives; and the hotel organizations objective will be to generate revenue through online reservations within these online social networks. Further, we can count as an expense the money that these companies invest in online advertisements through these online social networks.

However, since all these three concepts are basically different representations of an investment, we should merge these values into one input. Moreover, we should also count the hours that these people spend on using social media inside the organizations, e.g., to answer questions about certain products and services to the customers. This constitutes the second input.

Our output can be simply defined by taking the revenue that this social network generates through online booking from the online profiles. Our factors are resumed as follows:

Input factors:

- Time spent to create and maintain the respective pages in each social network, including the time to answer the requests via email, telephone etc.
- Money invested in salaries, equipment, and social media advertising.

Output factors:

- Revenue from online reservations through social media profiles.

As we said before, the DEA technique can handle multiple inputs and outputs. To evaluate online social network performances in hotel organizations we use in the following the input-oriented constant returns to scale (CRS) model by Charnes et al. (1978), which reads as follows:

$$\max_{v,u} E_m = \frac{\sum_{j=1}^{J} v_{jm} y_{jm}}{\sum_{i=1}^{I} u_{im} x_{im}}$$

subject to:

$$\frac{\sum_{j=1}^{J} v_{jm} y_{jm}}{\sum_{i=1}^{I} u_{im} x_m} \leq 1; \quad m = 1, \ldots, M$$

$$v_{jm}, u_{im} \geq 0; \quad i = 1, \ldots, I; \quad j = 1, \ldots, J,$$

where

E_m is the efficiency of the m-th Decision Making Unit (DMU), $m = 1, \ldots, M$,
y_{jm} is j-th output of the m-th DMU,
v_{jm} is the weight of that output,
x_{im} is i-th input of the m-th DMU,
u_{im} is the weight of that input, and
y_{jm} and x_{im} are the j-th output and i-th input, respectively, of the m-th DMU, $m = 1, \ldots, M$.

In our example, the m-th DMU is given by the m-th hotel organization, and further we have M = 13 (number of DMUs), I = 1 (number of outputs), and J = 2 (number of inputs).

For the efficiency analysis we used a 3-month period in 2012 and 2013 including the months May, June, and July. This is the most important period since it comprises the transition from the end of the low season to the start of the high season. The data were collected directly from hotel managers and from owners of the hotel organizations. Information provided by the hotels used in this study comes from Facebook, which is the most widely used social network in this sector.

An isotonicity test (Chen 1997) was carried out for all DEA variables (Inputs, Outputs) to confirm the legitimacy of our model, this means that greater quantities of inputs should result in greater output increase and not a decrease in any output. The inter-correlations show that the Pearson's coefficients are positive so the inclusion of the variables in the model is justified (see Table 10).

4   Results

4.1   Relative Efficiency

Table 4 presents the DEA efficiency scores from the 3-month period of May, June, and July 2013, and Table 5 the respective inputs and output. The highest level of efficiency was obtained by the Hotel Playa Mazatlán, which uses an aggressive investment strategy for searching new markets using online social networks along with their traditional business base (Porter 2001). Similarly, their online network profiles are aligned with the marketing efforts made by email and the one made in search engines like Google and Yahoo; but each communication effort is made depending on what customers want. In this case, it is a comprehensive strategy based on user preferences.

Table 4. Efficiency scores in 2013 for 3-month period (May, June and July). Source: own elaboration.

DMU	Hotel	Score	Rank
H13	Playa Mazatlán	100	1
H4	Royal Villas	69.02	2
H9	Quijote Inn	54.69	3
H2	Costa de Oro	29.45	4
H6	Pueblo Bonito Emerald Bay	14.62	5
H7	Pueblo Bonito Mazatlán	11.9	6
H1	Ramada Resort Mazatlán	11.76	7
H10	Hacienda Blue Bay	11.65	8
H5	Coral Island	9.85	9
H3	Quality In	9.4	10
H11	Don Pelayo	7.43	11
H12	Los Arcos	6.05	12
H8	Las Flores Beach Resort	5.88	13

Table 5. Inputs and output for 3-month period (May, June and July 2013). Source: own elaboration with data provided by the hotel organizations.

Dmu	Hotel	Hours	Overall investment	Revenue
H13	Playa Mazatlán	480	143,000	2,060,184.96
H4	Royal Villas	60	7,000	69,606
H9	Quijote Inn	180	18,000	141,830
H2	Costa de Oro	540	28,000	118,800
H6	Pueblo Bonito Emerald Bay	216	19,781.97	41,677.43
H7	Pueblo Bonito Mazatlán	216	19,781.97	33,904.17
H1	Ramada Resort Mazatlán	792	79,306	134,393.94
H10	Hacienda Blue Bay	60	40,023.6	30,000
H5	Coral Island	400	74,600	105,894
H3	Quality In	216	23,250	31,500
H11	Don Pelayo	60	7,000	7,497
H12	Los Arcos	2160	49,500	43,152
H8	Las Flores Beach Resort	360	25,000	21,187.95

Note: Investment and revenue are in Mexican pesos.

For example, if a potential customer types the words "Mazatlán vacations" into a search engine, Playa Mazatlán uses different types of offerings depending on the characteristics of their potential customers like sex, age, hobbies, etc. Before the deals are created, Playa Mazatlán utilizes a customer profiling process with all the available information primarily through social media, where they can identify new trends, new business opportunities and the right way to achieve this. They group individual profiles into market segments that they can use to develop the right offerings to their customer; with this strategy they make sure that they are serving the correct products and services according to their customer's needs. So one can assume from the given results that this might be the best strategy in online social networks.

Hotel Playa can be defined also as a fast innovation company because they immediately adapted their social media efforts to the technology trends, using the different options available, such as Facebook app for online reservations, PayPal online payments processing, etc.

Another important thing to remark is the well-structured social media team Hotel Playa Mazatlán has, the functions of each member of the team are defined so nobody can be confused on what to do when they interact with online customers.

Playa Mazatlán is making a huge investment on social media and they are getting high rates of return, they are achieving this using different set of strategies as follows:

- Dynamic Investment
- Fast mover
- Using social media as a pool of information to create market segments
- Custom offers depending on customer needs and habits
- Well-structured social media team
- Constant training.

On the other hand, we have the strategy used by the two hotels behind Playa Mazatlán, namely Quijote Inn and Royal Villas. Their social media strategy is based through direct care of their active customer base, keeping their existing customers happy rather than go out and find new markets as a cost saving strategy. Also, they only receive reservations via inbox messages, because for a long time they did not have promotional posts or offers through their Facebook profile.

Royal Villas have been publishing promotional posts and offering in recent days, because they are trying to compete with the hotel Playa Mazatlán, but Quijote Inn is staying with the same strategy, doing reservations only via inbox messages in Facebook and keeping their customer base.

Having said this, another type of strategy is presented; hotels like Pueblo Bonito and Quality Inn are using online social networks such as Facebook and Twitter only as a communication tool, to reach travel agencies, or to show information to customers, they don't think social media should be used to sell products or services directly to their customers, thought they are not closed to the idea, and they receive some reservations through this medium.

Some other hotels like Ramada, Hacienda Blue Bay, among others, are losing efficiency because they are paying relatively large salaries to their managers to create and maintain their online profiles, instead of reducing costs by hiring community managers to do this job. Hotel Playa has a team of community managers so they are not spending the money in the wrong way. Although Ramada and Pueblo Bonito have a higher number of Facebook fans (Table 6), Hotel Playa is assumed to have the best strategy given the results of the DEA analysis.

Table 6. Results of Facebook fan page profile analysis. Source: own elaboration from the Facebook profiles of the hotel organizations. Accessed 16 September 2014.

Dmu	Hotel	Rank	FB profile since	# FB fan page likes
H13	Playa Mazatlán	1	2009-10-02	31,795
H4	Royal Villas	2	2011-07-13	1,929
H9	Quijote Inn	3	2011-02-27	219
H2	Costa de Oro	4	2010-03-07	5,335
H6	Pueblo Bonito Emerald Bay	5	2008-07-10	57,991
H7	Pueblo Bonito Mazatlán	6	2008-07-10	57,991
H1	Ramada Resort Mazatlán	7	2011-01-13	42,122
H10	Hacienda Blue Bay	8	2011-03-17	9,341
H5	Coral Island	9	2011-03-01	3,703
H3	Quality In	10	2011-08-01	975
H11	Don Pelayo	11	2011-02-03	1,631
H12	Los Arcos	12	2010-04-16	15,454
H8	Las Flores Beach Resort	13	2010-08-20	3,869

The recompilation of bad strategies used by the poorer performers in the DEA analysis are shown as follows:

- Lack of interest of the managers to use social media in their business process.
- Lack of information on how to use social media.
- Ignoring which key performance indicators are needed to measure social media.
- Using highest-paid employees such as managers to create and maintain social media profiles.

With the current data it can be assumed that Hotel Playa has the best strategy given the results in both years 2012 and 2013 (Table 7), and as long as they are the only ones investing huge amount of money into social media, it is logical to assume that they will continue to have the most efficient online business in the hospitality sector in Mazatlán.

Table 7. Most efficient hotels from 2012 and 2013. Source: own elaboration.

Dmu	Hotel	2012 score	2012 rank	2013 score	2013 rank
H13	Playa Mazatlán	100	1	100	1
H4	Royal Villas	27	2	69.02	2

Even though Royal Villas made a significant evolution from one year to another, their revenue (Table 8) is still not compared to the market share owned by Playa Mazatlán in both years.

Table 8. Revenue evolution from 2012 to 2013. Source: own elaboration.

Dmu	Hotel	2012 revenue	2012 rank	2013 revenue	2013 rank
H13	Playa Mazatlán	2,245,509.11	1	2,060,184.96	1
H4	Royal Villas	74,884	2	69,606	2

Note: Revenue is in Mexican pesos.

The significant difference relies on the amount invested by both hotels, while Royal Villas decreased their investment from 2012 to 2013 (Table 9). Playa Mazatlán increased their budget year by year to gain more market for the years to come.

Table 9. Investment evolution from 2012 to 2013. Source: own elaboration.

Dmu	Hotel	2012 investment	2012 rank	2013 investment	2013 rank
H13	Playa Mazatlán	134,250	1	143,000	1
H4	Royal Villas	17,000	2	7,000	2

Note: Investment is in Mexican pesos.

4.2 Limitations of the Study

With the current data set, we compared hotels with different dimensions, rating, and infrastructure, at first sight they might not be directly comparable, but all the hotels in the data have in common the usage of the online social networks in their business activities, mostly Facebook. The hotels in current data set represents the early adopters, some others in Mazatlán are in recent days adapting their processes to this new technology so they do not have enough data to be considered, and the rest does not use any social media.

Another important limitation in the study was that although many organizations use online social networks in their strategic processes, they do not keep records of the investment and the financial costs of the same, i.e., they do not have accurate data on how much they have spent in that medium, this severely limited the sample, whereas many of these organizations agreed to give interviews, could not have reliable data due to the lack of records and financial information, so they were discarded from the study.

Table 10. Pearson correlations from 2013 DEA data. Source: own elaboration.

DEA variables	Hours	Overall investment	Revenue
Hours	1.00	0.30	0.03
Overall investment	0.30	1.00	0.82
Revenue	0.03	0.82	1.00

5 Conclusions

The main objective of this study was to propose a simple framework to evaluate the performance of the online social networks usage in organizations, to know which are the best practices that serve as a base to organizations to generate competitive strategies on this medium. The framework is based on Data Envelopment Analysis (Charnes et al. 1978) that can handle multiple inputs and outputs and can tell us how an organization can improve their efficiency changing their resource allocation.

The information shown in this study provides useful insight to know more how organizations can measure economic performance of their efforts in online social networks, and for future research on this topic.

With the results provided through Data Envelopment Analysis we could clarify how the relative efficiency is grouped in Mazatlán hotel organizations, also we could know that only one organization is leading the social media market through an aggressive and personalized strategy, allocating a lot of resources in order to achieve this goal. In this case the effective strategies can be summarized as follows: the best performer hotel uses an aggressive investment in social media campaigns, also in hiring and training qualified employees. Second, they use all the available information contained in their social media profiles to create unique and personalized products and services, the main key is to innovate their service offerings through their client's needs and opinions.

It is important to remark that with this study we found out that there are some practices that all managers need to avoid in order to improve their performance in social media in order to create better strategies, we can include in those bad practices, the managers lack of interest to use social media in their business, by doing this, managers are losing emerging market segments and we can assume that such businesses will lose competitiveness if they don't make a move into social media. We can also note that using high-paid employees such as managers to create and maintain social media profiles is a bad idea and makes a huge impact on the organization performance.

On other hand, we can also show that the lack of information systems that enables organizations to store critical data, such as hobbies or customer preferences (retrieved from social media), makes it impossible for these organizations to evaluate their performance at the most minimal level.

With the available data we have discovered some of the key points that makes an organization a good performer using social media, with broader data we will expect in future work to obtain more information about the good and the bad practices in social media (Table 10).

Acknowledgements. Manuel Cázares acknowledges support from CONACyT.
Oliver Schütze acknowledges support from CONACyT project no. 285599.

References

Charnes, A., Cooper, W., Rhodes, E.: Measuring the efficiency of decision making units. Eur. J. Oper. Res. **2**(6), 429–444 (1978)

Boyd, D., Ellison, N.: Social network sites: definition, history, and scholarship (2007). http://jcmc.indiana.edu/vol13/issue1/boyd.ellison.html

Schwartz, M.: Impact of IT on hotel industry and Hilton hotel corporation. Positive Concepts Inc., Troy (2012)

Witham, G.: Cornell hospitality research summit proceedings. In: Social Media and the Hospitality Industry: Holding the Tiger by the Tail, Nueva York (2011)

INEGI: Perspectiva estadística Sinaloa. INEGI, Ciudad de México (2011)

Bojórquez, H.: El flujo de turistas se dispara con la Mazatlán-Durango. Empresas El Debate, S.A. de C.V., 21 Octubre 2013. http://www.debate.com.mx/eldebate/noticias/default.asp?IdArt=13692957&IdCat=6097. Accessed 2 Dec 2013

Sectur Sinaloa, A.: Secretaría de Turismo - Crece 21% el turismo nacional hacia Mazatlán por apertura de carretera. 25 Octubre 2013. http://turismo.sinaloa.gob.mx/index.php?option=com_content&view=article&id=245:crece-21-el-turismo-nacional-hacia-mazatlan-por-apertura-de-carretera&catid=39&Itemid=271. Accessed 28 Apr 2014

Sectur: Secretaría de Turismo. Secretaría de Turismo (2013). http://www.datatur.beta.sectur.gob.mx/Documentos%20Publicaciones/sem302013.pdf. Accessed 1 Jan 2014

Hoffman, D., Fodor, M.: Can you measure the ROI of your social media marketing? MIT Sloan Manag. Rev. **52**(1), 40–48 (2010)

Cook, W., Zhu, J.: Data envelopment analysis: modeling operational processes and measuring productivity (2008)

Alonso de Magdaleno, M., Fernández, M., González, M.: Análisis de Eficiencia en el Sistema Hotelero Español: Una Aplicación al Caso De Sol Meliá. Investigaciones Europeas de Dirección y Economía de la Empresa **15**(3), 83–99 (2009)

Porter, M.: Strategy and the internet. Harvard Bus. Rev. 62–78 (2001)

Cayón, M.: Análisis Comparativo De La Eficiencia De La Empresa Pública Respecto La Empresa Privada; Aplicado a Empresas Hoteleras En España. Universidad Autónoma de Barcelona, Barcelona (2007)

Alberca, P., Parte, L.: Evaluación de la eficiencia y la productividad en el sector hotelero espa˜nol: un análisis regional. Investigaciones Europeas de Dirección y Economía de la Empresa, p. 10 (2012)

Hwang, S.-N., Chang, T.-Y.: Using data envelopment analysis to measure hotel managerial efficiency change in Taiwan. Tourism Manag. **23**, 10 (2002)

Pestana, C., Almeida, C.: The measurement of efficiency in Portuguese hotels using data envelopment analysis. J. Hosp. Tourism Res. **30**(3), 378–400 (2006)

O'Connor, P.: Managing social media (2011). http://hotelexecutive.com/business_review/2500/managing-social-media

Chen, T.: An evaluation of the relative performance of university libraries in Taipei. Asian Libraries, pp. 39–50 (1997)

Wilson, A., Murphy, H., Cambra, J.: The nature and implications of user-generated content. Cornell Hospitality Q. **53**, 220–228 (2012)

Mintzberg, H., Quinn, B.: El proceso estratégico: conceptos, contextos y casos. Prentice Hall, México (1993)

Kim, W., Lee, C., Hiemstra, S.: Effects of an online virtual community on customer loyalty and travel product purchases. Tourism Manag. **25**, 343–355 (2004)

Author Index

© Springer International Publishing AG, part of Springer Nature 2019
L. Trujillo et al. (Eds.): NEO 2017, SCI 785, pp. 311–312, 2019.
https://doi.org/10.1007/978-3-319-96104-0

Printed in the United States
By Bookmasters